AF684244

SLEEP APNEA: WHAT EVERY CLINICIAN (AND PATIENT) SHOULD KNOW

This volume includes the proceedings of the
Forty-Fourth Annual Moyers Symposium and the Forty-Second
Annual International Conference on Craniofacial Research
March 4-5, 2017
Ann Arbor, Michigan

Editors
James A. McNamara, Jr., DDS, PhD
Anita Valanju Shelgikar, MD

Associate Editor
Kristin Y. De Koster

Volume 54
Craniofacial Growth Series

Department of Orthodontics and Pediatric Dentistry
School of Dentistry; and
Center for Human Growth and Development

The University of Michigan
Ann Arbor, Michigan

©2018 by the Department of Orthodontics and Pediatric Dentistry,
School of Dentistry and
Center for Human Growth and Development
The University of Michigan, Ann Arbor, MI 48109

Publisher's Cataloguing in Publication Data

Department of Orthodontics and Pediatric Dentistry and
Center for Human Growth and Development
Craniofacial Growth Series
Sleep Apnea: What Every Clinician (and Patient) Should Know
Volume 54

ISSN 0162 7279
ISBN 0-929921-00-3
ISBN 0-929921-50-x
No part of this publication may be reproduced, stored in a retrieval system, or transmitted in any form by any means, electronic, mechanical, photocopying, recording, or otherwise, without the prior written permission of the Editor-in-Chief of the Craniofacial Growth Series or designate.

CONTRIBUTORS

BHOOMIKA AHUJA, Assistant Professor in Orthodontics, School of Dentistry, Department of Developmental Sciences, Marquette University, Marquette, MI.

SHARON ARONOVICH, Clinical Assistant Professor, Department of Oral and Maxillofacial Surgery, The University of Michigan, Ann Arbor, MI.

SUELEN W. BORGES, Assistant Professor, School of Dentistry, University Positivo, Curitiba, Paraná, Brazil.

MARY E. BURNS, private practice, New Hope, PA.

SEAN K. CARLSON, Associate Professor, School of Dentistry, Department of Orthodontics, University of the Pacific, San Francisco, CA; private practice, Mill Valley, CA.

JOHN C. CARTER, Assistant Professor, Division of Pulmonary, Critical Care and Sleep Medicine, Case Western Reserve University School of Medicine, Cleveland, OH; Assistant Professor, Divisions of Pulmonary and Sleep Medicine, Pediatric Neurology, Seattle Children's Hospital, Seattle, WA.

PAOLO M. CATTANEO, Associate Professor, Section of Orthodontics, Department of Dentistry and Oral Health, Faculty of Health, Aarhus University, Aarhus, Denmark.

LUCIA H.S. CEVIDANES, Associate Professor of Orthodontics, School of Dentistry, Department of Orthodontics and Pediatric Dentistry, The University of Michigan, Ann Arbor, MI.

CAUBY M. CHAVEZ, JR., Associate Professor of Orthodontics, School of Dentistry, Federal University of Ceará, Fortaleza/Ceará, Brazil.

LOUIS CHMURA, private practice, Marshall, MI.

R. SCOTT CONLEY, LB Badgero Endowed Associate Professor and Chair, School of Dental Medicine, University of Buffalo, Buffalo, NY.

GABRIELE DI CARLO, PhD Fellow, Department of Oral and Maxillofacial Science, Sapienza University of Rome, Rome, Italy; Section of Orthodontics, Department of Dentistry and Oral Health, Faculty of Health Science, Aarhus University, Aarhus, Denmark.

SIRWAN FERNANDEZ GURANI, PhD Fellow, University of Southern Denmark, Odense, Denmark; Faculty of Health Sciences, Institute of Regional

Health Services Sciences and Department of Oral and Maxillofacial Surgery, Hospital of South West Denmark, Esbjerg, Denmark.

CARLOS FLORES-MIR, Professor and Division Head, School of Dentistry, Division of Orthodontics, University of Alberta, Edmonton, Alberta, Canada.

JAMES K. HARTSFIELD, JR., E. Preston Hicks Endowed Professor of Orthodontics and Oral Health Research, College of Dentistry, University of Kentucky, Lexington, KY.

JOSEPH I. HELMAN, C.J. Lyons Endowed Professor, Department of Oral and Maxillofacial Surgery, The University of Michigan, Ann Arbor, MI.

TIMOTHY F. HOBAN, Director of Pediatric Sleep Medicine and Clinical Neurophysiology, Professor of Pediatrics and Communicable Diseases, Professor of Neurology, Department of Neurology, The University of Michigan, Ann Arbor, MI.

MARCOS IOSHIDA, Research Fellow, School of Dentistry, Department of Orthodontics and Pediatric Dentistry, The University of Michigan, Ann Arbor, MI.

GEORGE JERYN JACOB, Craniofacial Genetics Extern, College of Dentistry, University of Kentucky, Lexington, KY.

VISHESH K. KAPUR, Professor, Division of Pulmonary, Critical Care and Sleep Medicine, University of Washington, Seattle, WA.

LAEHYUN KIM, Principal Researcher, Center for Bionics, Korea Institute of Science and Technology, Seoul, South Korea.

YOON-JI KIM, Associate Professor, Department of Orthodontics, Korea University Anam Hospital, Seoul, South Korea.

YOUNGJUN KIM, Senior Researcher, Center for Bionics, Korea Institute of Science and Technology, Seoul, South Korea.

DOUGLAS KIRSCH, Medical Director, CHS Sleep Medicine Associate Professor, UNC School of Medicine, Carolinas HealthCare System, Charlotte, NC.

G. THOMAS KLUEMPER, Professor and Division Chief of Orthodontics, College of Dentistry, University of Kentucky, Lexington, KY.

BRIAN LUONG, Former Resident, College of Dentistry, Orthodontic Section, University of Nebraska Medical Center, Lincoln, NE.

LORRI ANN MORFORD, Research Assistant Professor, College of Dentistry, University of Kentucky, Lexington, KY.

ALEXANDRE MORO, Associate Professor, Department of Orthodontics, Federal University of Paraná and Positivo University, Curitiba/Paraná, Brazil.

ALEXANDRE T. MOTTA, Associate Professor, School of Dentistry, Department of Orthodontics, Fluminense Federal University, Rio de Janeiro, Brazil.

TUNG NGUYEN, Associate Professor, School of Dentistry, Department of Orthodontics, University of North Carolina, Chapel Hill, NC.

EUNG-KWON PAE, Chair and Program Director, School of Dentistry, Department of Orthodontics and Pediatric Dentistry, University of Maryland, Baltimore, MD.

JOSEANE PIZZATTO, Research Associate, Department of Orthodontics, University of Alberta, Vancouver, British Columbia, Canada.

BENJAMIN PLISKA, Assistant Professor, Department of Oral Health Sciences, Faculty of Dentistry, University of British Columbia, Vancouver, British Columbia, Canada.

SUNDARALINGAM PREM PREMARAJ, Associate Professor and Graduate Program Director, College of Dentistry, Orthodontic Section, University of Nebraska Medical Center, Lincoln, NE.

THYAGASEELY SHEELA PREMARAJ, Assistant Professor, College of Dentistry, University of Nebraska Medical Center, Lincoln, NE.

ANTONIO C.O. RUELLAS, Associate Professor of Orthodontics, Federal University of Rio de Janeiro, Rio de Janeiro, Brazil.

JAE-JUN RYU, Professor, Department of Prosthodontics, Korea University Anam Hospital, Seoul, South Korea.

JASON ROBERT SCOTT, private practice, Spokane, WA.

ANITA VALANJU SHELGIKAR, Director, Sleep Medicine Fellowship; Assistant Professor, Department of Neurology, The University of Michigan, Ann Arbor, MI.

PAULA P. SPADA, Associate Professor, School of Dentistry, University Positivo, Curitiba/Paraná, Brazil.

KIM SUNG, Assistant Professor, College of Dentistry, Department of Oral Biology, University of Nebraska Medical Center, Lincoln, NE.

INGE ELLY KIEMLE TRINDADE, Professor, Bauru Dental School, Hospital for Rehabilitation of Craniofacial Anomolies, University of São Paulo, Bauru/São Paolo, Brazil.

IVY KIEMLE TRINDADE SUEDAM, Associate Professor, Bauru Dental School, Hospital for Rehabilitation of Craniofacial Anomolies, University of São Paulo, Bauru/São Paolo, Brazil.

MARILIA YATABE, Post-Doctoral Fellow, Hospital for Rehabilitation of Craniofacial Anomalies, University of São Paulo, Bauru/São Paolo, Brazil; Craniofacial Fellow, School of Dentistry, Department of Orthodontics, The University of Michigan, Ann Arbor, MI.

PREFACE

Obstructive sleep apnea (OSA) is a relatively common disorder characterized by the repeated collapse of the upper airway, resulting in sleep fragmentation and episodic hypoxemia. The consequences of untreated sleep apnea can be significant and include increased risk of cardiovascular events (e.g., heart attack, stroke), decreased quality of life and motor vehicle accidents. It is estimated that over 80% of individuals who have sleep apnea do not know that they have this medical problem, emphasizing the need for greater awareness among both clinicians and the lay population.

OSA has a complex multi-factorial etiology and is more common in older adults who are overweight, but it can affect individuals of any age and body type. Even children, especially those with enlarged tonsillar or constricted nasopharyngeal tissues, may have OSA. It requires long-term management; lifestyle changes, positive airway pressure, oral appliances and/or surgery can be used treat sleep apnea successfully once the condition is diagnosed.

The Moyers Symposium has had a long history of dealing with interdisciplinary topics and the *44th Annual Moyers Symposium* was no exception. We brought together ten healthcare providers in both medicine and dentistry who have expertise in sleep-disordered breathing to discuss the diagnosis and treatment of OSA patients, considering in detail the multiple treatment approaches that are available. Some of the attendees also may have discovered that this condition was relevant personally, as the signs, symptoms and history of OSA were presented during the meeting.

The *44th Annual Moyers Symposium* was held at The University of Michigan on Saturday and Sunday, March 4-5, 2017. The interdisciplinary audience was large, with approximately 550 individuals in attendance. The annual *Presymposium* meeting (*aka* the *42nd Annual International Conference on Craniofacial Research*) was held on Friday March 2, 2017 in the 4th Floor Amphitheater of the Horace H. Rackham School of Graduate Studies Building. Thirteen of the 16 papers presented by an international group of investigators were focused specifically on the topic of the 2017 Moyers Symposium. Many of the *Presymposium* papers are included as chapters in this volume.

As in previous years, the Symposium honored the late Dr. Robert Edison Moyers, Professor Emeritus of Dentistry and Fellow Emeritus and Founding Director of the Center for Human Growth and Development at The University of Michigan. This meeting was co-sponsored by the School of Dentistry and the Center for Human Growth and Development.

First, I would like to thank my co-editor of this volume, Anita V. Shelgikar, MD, for her tireless work in helping us plan and execute the 2017 Moyers Symposium. Anita was instrumental in identifying outstanding speakers in sleep medicine to participate, as well as delivering the keynote 21st Annual Robert E. Moyers Memorial Lecture entitled *Sleep Apnea: What Is It and Why Should We Care?* She then demonstrated excellent editing skills in critiquing chapters from both medical and dental disciplines, making many suggestions that have clarified and improved each chapter. Her participation was integral from beginning to end both for the Symposium and the subsequent volume.

We continue to recognize the enormous contribution of Kris De Koster, Associate Editor of the Craniofacial Growth Series, for her efforts on this book. For the past ten years, Kris has facilitated the publication of this annual volume through her interactions with the authors, editing, manipulating a variety of figure formats and formatting the layout of the book. It is always a challenge for us to produce such a volume in the time frame prior to the next Symposium; Kris has a stellar record of producing a high-quality book within this limited period. In addition, we recognize the work of Kathy Ribbens who provided assistance in the final formatting the book for publication. We also thank the contributors for sending us their material in a timely fashion.

We acknowledge and thank Dr. Nan Hatch, Chair of the Department of Orthodontics and Pediatric Dentistry, for providing the financial resources to underwrite partially the publication of this book. We also must thank Dr. Brenda Volling, the Director of the Center for Human Growth and Development, for the continued financial and moral support of the Moyers Symposium provided by the Center for the last 44 years.

We are fortunate to work with the staff from the Office of Continuing Dental Education, who organize and run the *Presymposium* and

Symposium. Michelle Jones, Karel Barton and Elizabeth Fee managed both meetings in an exceptionally smooth and efficient fashion.

We all are very pleased that the *44th Moyers Symposium* was so successful. The topic obviously was of heightened clinical relevance to clinicians and researchers in both medicine and dentistry. The chapters within this volume are written with both the clinician and patients in mind, so that we all can benefit from sharing our knowledge provided from a variety of perspectives.

James A. McNamara, Jr.
Editor-in-Chief, Craniofacial Growth Series
The University of Michigan
Ann Arbor, Michigan
December 2017

TABLE OF CONTENTS

OBSTRUCTIVE SLEEP APNEA: WHAT IS IT AND WHY SHOULD WE CARE?

Anita Valanju Shelgikar

ABSTRACT

Obstructive sleep apnea (OSA) is a chronic medical condition caused by repetitive collapse of the upper airway during sleep. This condition, in turn, can lead to sleep fragmentation and poor sleep quality. Excessive daytime sleepiness, decreased alertness, impaired mood and diminished quality of life may result from untreated OSA. Other consequences of untreated OSA include increased risk of hypertension, cardiac arrhythmia, myocardial infarction and stroke. Multiple other biological processes and organ systems are affected adversely by OSA as well. Untreated OSA can be dangerous, especially for those individuals who work in transportation. Treatment of OSA can yield improvement in quality of life, reduction in associated medical comorbidities and have a positive economic effect on a population level.

KEY WORDS: obstructive sleep apnea (OSA), physiology, treatment

---

PHYSIOLOGY OF OBSTRUCTIVE SLEEP APNEA

Obstructive sleep apnea (OSA) is a sleep-related breathing disorder characterized by repetitive collapse of the upper airway during sleep, which causes complete or partial cessation in airflow (Eckert and Malhotra, 2008). Respiratory effort persists despite the reduction in airflow. Apnea refers to complete cessation of airflow; hypopnea indicates limited airflow. Figure 1 shows an example of an obstructive apnea, in which airflow is absent, but respiratory effort continues with subsequent oxygen desaturation.

The pathogenesis of OSA is multi-factorial, affected by anatomic structure, neuromuscular tone and a number of other contributors. Table 1 lists factors that affect the development of OSA. While some factors are modifiable (e.g., presence of obesity), others are more difficult to change. The degree to which any one factor influences the development of OSA

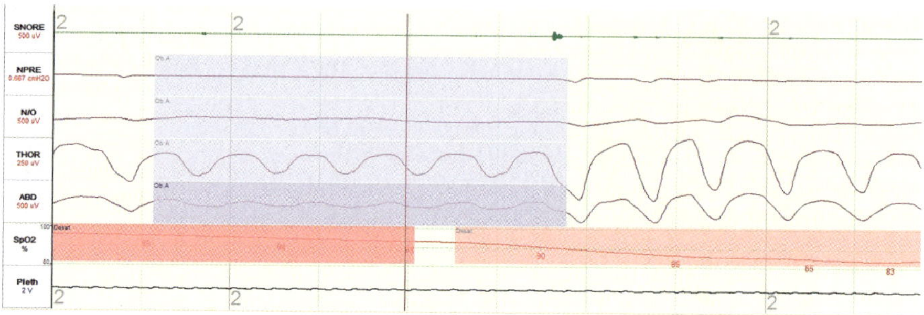

Figure 1. Image from a baseline polysomnogram showing oxygen desaturations following an obstructive apnea. NPRE = nasal pressure; N/O = nasal-oral thermistor; THOR = thoracic effort belt; ABD = abdominal effort belt; SpO2 = blood oxygen saturation; Pleth = plethysmography.

Table 1. Factors that contribute to pathogenesis of OSA.

Factors
Craniofacial anatomy
Loss of tone in upper airway dilator muscles
Body habitus
Gender
Age
Arousal threshold
Fluid shifts
Impaired loop gain in the neurologic control of breathing during sleep
Medical comorbidities

varies from one person to the next, as does the combination of risk factors in a given individual. This variation yields a heterogeneous patient population. This inherent diversity of disease adds to the scientific intrigue surrounding OSA and also shapes the multitude of treatment options currently available for OSA.

Upper airway anatomy plays a key role in the development of OSA. Cephalometric imaging gives a static view of the upper airway. Such imaging provides insight about areas along the upper airway that are narrow at baseline and may have a higher propensity to obstruct during sleep. An important caveat, however, is that cephalometric imaging is obtained while the patient is awake and upright; thus, it does not yield a true image of upper airway caliber during sleep in a recumbent position. Figure 2 shows a lateral cephalogram with a superimposed tracing of the

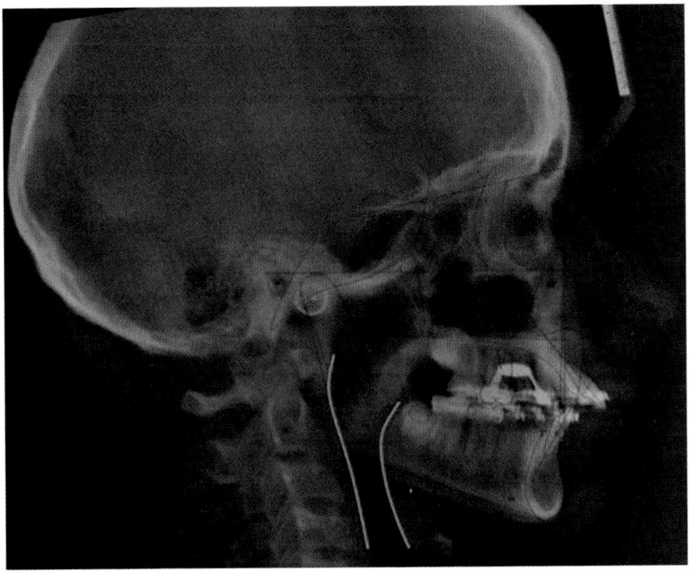

Figure 2. Lateral cephalogram with upper airway tracing.

upper airway. This particular image does not indicate significant narrowing at one particular level along the upper airway. However, the upper airway cross-sectional area is known to be reduced in individuals with OSA, compared with those without OSA (Schwab et al., 1995).

In addition to anatomic considerations, upper airway muscle tone is another important consideration. The role of neuromuscular tone in the pathogenesis of OSA should not be forgotten; OSA is not simply a "structural problem." As the brain transitions from wake to sleep, upper airway dilator muscle tone decreases and resistance within the upper airway increases. Understanding this physiology helps explain changes in upper airway caliber in different states (e.g., wake *versus* sleep).

As sleep is sustained, lung volumes decrease and chemoreceptors become less sensitive. Decreased chemoreceptor sensitivity affects the input provided for neural control of breathing. As a result of reduced lung volumes, relative hypoventilation ensues. As the upper airway dilator muscles become even less active, the upper airway caliber narrows further, either to partial or complete obstruction. This condition leads to an obstructive hypopnea (persistent work of breathing in the setting of a partially occluded airway) or an obstructive apnea (persistent work of breathing in the setting of a fully occluded airway).

When airflow is restricted, blood oxygen saturation may drop while carbon dioxide levels may rise. Respiratory effort increases as the respiratory apparatus continues to work against a partially or fully obstructed airway. The increased work of breathing causes an arousal, in which the brain wakes from sleep. During this transient period of wakefulness, upper airway muscle activity increases and chemosensitivity improves. This situation is followed quickly by a brisk opening of the upper airway. Once airflow is restored, the brain again transitions back from wake to sleep and the cycle resumes (Eckert and Malhotra, 2008).

Sleep-wake instability results from untreated OSA. Increased respiratory effort leads to frequent arousals from sleep, which last only a few seconds and then the brain changes from wake to sleep. In patients with frequent respiratory disturbances, sleep can become highly fragmented as a function of endless transitions from wake to sleep and back again.

In addition to regulation of sleep-wake transitions, the brain also modulates ventilatory control. Neural control of breathing is an intricate process with a number of variable inputs. Loop gain refers to sensitivity of a variable system (Naughton, 2010). In OSA, the variable system is a feedback cycle between peripheral inputs and central regulation of breathing. Components include:

1. The blood gas response to a change in ventilation;
2. The speed at which blood carbon dioxide level is signaled to the central controller; and
3. The central ventilatory response to carbon dioxide.

In patients with untreated OSA, the loop gain is impaired; chemoreceptors are less sensitive to small changes in blood carbon dioxide levels and ventilation becomes more unstable (Kapur, 2010).

Extracellular fluid and body positioning also may have a role in the pathogenesis of OSA. When recumbent, extracellular fluid in the lower extremities can shift rostrally and accentuate collapse of the upper airway (Chiu et. al., 2006). This fluid shift has been correlated with increased in neck circumference (Chiu et. al., 2006) and frequency of apneas and hypopneas per hour of sleep in non-obese healthy men in whom there was a suspicion of OSA (Kapur, 2010). The finding that fluid shifts also may contribute to the pathogenesis of OSA is gaining more attention and may allow for development of new therapeutic pathways for OSA patients.

COMORBIDITIES ASSOCIATED WITH OSA

The heart and brain are the two end organs most studied in untreated OSA. The aforementioned fluctuations in blood levels of oxygen and carbon dioxide and sleep-wake instability can lead to increased sympathetic neural activity. The increased sympathetic tone causes acute increases in heart rate, blood pressure and cardiac stress, particularly left ventricular stress. The accumulation of these sympathetic surges and related consequences can lead to clinical manifestations of hypertension, cardiac arrhythmias, stroke and sudden cardiac death (Gami and Somers, 2007).

Untreated OSA is an independent risk factor for hypertension; the risk increases as the severity of OSA increases (Peppard et al., 2000). In a prospective study of men with OSA, odds of fatal and non-fatal cardiovascular increased with severe untreated OSA (Marin et al., 2005). The cumulative frequency of atrial fibrillation increases over time for individuals with OSA compared with those without (Gami et al., 2007). Just as the risk of cardiovascular events increases with untreated OSA, the risk of stroke or death rises as the severity of OSA increases (Yaggi et al., 2005). The risk of these adverse outcomes is mitigated with continuous positive airway pressure (CPAP) for treatment of OSA (Park et al., 2011).

Ongoing research indicates many other organs and homeostatic processes are affected adversely by untreated OSA. For example, blood glucose regulation can be impaired as patients with untreated OSA have increased insulin resistance. Frequent, repetitive sympathetic surges can lead to increased oxidative stress, endothelial dysfunction, hypercoagulability and systemic inflammation. These cellular-level changes not only increase risk for cardiovascular and cerebrovascular morbidities, but also may contribute to development of certain cancers.

Table 2 lists conditions affected by untreated OSA (Chihorek et al., 2007; Knauert et al., 2015). This list undoubtedly will continue to grow as more work is completed on cellular mechanisms of disease in patients with untreated OSA. Given that risk of cardiovascular and cerebrovascular adverse events improves with OSA treatment, the same may be true of other medical comorbidities associated with untreated OSA.

Table 2. Examples of conditions that can be exacerbated by untreated OSA.

Hypertension
Atrial fibrillation
Depression
Myocardial infarction
Stroke
Cancer
Adverse perioperative outcomes
Epilepsy
Diabetes mellitus

For some patients, discussion of increased risk of myriad medical comorbidities may seem abstract and depersonalized. Untreated OSA also can have a very personal, real-life impact for an affected individual, as exemplified in Figure 3. Excessive daytime sleepiness and decreased quality of life are two of the most commonly discussed consequences of untreated OSA (Knauert et al., 2015). For symptomatic individuals, excessive daytime sleepiness can have implications on employment options, social engagements and interpersonal relationships. Impaired academic and/or professional performance and restricted social interactions can have strong bearing on quality of life for those with untreated OSA. Effective OSA treatment can help to reverse these negative symptoms and can yield improved quality of life.

CLINICAL EVALUATION OF OSA

The most critical component of the clinical evaluation for OSA is a detailed history. The examiner should inquire about sleep-related symptoms and consequences on daytime functioning. Sleep-related symptoms suggestive of OSA include snoring, witnessed apneas, frequent nocturnal awakenings and non-restorative sleep. Patients also may endorse frequent mouth breathing and waking with a dry mouth or sore throat. Other symptoms experienced by untreated OSA patients include acid reflux, nocturnal urination and morning headaches. History of weight gain also may provide insight in the context of the patient's other symptoms.

Presence of excessive daytime sleepiness, the situations in which it occurs and the extent to which it exists are important to gauge during the evaluation for possible OSA. The clinician should counsel patients to

Clinical History: A 36-year-old gentleman presents to the Sleep Disorders Clinic for evaluation of daytime sleepiness. He has been experiencing excessive daytime sleepiness for years. He often nods off during conversations and finds that his daytime sleepiness makes it difficult for him to interact with people for longer than an hour. He recounts a few instances during which he dozed while waiting at a stoplight; he woke to the sound of another driver honking the horn. He denies any motor vehicle collisions due to sleepiness. He sleeps from 10:30 p.m.-5:30 a.m. during the week and until 7:30 a.m. on weekends. He feels refreshed in the morning, but feels sleepy after a couple hours. He has three cups of coffee during the day and does not take naps. He snores when he sleeps on his back and rarely awakens gasping for air. He has been witnessed to have apneas during sleep. He is a university faculty member and has been teaching for about ten years. He reports that he has difficulty keeping up with the demands of his job because his daytime sleepiness interferes with his ability to concentrate and work efficiently. He decided to pursue medical evaluation of his sleepiness because he is concerned that he may be placed on probationary status at work.
Diagnostic Evaluation: A baseline polysomnogram showed that he had 38 apneas + hypopneas/hour and oxygen desaturations to 78%. He was diagnosed with OSA and was prescribed continuous positive airway pressure (CPAP).
Treatment Response: He committed to using CPAP nightly. He had resolution of his excessive daytime sleepiness and no longer felt sleepy while driving. Due to improved attention and focus, he was able to improve the quality of his work and be considered for promotion.

Figure 3. Case example of a how untreated OSA can affect quality of life.

avoid high-risk situations (e.g., driving) when experiencing excessive daytime sleepiness. Occupational considerations also should be discussed to ensure that excessive daytime sleepiness does not impair the patient's ability to perform his/her job safely.

The STOP-Bang questionnaire (Chung et al., 2013) is a quick, effective screening tool that can help identify patients at risk for severe OSA. The screening items are:

Snoring
Tiredness
Observed apnea
high blood **P**ressure *(STOP)*
Body mass index (BMI) above 35 kg/m^2
Age above 50 years
Neck circumference above 40 cm and
male **G**ender *(Bang)*.

One point can be assigned for each category, for a maximum of 8 points. A validation study in obese and morbidly obese surgical patients (Chung et al., 2013) showed that a STOP-Bang score of 4 has 88% sensitivity for severe OSA. A STOP-Bang score of 6 is more specific for severe OSA in obese and morbidly obese surgical patients.

Even the STOP questionnaire, which contains the first four elements (Snoring, Tiredness, Observed apnea and high blood Pressure), is 79.5% sensitive in pre-surgical patients with severe OSA (Chung et al., 2008). For clinicians who are unable to obtain measurements of a patient's BMI and neck circumference, the STOP questionnaire may be a useful screening tool to incorporate into the clinical evaluation for OSA.

EPIDEMIOLOGY OF OSA

The estimated prevalence of OSA in the U.S. often is cited as 3-7% in men and 2-5% in women (Park et al., 2011). Other studied populations, including those in Australia, India, China and Korea also have a higher prevalence in men compared to women (Punjabi, 2008). The prevalence of OSA is higher among adults aged 65 years and older than it is for adults aged 30 to 64 (Bixler et al., 1998).

The OSA prevalence estimate is higher in other subgroups of the general population. In patients with obesity, OSA prevalence rises as BMI increases. A retrospective study of morbidly obese patients who presented for weight loss surgery showed a total OSA prevalence of 78% across all BMI categories. In patients with a BMI above 60 kg/m^2, the prevalence measured 95% (Lopez et al., 2008). The prevalence of OSA can be expected to rise as the obesity epidemic continues in the U.S. and other countries. As a result, estimates from earlier landmark epidemiologic studies on OSA may underrepresent the current prevalence of OSA in the general population.

PUBLIC HEALTH CONCERNS RELATED TO OSA

Each comorbidity associated with untreated OSA comes with its own impact to the individual patient. The most apparent and worrisome public health concern related to untreated OSA is the risk of drowsy driving and subsequent motor vehicle collision. OSA patients are nearly 2.5x more likely to be the driver in a motor vehicle accident when compared with individuals who do not have OSA (Karimi et al., 2015). This risk is reduced by 70% in OSA patients who use CPAP for at least four hours/ night.

The increased accident risk for untreated OSA patients is dangerous, especially in those employed in transportation: truck drivers, train operators and pilots. News reports can be found to describe tragic accidents in each of these settings, with the incidents attributable at least partly to the operator's/driver's untreated OSA. These accidents occur around the globe. In the U.S., federal agencies are reviewing published data on OSA prevalence in transportation industries to inform policymaking about OSA screening, evaluation and management.

The prevalence of OSA in commercial drivers has been studied in multiple populations, with most studies showing a higher OSA prevalence compared with the general population. For instance, a study of 388 commercial drivers showed a 78% OSA prevalence (Stoohs et al., 1995). In light of this high clinical risk, coupled with the risk of injury to self and others in the event of a crash, commercial drivers, pilots and train operators may be subject to OSA screening, evaluation and treatment as indicated.

The American Academy of Sleep Medicine has outlined detailed recommendations for transportation workers with safety-sensitive duties who are diagnosed with OSA (http://www.aasmnet.org/resources/pdf/AASMResponseFMCSAFRA.pdf). These recommendations include considerations for "No Restrictions," "Conditional Restrictions" and "Immediate Disqualification" from occupational responsibilities. The different designations are conferred based on clinical symptoms, severity of OSA and adherence to OSA treatment. These individuals require longitudinal follow-up to ensure that OSA is treated fully and that the management plan is reassessed and adjusted if the patient's clinical picture changes.

ECONOMIC IMPACT OF OSA

A number of studies and reviews have examined the economics of OSA with regard to the cost of untreated OSA and the cost saving that results from OSA treatment. Knauert and colleagues (2015) conducted a review of 106 studies that examined "physiologic, clinical and societal consequences of obstructive sleep apnea syndrome as well as the costs associated with these consequences." This review showed that in terms of healthcare costs, undiagnosed OSA patients cost $1,950 to $3,899 per year more than patients without OSA. Cost is increased even when patients are diagnosed with OSA, but remain untreated. One study estimated that the annual cost of treating the medical consequences of OSA in the U.S. is $3.4 billion (Kapur et al., 1999).

The aforementioned review of 106 studies showed that in terms of healthcare utilization, OSA patients treated with CPAP cost $2,700-$5,200 less per year, compared with OSA patients who are not treated (Knauert et al., 2015). Incremental cost-effectiveness ratio (ICER)—the ratio of the incremental cost and incremental change in quality adjusted life years (QALY) that stems from use of a specified treatment—is a measure of cost effectiveness (Tan et al., 2008). An ICER/QALY value of $50,000 typically is considered to be acceptable or beneficial (Ubel et al., 2003). Economic analyses with ICER/QALY values have shown the cost effectiveness of CPAP therapy in OSA patients. Use of CPAP therapy compared with no treatment in moderate-to-severe OSA patients had an ICER/QALY value of $15,915, indicative of significant cost-effectiveness of CPAP therapy in these patients (Pietzsch et al., 2011).

CONCLUSIONS

OSA is an intriguing, multi-faceted disease process. Craniofacial anatomy and upper airway neuromuscular tone are just two of many factors that contribute to development of OSA in an individual. The consequences of OSA affect not only the patient, but society at large. Untreated OSA patients are at increased risk for a number of cardiovascular comorbidities; risk of stroke, endocrinologic disorders and cancer also increase when OSA is not treated. Patients with untreated OSA may experience excessive daytime sleepiness and have a higher risk of motor vehicle accidents compared with those who do not have OSA.

The bright side is that OSA is a treatable condition. Many treatment options exist for patients with OSA and the list of available therapies will continue to expand as the pathophysiology of OSA is understood better. Effective treatment of OSA can lead to increased quality of life for the affected individual and cost savings for society, which are two compelling reasons to evaluate and treat patients at high risk for having OSA.

REFERENCES

American Academy of Sleep Medicine. http://www.aasmnet.org/resources/pdf/AASMResponseFMCSAFRA.pdf. Accessed May 8, 2017.

Bixler EO, Vgontzas AN, Ten Have T, Tyson K, Kales A. Effects of age on sleep apnea in men: I. Prevalence and severity. Am J Respir Crit Care Med 1998;157(1):144-148.

Chihorek AM, Abou-Khalil B, Malow BA. Obstructive sleep apnea is associated with seizure occurrence in older adults with epilepsy. Neurology 2007;69(19):1823-1827.

Chiu KL, Ryan CM, Shiota S, Ruttanaumpawan P, Arzt M, Haight JS, Chan CT, Floras JS, Bradley TD. Fluid shift by lower body positive pressure increases pharyngeal resistance in healthy subjects. Am J Respir Crit Care Med 2006;174(12):1378-1383.

Chung F, Yang Y, Liao P. Predictive performance of the STOP-Bang score for identifying obstructive sleep apnea in obese patients. Obes Surg 2013;23(12):2050-2057.

Chung F, Yegneswaran B, Liao P, Chung SA, Vairavanathan S, Islam S, Khajehdehi A, Shapiro CM. STOP questionnaire: A tool to screen patients for obstructive sleep apnea. Anesthesiology 2008;108(5):812-821.

Eckert DJ, Malhotra A. Pathophysiology of adult obstructive sleep apnea. Proc Am Thorac Soc 2008;5(2):144-153.

Gami AS, Hodge DO, Herges RM, Olson EJ, Nykodym J, Kara T, Somers VK. Obstructive sleep apnea, obesity, and the risk of incident atrial fibrillation. J Am Coll Cardiol 2007;49(5):565-571.

Gami AS, Somers VK. Sleep apnea and cardiovascular disease. In: Libby P, Bonow RO, Mann DL, Zipes DP, eds. *Braunwald's Heart Disease: A Textbook of Cardiovascular Medicine.* 8th ed. Philadelphia: Saunders 2007.

Kapur V, Blough DK, Sandblom RE, Hert R, de Maine JB, Sullivan SD, Psaty BM. The medical cost of undiagnosed sleep apnea. Sleep 1999;22(6):749-755.

Kapur VK. Obstructive sleep apnea: Diagnosis, epidemiology, and economics. Respir Care 2010;55(9):1155-1167.

Karimi M, Hedner J, Häbel H, Nerman O, Grote L. Sleep apnea-related risk of motor vehicle accidents is reduced by continuous positive airway pressure: Swedish Traffic Accident Registry data. Sleep 2015;38(3): 341-349.

Knauert M, Naik S, Gillespie MB, Kryger M. Clinical consequences and economic costs of untreated obstructive sleep apnea syndrome. World J Otorhinolaryngol Head Neck Surg 2015;1(1):17-27.

Lopez PP, Stefan B, Schulman CI, Byers PM. Prevalence of sleep apnea in morbidly obese patients who presented for weight loss surgery evaluation: More evidence for routine screening for obstructive sleep apnea before weight loss surgery. Am Surg 2008;74(9):834-838.

Marin JM, Carrizo SJ, Vicente E, Agusti AG. Long-term cardiovascular outcomes in men with obstructive sleep apnoea-hypopnoea with or without treatment with continuous positive airway pressure: An observational study. Lancet 2005;365(9464):1046-1053.

Naughton MT. Loop gain in apnea: Gaining control or controlling the gain? Am J Respir Crit Care Med 2010;181(2):103-105.

Park JG, Ramar K, Olson EJ. Updates on definition, consequences, and management of obstructive sleep apnea. Mayo Clin Proc 2011;86(6): 549-555.

Peppard PE, Young T, Palta M, Skatrud J. Prospective study of the association between sleep-disordered breathing and hypertension. N Engl J Med 2000;342(19):1378-1384.

Pietzsch JB, Garner A, Cipriano LE, Linehan JH. An integrated health-economic analysis of diagnostic and therapeutic strategies in the treatment of moderate-to-severe obstructive sleep apnea. Sleep 2011; 34(6):695-709.

Punjabi NM. The epidemiology of adult obstructive sleep apnea. Proc Am Thorac Soc 2008;5(2):136-143.

Schwab RJ, Gupta KB, Gefter WB, Metzger LJ, Hoffman EA, Pack AI. Upper airway and soft tissue anatomy in normal subjects and patients with sleep-disordered breathing: Significance of the lateral pharyngeal walls. Am J Respir Crit Care Med 1995;152(5 Pt 1):1673-1689.

Stoohs RA, Bingham LA, Itoi A, Guilleminault C, Dement WC. Sleep and sleep-disordered breathing in commercial long-haul truck drivers. Chest 1995;107(5):1275-1282.

Tan MC, Ayas NT, Mulgrew A, Cortes L, FitzGerald JM, Fleetham JA, Schulzer M, Ryan CF, Ghaeli R, Cooper P, Marra CA. Cost-effectiveness of continuous positive airway pressure therapy for patients with obstructive sleep apnea-hypopnea in British Columbia. Can Respir J 2008; 15(3):159-165.

Ubel PA, Hirth RA, Chernew ME, Fendrick AM. What is the price of life and why doesn't it increase at the rate of inflation? Arch Intern Med 2003;163(14):1637-1641.

Yaggi HK, Concato J, Kernan WN, Lichtman JH, Brass LM, Mohsenin V. Obstructive sleep apnea as a risk factor for stroke and death. N Engl J Med 2005;353(19):2034-2041.

WHAT'S YOUR NUMBER?
DIAGNOSIS OF OBSTRUCTIVE SLEEP APNEA

Douglas Kirsch

ABSTRACT

The understanding of obstructive sleep apnea (OSA) has advanced significantly over the last several decades, but there still is much to discover. The diagnostic testing for OSA initially was limited to specialized sleep laboratories, but now often is performed in the patient's home with more limited equipment.

The primary severity marker for OSA is a work in progress, given disagreement about the definition of abnormal respiratory events. This chapter will review the recent evolution in testing and diagnostic criteria for OSA, while also forecasting what the near future might bring in terms of OSA testing.

KEY WORDS: polysomnography, home sleep apnea test, obstructive sleep apnea, Apnea-Hypopnea Index, respiratory event index

---

INTRODUCTION

Obstructive sleep apnea (OSA) is a relatively modern diagnosis within the long history of the field of medicine. In past generations, the primary sign of OSA—snoring—was viewed as a sign of a deep and restful sleep. Sleep medicine clinicians currently understand that the loud rumbles emanating from a person's upper airway are not likely to be a sign of high-quality sleep, but a potential harbinger of a significant sleep disorder and a potential long-term health risk. This chapter will review the history of the diagnosis of OSA, the evolution in sleep apnea testing that has occurred in the last few decades and a glimpse of the possible future for evaluation of sleep-disordered breathing (SDB).

THE HISTORY OF POLYSOMNOGRAPHY

Reports of sleep apnea-like phenomena date back to ancient Greece, in which the obese Greek tyrant Dionysius (after nights of merriment)

found himself being poked by needles to restart his breathing while sleeping (Kryger, 1983). A few thousand years later, Charles Dickens created the character Fat Joe in the Posthumous Papers of the Pickwick Club, noting that this corpulent character would fall asleep while chewing food and talking to others, suggestive of symptoms of severe SDB. Even more eloquently, H. V. Morton (1930) described the "Snorer of Kilkenny" thusly:

> *How that man made me suffer. His ghastly organ recital was as regular in its devilish rhythm as a sawmill. Once every half hour he was seized with a kind of convulsion. I hoped that he was dying. The debasing sounds shuddered to a pianissimo and ceased, then he gave a violent gasp, a snort, appeared to be choking, grunted, gasped, and got into top gear again.*

These examples, spread over millennia, are recognition of serious snoring, apneic episodes and significant daytime sleepiness. The physiological evaluation of sleep, however, was stimulated in part by the discovery of Richard Caton, a Scottish scientist, who was the first to record electrical rhythms originating from the brains of rabbits and monkeys (Caton, 1875). It was not until decades later, however, that Hans Berger (1929), a German psychiatrist, was able to record human brain activity, including pattern changes with sleep, with his newly named "electroencephalogram."

The study of sleep on American soil began with two groups, one at Harvard and another at the University of Chicago. Loomis and the Harvard group began to describe different sleep stages (A-E) while studying sleeping patients with electroencephalography (EEG). Loomis (1937), Aserinsky and Kleitman (1953) and other researchers at the University of Chicago took another step forward by studying eye movements during sleep, discovering the presence of rapid eye movements during sleep. Finally, Dement and Kleitman (1957) recognized the patterns of sleep stages, including that of rapid eye movement (REM) sleep and its relationship to dreaming, through the use of entire night recordings with EEG, electromyography (EMG) and electrooculography (EOG). Thus, the full-night sleep recording was born.

Focus on the respiratory aspects of sleep and resultant abnormalities occurred in a similar timeframe as the discovery of sleep stages.

Bickelmann and colleagues (1956) described a patient with obesity, hypoventilation and sleepiness, naming the symptom set "Pickwickian syndrome" after the previously referenced Dickens story. Shortly thereafter, research groups in Europe evaluated patients with Pickwickian syndrome, discovering the presence of apneic spells with continued respiratory effort (Gastaut et al., 1965). After postulating that this finding suggested obstruction at the upper airway, physicians treated these patients with tracheostomy, eliminating the symptoms and the apneas (Kuhlo et al., 1969).

Further research on respiration during sleep led to differentiation between obstructive, mixed and central apneas. Moving beyond respiration, scientists in Europe and the United States published data leading to the addition of cardiac and leg movement leads for a sleep study. Finally, the components were in place for the modern polysomnogram (PSG) that we use clinically today. Figure 1 demonstrates a few epochs of a typical PSG, including sleep staging, respiratory assessment and, in this case, a diagnosis of OSA.

MEASURING OSA

Once OSA was recognized as a disorder of sleep, more clear guidelines for making a diagnosis were required. Early studies of the airway with oral/nasal thermistor and effort bands identified the range of normal and abnormal with clear obstructive apneas. Over time, however, scientific study clarified that airway closure was not an all-or-none phenomenon, but was variable continuously. Partial obstruction of the airway, known as a hypopnea, was demonstrated to be relevant equally as an apnea to sleep disruption.

A hypopnea initially was defined polysomnographically as a 50% decrement in airflow along with some amount of oxygen desaturation; between 2% and 4% desaturations were used in different studies. Other research examined the impact of the respiratory effort-related arousal (RERA), a respiratory event in which smaller limitations in airflow than with a hypopnea were associated with an EEG-based arousal from sleep, leading to sleep disruption and neuro-cognitive effects. Even trying to define the length of an impactful EEG arousal was a scientific process, with a three-second change in EEG found to be most relevant to predicting daytime performance (Stradling and Davies, 2004).

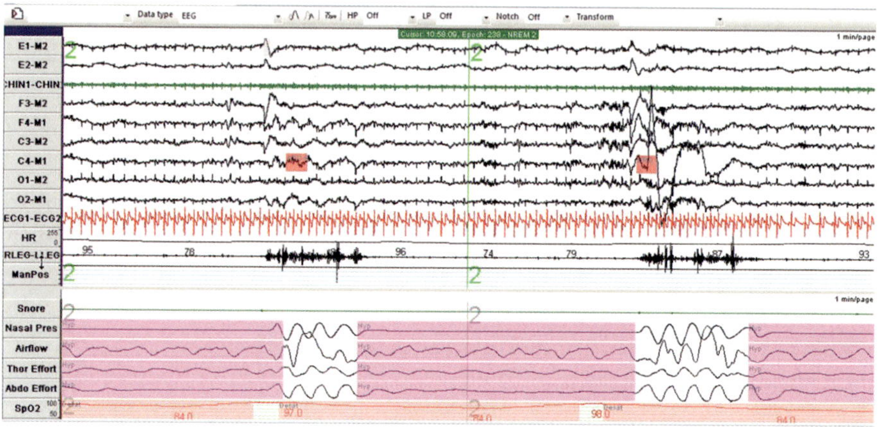

Figure 1. This image is a one-minute epoch of a laboratory-based PSG (Compumedics, Charlotte, NC). The leads on the top half of the image include EOG (E1, E2); chin EMG (Chin), EEG (F3, F4, C3, C4, O1, O2), electrocardiogram (ECG and heart rate) and leg EMG (RLEG-LLEG); Body Position (ManPos). The lower half of the image includes snoring, nasal pressure (Nasal Pres), nasal-oral thermistor (Airflow), thoracic and abdominal effort (Thor effort, Abd effort) and oximetry (SpO2). This epoch demonstrates two epochs of stage N2 sleep with multiple hypopneas (shaded in airflow/effort leads) associated with oxygen desaturations (in SpO2 lead) and arousals (shaded area in EEG leads).

Having multiple respiratory events definitions led to complications in scientific research and clinical activities; an effort then was made to standardize these definitions. Multiple committee and task forces were convened over the course of several years to clarify the definitions of respiratory events. An American Academy of Sleep Medicine (AASM; 1999) consensus report recommended standardized scoring criteria for a range of respiratory events in clinical research (known as the Chicago criteria). This report described two types of hypopneas: 1) respiratory events with a > 50% decrease in a valid measure of airflow without a requirement for associated oxygen desaturation or arousal; and 2) those with a lesser airflow reduction in association with oxygen desaturation of > 3% or an arousal (Sleep, 1999).

Soon thereafter, a clinically-focused paper from the AASM Clinical Practices Review Committee used the hypopnea definition from the Sleep Heart Health Study as an abnormal respiratory event lasting ≥ 10 seconds in length with ≥ 30% reduction in airflow or chest wall movement and with ≥ 4% oxygen desaturation (Meoli et al., 2001). This particular definition has been used by the Centers for Medicare and Medicaid Services (CMS) to determine treatment eligibility for patients with OSA diagnosed according to this definition.

In order to better codify the variety of available scoring rules and improve clinical standardization, a scoring manual was created by the AASM (Iber et al., 2007). The first edition of the Manual for the Scoring of Sleep and Associated Events was published in 2007 with two definitions for hypopneas. The recommended definition in this edition was the same as the CMS definition: hypopnea scoring requires ≥ 30% reduction in nasal pressure signal excursions from baseline and associated ≥ 4% desaturation from pre-event baseline.

The alternative hypopneas definition includes a ≥ 50% reduction in nasal pressure signal excursions and associated ≥ 3% desaturation or arousal. Later editions of the scoring manual have reversed the recommended and alternative definitions; the recommended definition now is hypopneas associated with 3% oxygen desaturation and/or arousal.

Using the different AASM definitions for respiratory events can have dramatic differences on determining OSA severity. For instance, in Ruehland and associates' study (2009), applying the Chicago criteria to patients led to a median Apnea-Hypopnea Index (AHI) of 25.1/hour, 3% oxygen desaturation and/or arousal criteria led to a median AHI of 14.9/hour and 4% oxygen desaturation criteria led to a median AHI of 8.3 events/hour. These variations would lead on average to the same patient being diagnosed with mild OSA using one definition and high-to-moderate OSA with another, thus leading to potentially different treatment algorithms for the same patient. Debate on a "true" definition for hypopneas is ongoing and unlikely to be resolved in the near future.

THE AHI: IMPORTANT, BUT PROBLEMATIC

Once respiratory events were defined more clearly, the AHI became the primary method of assessing OSA severity. The AHI is defined as the total number of apneas added to the total number of hypopneas divided by the total hours of sleep. The AHI often is used to stratify OSA severity (Table 1); severity level impacts the determination of which treatment choices are most appropriate for a given patient. The AHI often is confused with the respiratory disturbance index (RDI), which adds RERAs to the apneas and hypopneas before dividing by

Table 1. OSA severity by AHI criteria.

Severity	Total AHI/hour of sleep
No OSA	0-4.99
Mild OSA	5-14.99
Moderate OSA	15-29.99
Severe OSA	≥ 30

total sleep hours. Another potential metric used to evaluate SDB is the oxygen desaturation index (ODI), which is the total number of oxygen desaturations divided by the hours of sleep.

The reason for the predominant use of the AHI in current clinical practice is that crucial research comprising the backbone of clinical sleep medicine used the AHI as the primary determinant. In the 1993 Wisconsin Sleep Cohort epidemiological study, the AHI was the primary metric used to define the prevalence of OSA; this large cohort study found in middle-aged adults that 4% of men and 2% of women had an AHI > 5 along with excessive daytime sleepiness. Sleepiness is one of the most frequent symptoms of OSA; for instance, patients with OSA have as much as a 5x higher risk of motor vehicle crash than control subjects (Tregear et al., 2009).

When evaluating the health risks to patients who suffer from OSA, the AHI typically has been used as the severity determinant. Multiple papers have demonstrated that AHI-determined OSA leads to hypertension (HTN) and that treatment of OSA will improve blood pressure. Epidemiological studies have linked OSA (based on AHI) with myocardial infarction, congestive heart failure and stroke (Gottlieb et al., 2010). However, a single definition for the respiratory events comprising the AHI may not elicit all the possible disease associations, as OSA may have different effects on different organ systems. For instance, 4% oxygen desaturations better predict hypertension, while 2% oxygen desaturations better predict fasting hyperglycemia and the arousal frequency predicts impaired memory consolidation (Jordan et al., 2014).

Having a mostly formalized definition for the AHI has some advantages in that it encapsulates a complex disorder into a single metric. However, as noted above, there are aspects of the AHI that raise concern; in particular, the AHI may not tell the entire story about OSA severity.

Questions about the metric are encapsulated in Table 2 and will be discussed in further detail.

First, while research suggests that apneas and hypopneas are similar in physiological terms, it does not define them necessarily as biologically equivalent or even that all hypopneas have the same biological impact. This non-equivalence is true, particularly when a scored hypopnea is associated with an arousal rather than an oxygen desaturation. While that arousal from sleep may have downstream impact on quality of life, wellbeing and medical conditions, it likely has a different impact than those events associated with oxygen desaturation, whether that desaturation is 3%, 4% or other. Thus, the AHI—regardless of the definition used—is unlikely to encapsulate completely the variety of effects that OSA causes because of the variety of events that comprise the metric.

Second, respiratory events—whether apneas or hypopneas—are associated with different levels of oxygen desaturation. Thus, a hypopnea with a 4% desaturation is considered of equivalent severity to that associated with a 10% desaturation, though knowledge of the human body's response to hypoxia would suggest that these hypopneas likely have a differential effect. While some studies suggest that increasing depth of oxygen desaturation during sleep has a negative impact on health and increases mortality (**Muraja-Murro et al., 2013)**, there are limited data to inform whether depth or duration of oxygen desaturation may lead to an increased impact.

Third, by definition, all respiratory events must occur for at least ten seconds. However, does a ten-second hypopnea have a similar effect to a 60-second apnea? And could a shorter than ten seconds respiratory event have a health-related impact? In children, respiratory events of shorter length (two-breath minimum) are deemed to be important physiologically; however, currently only longer (more than ten seconds) respiratory events have been defined as relevant in adults.

Table 2. Questions about the AHI.

1. Are apneas and hypopneas equivalent?
2. Does the depth of oxygen desaturation associated with each individual respiratory event matter?
3. Does the temporal length of a respiratory event matter?
4. Does the temporal dispersion of the events during the night matter?

Fourth, respiratory events typically may not distribute evenly over the course of a sleep study. In some cases, respiratory events may occur on a positional basis (e.g., supine *versus* non-supine) or may occur on a sleep stage basis (e.g., REM sleep *versus* non-REM sleep *versus* during wake-sleep transitions). Thus, we can compare two patients who have the same overall AHI. Patient 1 has a high supine-position AHI, but only sleeps on his/her back for an hour during a PSG. Patient 2 has a high AHI only during REM sleep, but limited respiratory events during NREM sleep. Technically these patients have an equivalent severity of apnea based on their total AHI, though whether they truly are equivalent disease severity or are likely to have dissimilar symptoms or disease-specific outcomes currently is unclear.

BEYOND THE LABORATORY-BASED PSG

For many years, patients questioned whether their sleep tested in a sleep laboratory was reflective of their typical sleep at home. With the advent of home sleep apnea testing (HSAT), patients now are able to be tested in their usual sleeping environment. The home testing devices are designed to be used by a patient at nighttime without significant technical assistance to evaluate them for OSA.

The unattended nature of the test is one of several differences between testing with HSAT devices and in-laboratory PSG. Other significant differences include: the absence of measurement of the patient's total sleep time and EEG-related arousals from sleep from most HSAT devices; HSAT devices do not assess appropriately for disorders other than OSA; and there is a lower cost of testing with use of a HSAT device compared to in-laboratory PSG. Table 3 compares and contrasts HSAT devices and in-laboratory PSG.

Though HSATs have been available for decades, the common use of these devices was discouraged due to insufficient data regarding their use in clinical practice. As research study publications increased, national medical societies and the CMS re-evaluated and eventually approved use of HSATs for diagnosis of OSA in 2007-2008. The AASM released *Clinical Guidelines for the Use of Unattended Portable Monitors in the Diagnosis of Obstructive Sleep Apnea in Adult Patients* (Collop et al., 2007).

Based on the 2007 AASM guidelines, the HSAT devices should be used in patients who have a high pre-test probability of moderate to

Table 3. Differences between HSAT devices and in-laboratory PSG.

Differences	HSAT device	In-lab PSG
Location	Home	Sleep laboratory
Attended with a technologist	No	Yes
Measures sleep time	Rarely	Always
Typical measurement leads #	3-4	16-20
Measures EEG arousals	No	Yes
Measures EEG-based sleep time	No	Yes
Can diagnose	OSA, occasionally CSA	Many sleep disorders
Expense per test	Low ($200-$400)	High ($800 to multiple thousands)
Population	Adults only	Children and adults
Suggested restrictions	Patients without serious co-morbid medical or sleep disorders	None

severe OSA. HSAT use in this population improves the specificity and sensitivity of the test. The guidelines also suggest that the devices not be used in patients with significant sleep co-morbidities (e.g., periodic limb movements of sleep, insomnia, parasomnias) and severe medical co-morbidities that may affect breathing (e.g., chronic obstructive pulmonary disease, congestive heart failure, neuromuscular conditions). On a practical level, however, local insurance regulations may affect the strict application of these guidelines to clinical care.

The use of HSATs now applies a potentially different structure to the assessment of respiratory event frequency. While laboratory testing includes both nasal pressure and oronasal thermistor leads, HSATs generally include one of those two leads, typically nasal pressure. EEG-staged sleep also is not measured on most HSAT devices, thus the breathing index is determined by the denominator of total recording time, rather than total sleep time. This change may lead to underestimation of OSA severity, given that most patients do not sleep for the entire recording of the home test.

The absence of EEG leads on the HSAT also means that arousals from sleep cannot be scored. The fallout from this absence is that only the oxygen desaturation hypopnea criteria can be utilized. Thus, the AASM has opted to call a HSAT breathing index the "respiratory event index" or "REI." Though similar, the REI truly is not synonymous with the AHI, thus further muddying the already cloudy waters of OSA severity. A sample image of a HSAT from a commonly used device can be seen in Figure 2.

SMARTPHONES AND BEYOND

Less data is available about Smartphone apps and other consumer devices that evaluate SDB. Fitbits™ and other personal trackers evaluate sleep *via* privately designed algorithms that have mixed data when compared to PSG (Mantua et al., 2016). To this point, however, while the devices may detect movement-based arousals, they do not have a clear method for assessing respiratory-based sleep disorders.

Smartphone application is another method of assessment of sleep and SDB breathing at home. These applications typically use gyroscopic movement and noises to determine sleep quality. Limited data exists comparing these applications to PSG, but what does exist suggests that the sensitivity and specificity of the applications may not be high (Bhat et al., 2015). A sample patient image from the Smart Alarm application can be seen in Figure 3, demonstrating reported depth of sleep and areas of user re-playable noise during the night (at the circles).

However, better data exist regarding Smartphone assessment of sleep-related breathing. Researchers in Japan were able to use the microphone from a Sharp Smartphone attached to the sternum to estimate a research subject's AHI compared to a PSG with a high level of correlation (Nakano et al., 2014). There also has been a growth in the area of snoring applications in which the Smartphone microphone is used to record and quantify snoring sounds. A recent review rated fourteen snoring applications, selecting one for a detailed analysis (Camacho et al., 2015). This application was compared to PSG and gathered snoring data that was significantly similar. In 2015, an abstract was published in the journal, Sleep, demonstrating that ApneaApp, a Smartphone application from the University of Washington, correlated well with the AHI from a PSG in a contactless manner (Nandakumar et al., 2015). Identification of OSA

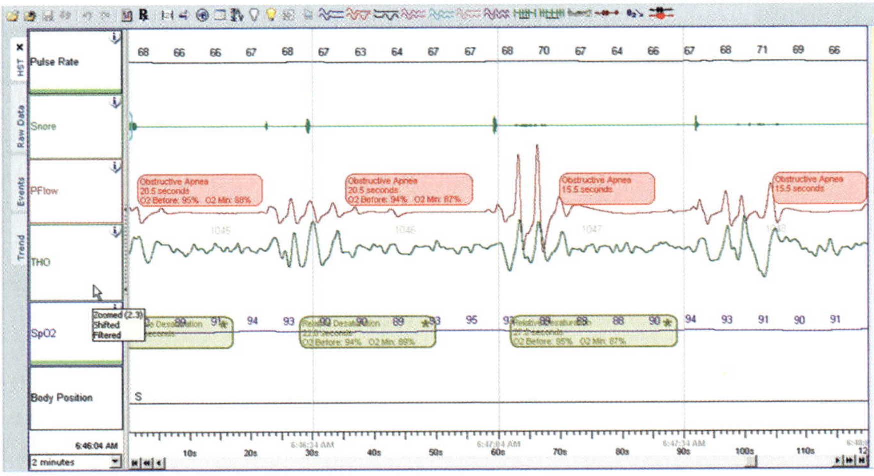

Figure 2. This two-minute epoch from a Philips Alice Night One HSAT demonstrates OSA (Philips Respironics, Pittsburgh, PA). From top to bottom, the leads are: pulse rate, snoring, nasal pressure (PFlow), thoracic effort (THO), oxygen saturation (SpO2) and body position (S=supine). The red shaded images in the nasal pressure signal demonstrate the absence of air flow in the presence of continued effort and are obstructive apneas. Oxygen desaturations associated with the respiratory events are seen in the tan shaded areas in the SpO2 lead.

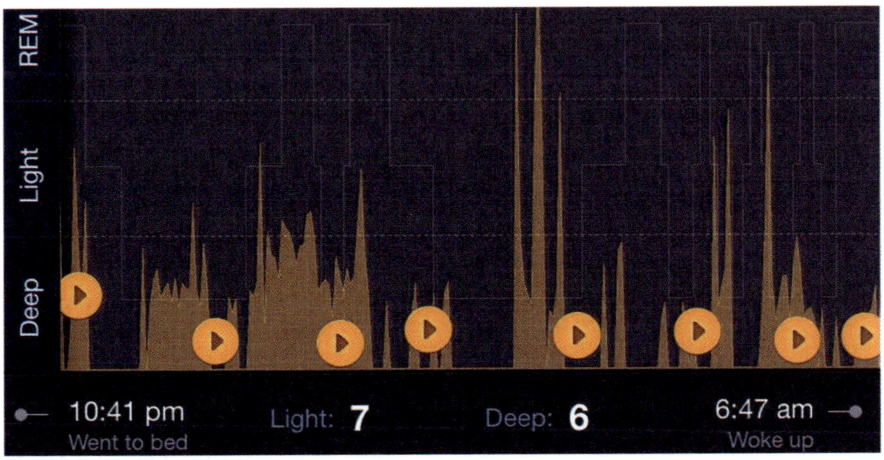

Figure 3. This image demonstrates overnight sleep tracking from the Smartphone application Smart Alarm Clock. Sleep levels move from deep when the orange line is the lower level to light sleep to REM sleep at the top of the image. The circles with the triangles are periods in which enough noise is audible to record and can be played back later; these areas represent snoring in some cases.

via Smartphone appears to be moving forward quickly, using similar metrics to those being used currently in the laboratory.

FUTURE THOUGHTS

Though the AHI has much data supporting its use, other metrics have been proposed. Some of the areas that are being explored actively are shown in Table 4. These metrics take into account some of the issues raised with the AHI above. At this point, these options are theoretical constructs; little clinical data are able to support the use of these metrics.

In a paper evaluating interactions between event duration and oxygen desaturation, however, it is clear that two patients with similar AHIs can have different findings when looking at these novel metrics (Kulkas, 2013). These differences are just one of several possible explanations for why patients with different AHIs have different clinical symptoms and health outcomes. Another study evaluated patients with mild-to-moderate OSA retrospectively, using the integrated area of desaturation curve (IAD) to re-evaluate them. Patients with high IADs were more likely to have had a cardiovascular event compared to those with lower IADs, when AHI alone did not distinguish these groups (Asano et al., 2009). Punjabi (2016) remarks about the breadth of available data on the sleep study in a recent article:

> *The field of sleep disorders medicine should not settle for taking a large series of time varying signals from the PSG that often exceed one gigabyte of data and just use one metric to summarize all key features. Certainly, a "one size fits all" solution will never exist that can embody the physiologic diversity of apneas and hypopneas.*

Table 4. Potential new metrics in assessment of OSA.

Potential new metrics
Quantification of depth of desaturation
Interaction of event length and desaturation
Heart-rate variability
Sleep stage transitions
Quantifying amount of flow-limited breathing

CONCLUSIONS

The young field of sleep medicine has come a long way in a short time, medically speaking. As sleep medicine continues to evolve in ways both predictable and not, practitioners should embrace the idea that while our current metrics have value based on our history, new metrics may evaluate OSA better severity by linking more strongly with patient symptoms and important clinical outcomes. This knowledge should aid in the shared decision-making processes around treatments and lead to better overall patient health outcomes.

REFERENCES

Asano K, Takata Y, Usui Y, Shiina K, Hashimura Y, Kato K, Saruhara H, Yamashina A. New index for analysis of polysomnography, 'integrated area of desaturation,' is associated with high cardiovascular risk in patients with mild to moderate obstructive sleep apnea. Respiration 2009;78(3):278-284.

Aserinsky E, Kleitman N. Regularly occurring periods of eye motility, and concomitant phenomena, during sleep. Science 1953;118(3062):273-274.

Berger H. Ueber das Elektroenkephalogramm des Menschen [On the human electroencephalogram]. Archiv Psychiatr Nervenkr 1929;87:527-570. [In German.]

Bhat S, Ferraris A, Gupta D, Mozafarian M, DeBari VA, Gushway-Henry N, Gowda SP, Polos PG, Rubinstein M, Seidu H, Chokroverty S. Is there a clinical role for Smartphone sleep apps? Comparison of sleep cycle detection by a Smartphone application to polysomnography. J Clin Sleep Med 2015;11(7):709-715.

Bickelmann AG, Burwell CS, Robin ED, Whaley RD. Extreme obesity associated with alveolar hypoventilation: A Pickwickian syndrome. Am J Med 1956;21(5):811-818.

Camacho M, Robertson M, Abdullatif J, Certal V, Kram YA, Ruoff CM, Brietzke SE, Capasso R. Smartphone apps or snoring. J Laryngol Otol 2015;129(10):974-979.

Caton R. The electric currents of the brain. Br Med J 1875;2:278.

Collop NA, Anderson WM, Boehlecke B, Claman D, Goldberg R, Gottlieb DJ, Hudgel D, Sateia M, Schwab R; Portable Monitoring Task Force of the American Academy of Sleep Medicine. Clinical guidelines for the use of unattended portable monitors in the diagnosis of obstructive sleep apnea in adult patients. J Clin Sleep Med 2007;3(7):737-747.

Dement W, Kleitman N. Cyclic variations in EEG during sleep and their relation to eye movements, body motility, and dreaming. Electroencephalogr Clin Neurophysiol 1957;9(4):673-690.

Gastaut H, Tassinari CA, Duron B. [Polygraphic study of diurnal and nocturnal (hypnic and respiratory) episodal manifestations of Pickwick syndrome]. Rev Neurol (Paris) 1965;112(6):568-579. [In French.]

Gottlieb DJ, Yenokyan G, Newman AB, O'Connor GT, Punjabi NM, Quan SF, Redline S, Resnick HE, Tong EK, Diener-West M, Shahar E. Prospective study of obstructive sleep apnea and incident coronary heart disease and heart failure: The sleep heart health study. Circulation 2010;122(4):352-360.

Iber C, Ancoli-Israel S, Chesson AL Jr, Quan S; for the American Academy of Sleep Medicine. *The AASM Manual for the Scoring of Sleep and Associated Events: Rules, Terminology and Technical Specifications.* 1st ed. Westchester, IL: American Academy of Sleep Medicine 2007.

Jordan AS, McSharry DG, Malhotra A. Adult obstructive sleep apnoea. Lancet 2014;383(9918):736-747.

Kryger MH. Sleep apnea: From the needles of Dionysius to continuous positive airway pressure. Arch Intern Med 1983;143(12):2301-2303.

Kuhlo W, Doll E, Franck MC. [Successful management of Pickwickian syndrome using long-term tracheostomy]. Dtsch Med Wochenschr 1969;94(24):1286-1290. [In German.]

Kulkas A, Tiihonen P, Eskola K, Julkunen P, Mervaala E, Töyräs J. Novel parameters for evaluating severity of sleep disordered breathing and for supporting diagnosis of sleep apnea-hypopnea syndrome. J Med Eng Technol 2013;37(2):135-143.

Loomis AL, Harvey EN, Hobart GA. Cerebral states during sleep, as studied by human brain potentials. J Exp Psychol 1937;21(2):127-144.

Mantua J, Gravel N, Spencer RM. Reliability of sleep measures from four personal health monitoring devices compared to research-based actigraphy and polysomnography. Sensors (Basel) 2016;16(5).

Meoli AL, Casey KR, Clark RW, Coleman JA Jr, Fayle RW, Troell RJ, Iber C; Clinical Practice Review Committee. Hypopnea in sleep-disordered breathing in adults. Sleep 2001;24(4):469-470.

Morton HV. *The Search for Ireland.* Methuen, London:1930;66-67.

Muraja-Murro A, Kulkas A, Hiltunen M, Kupari S, Hukkanen T, Tiihonen P, Mervaala E, Töyräs J. The severity of individual obstruction events is related to increased mortality rate in severe obstructive sleep apnea. J Sleep Res 2013;22(6):663-669.

Nakano H, Hirayama K, Sadamitsu Y, Toshimitsu A, Fujita H, Shin S, Tanigawa T. Monitoring sound to quantify snoring and sleep apnea severity using a Smartphone: Proof of concept. J Clin Sleep Med 2014;10(1): 73-78.

Nandakumar R, Gollakota S, Watson NF. A contactless system for respiratory event identification. Sleep 2015;38:A163 (abstract supplement).

Punjabi NM. Counterpoint: Is the apnea-hypopnea index the best way to quantify the severity of sleep disordered breathing? No. Chest 2016; 149(1):16-19.

Ruehland WR, Rochford PD, O'Donoghue FJ, Pierce RJ, Singh P, Thornton AT. The new AASM criteria for scoring hypopneas: Impact on the apnea hypopnea index. Sleep 2009;32(2):150-157.

Sleep-related breathing disorders in adults: Recommendations for syndrome definition and measurement techniques in clinical research. The Report of an American Academy of Sleep Medicine Task Force. (No authors listed) Sleep 1999;22(5):667-689.

Stradling JR, Davies RJ. Sleep. 1: Obstructive sleep apnoea/hypopnoea syndrome: Definitions, epidemiology, and natural history. Thorax 2004;59(1):73-78.

Tregear S, Reston J, Schoelles K, Phillips B. Obstructive sleep apnea and risk of motor vehicle crash: Systematic review and meta-analysis. J Clin Sleep Med 2009;5(6):573-581.

CPAP IN THE TREATMENT OF SLEEP APNEA

John C. Carter and Vishesh K. Kapur

ABSTRACT

Continuous positive airway pressure (CPAP) is a first-line therapy for adult patients with obstructive sleep apnea (OSA), particularly those with daytime symptoms. CPAP effectively resolves upper airway obstruction and is associated with improvements in daytime sleepiness and hypertension in patients with OSA. Adherence to CPAP is a key consideration in the effectiveness of this therapy and remains suboptimal for many patients, despite technological advancements in CPAP aimed at improving patient comfort. Educational and behavioral interventions have been associated with improvements in adherence.

KEY WORDS: CPAP, sleep apnea, effectiveness, efficacy, adherence

---

INTRODUCTION

Continuous positive airway pressure (CPAP) is the most efficacious treatment for obstructive sleep apnea (OSA) and, therefore, the most frequently prescribed. Regardless of which of several potential pathogenic causes is at play, application of positive pressure inside the upper airway typically relieves the upper airway obstruction that defines the disorder. Since the discovery of CPAP, a number of technological innovations have improved its utility. Despite these advancements, suboptimal adherence remains a factor that limits effectiveness in a significant proportion of patients who are prescribed CPAP. The scientific literature regarding the effectiveness of CPAP reflects the interplay of the application of a highly efficacious therapy with suboptimal adherence. Research also reveals avenues to realize the potential of CPAP more fully by improving adherence.

HISTORY OF CPAP

CPAP was used to treat OSA first in 1980 by the Australian physician Colin Sullivan (Sullivan et al., 1981). The first successful demonstration occurred in a patient with severe sleep apnea who had refused placement of a tracheostomy. This first CPAP unit consisted of little more than a commercially available blower fan with an attached pressure regulator that delivered positive air pressure to the patient *via* a jury-rigged nasal mask fashioned with rapid-setting silicone sealant. As the air pressure delivered was increased, the patient's repetitive apneas quickly gave way to normal respiratory flow patterns. When pressure was decreased, apneas recurred. A polysomnographic recording indicated a fairly rapid entry into REM sleep with seven subsequent hours of stable sleep. The next morning, the patient reported feeling more awake and alert than he had in years (Sullivan et al., 1981).

MECHANISMS OF ACTION

The mechanism by which CPAP treats OSA is straightforward: positive pressure generated by an air pump is transmitted to the upper airway to maintain a constant intraluminal positive pressure sufficient to prevent airway collapse (Strohl and Redline, 1986). CPAP is highly efficacious; if positive airway pressure (PAP) is applied at the level required to overcome the forces favoring airway collapse, OSA typically is resolved. This phenomenon is documented regularly in patients undergoing manual CPAP titration in the sleep laboratory (Sanders et al., 1993; Kushida et al., 2006; Fig. 1). CPAP may not be efficacious fully in a small minority of OSA patients for whom instability of breathing control is a prominent factor; these patients may develop central apneas with application of CPAP (Javaheri et al., 2009).

CPAP has another important physiological effect, namely an increase in intra-thoracic pressure that has effects on the heart. Positive pressure surrounding the heart impacts transluminal pressures in the cardiac chambers, which results in reduced pre- and after-load (Gay et al., 2006; Sun et al., 2013). This effect can have a positive or negative impact on cardiac output in patients with significant heart failure depending on volume status (Kaneko et al., 2003; Mansfield et al., 2004; Cowie et al., 2013; Fig. 2).

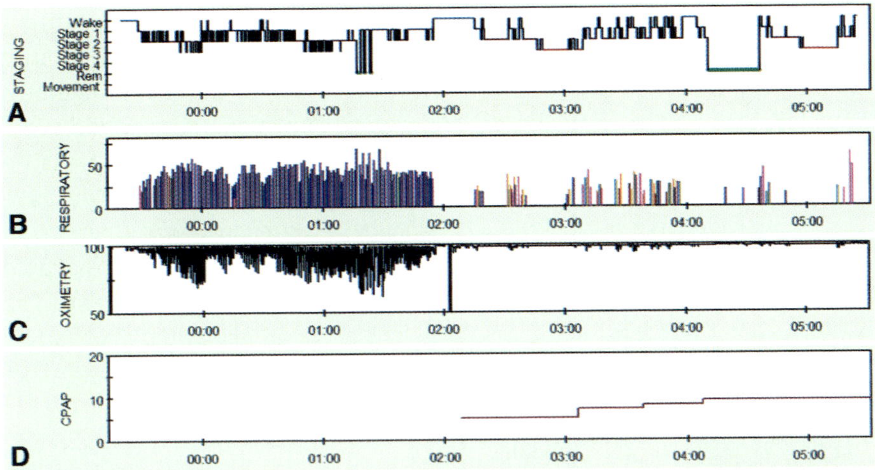

Figure 1. Split-night polysomnogram (PSG) showing response of OSA to CPAP. *A:* Sleep architecture and stages. *B:* Respiratory events including obstructive apneas and hypopneas as well as central apneas. *C:* Oxyhemoglobin saturations measured by oximetry. *D:* CPAP delivered to the patient. Note fragmentation in sleep architecture (*A*) prior to the initiation of CPAP, accompanied by frequent respiratory events and severe oxyhemoglobin desaturations that improve after CPAP is applied beginning at 5 cmH2O and titrating up to 10 cmH2O.

ADVANCEMENTS IN PAP TECHNOLOGY

PAP technology has advanced significantly over time in several respects. Modern devices are compact, quiet and typically incorporate humidification and monitoring capability that can be accessed remotely by care providers. In addition, devices have been developed that deliver variable pressure during individual respiratory cycles, as well as over successive respiratory cycles (Johnson and Johnson, 2015).

Bi-level PAP therapy provides two different levels of PAP: a lower level during expiration (EPAP) and a higher one during inspiration (IPAP; Fig. 3). The use of bi-level PAP for the treatment of OSA was described first by Sanders and Kern (1990) and currently is used in several circumstances. First, when high PAP settings are required for OSA therapy,

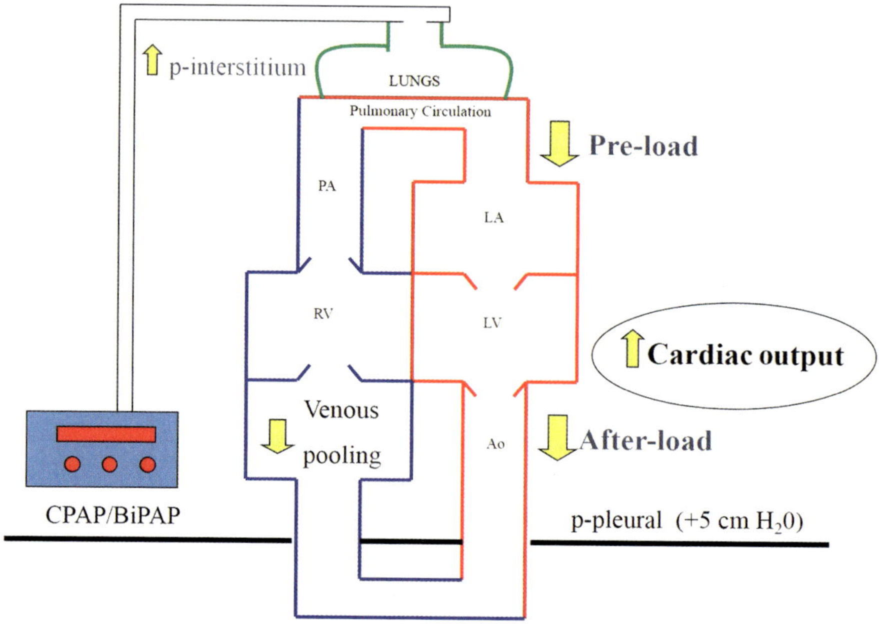

Figure 2. Cardiovascular consequences of positive pressure breathing in heart failure (HF). The application of CPAP increases intra-thoracic pressure, thereby reducing the pressure gradient across the cardiac wall. This increased pressure reduces cardiac pre- and after-load that may increase cardiac output in some HF patients. Ao = aorta; LA = left atrium; LV = left ventricle; RV = right ventricle; PA = pulmonary artery. Image courtesy of William DePaso; reproduced with permission.

bi-level PAP may relieve the sensation of difficulty with exhalation (Reeves-Hoche et al., 1995; Powell et al., 2012; Carlucci et al., 2015).

Second, in the case of hypoventilation, bi-level PAP provides ventilatory support by way of additional pressure (pressure support) delivered during inspiration relative to expiration. When used to treat hypoventilation, the EPAP is used to maintain patency of the airway, while the IPAP is set to maintain adequate ventilation. Bi-level devices also may include a backup rate, which prompts the device to provide inspiratory pressure automatically if a sufficient inspiratory effort is not generated by the patient to trigger the device to do so.

Conditions in which bi-level PAP with a back-up rate is indicated include central sleep apnea (CSA) when no inspiratory effort is made and severe respiratory muscle weakness when adequate inspiratory effort to

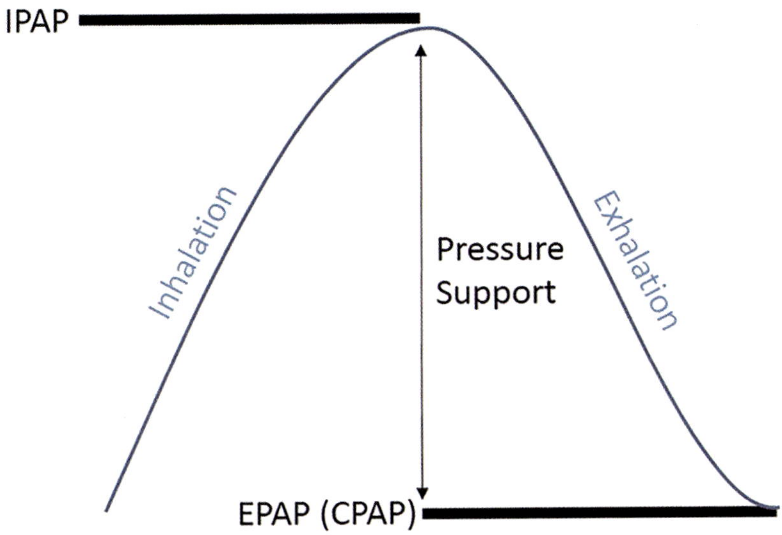

Figure 3. Schematic representing the pressure cycle provided a bi-level PAP device. EPAP maintains airway patency at the end of exhalation. IPAP provides additional pressure during inhalation, allowing for greater lung expansion and ventilation. The pressure differential between IPAP and EPAP is called *pressure support*.

trigger bi-level PAP may not occur consistently (Loube et al., 1999; Epstein et al., 2009). More advanced bi-level PAP devices have been developed that can adjust inspiratory and/or expiratory pressures during the night to maintain minimum levels of ventilation (for hypoventilation) or to limit variability of ventilation (for CSA).

Auto-titrating positive airway pressure (APAP) is designed to adjust CPAP settings continuously over the sleep period to provide the minimum pressure needed to relieve OSA (Roux and Hilbert, 2003). Proprietary algorithms adjust pressure on a continuous basis. The algorithms function based upon two basic mechanisms of action: 1) episodically monitor markers of airway patency; and 2) adjust pressure to achieve desired airway patency at the lowest pressure within pre-set pressure boundaries. The proprietary algorithms differ in terms of their aggressiveness in changing pressures and the breathing-related parameters used to monitor adequacy of pressures.

Potential advantages of APAP over CPAP include the ability to initiate PAP therapy without a laboratory-based CPAP titration study and minimization of PAP delivered to relieve OSA (Ayas et al., 2004; Aloia et al., 2005; Morgenthaler et al., 2008). By minimizing PAP pressures, pressure-related side effects of PAP (pressure intolerance, mask leak and aerophagia—excessive swallowing of air that goes to the stomach) may be ameliorated, leading to improved patient comfort. A potential disadvantage of APAP is failure to provide appropriate pressures (either too high/low) to relieve OSA (Ip et al., 2012), due to algorithmic miscalculation or mask/oral leak.

Expiratory pressure relief (EPR) is a technology that varies pressure delivered over the respiratory cycle (inspiration and expiration). Lower PAP is needed to maintain airway patency at the start of exhalation because lung volumes are larger at end-inhalation (which stiffens the upper airway *via* tracheal tug) and intra-luminal airway pressures are positive during exhalation. When pressure relief is activated, PAP is reduced during the beginning of exhalation in a manner that ideally should not affect efficacy.

Reduction of pressure at the start of exhalation can improve comfort for some patients. The risk associated with EPR use is that efficacy may be reduced unintentionally (Aloia et al., 2005; Bakker and Marshall, 2011). Most CPAP devices also incorporate a ramp function that can be set to start at a more comfortable sub-therapeutic pressure and increase over a variable period of time (10 to 45 minutes) to the optimum pressure. This is useful for patients who find the optimum pressure difficult to tolerate while falling asleep.

Modern CPAP devices provide the clinician and respiratory technologist with data on usage, pressure delivery, leak and efficacy. Such data now are available *via* remote download, streamlining the process of following up on CPAP therapy. Device settings can be viewed and modified remotely. This remote management represents a significant step forward in the convenience of managing patients on PAP therapy.

MASKS

CPAP interfaces have evolved significantly since the first makeshift one used by Dr. Sullivan. This is fortunate since the interface most acutely impacts patient comfort and tolerance, as it is the equipment that directly contacts the patient. With over 150 varieties of interfaces currently on the market, users of PAP have an abundance of choice.

Interfaces can be classified by type, the most common being nasal masks, intra-nasal pillows and full-face masks. Evaluation of a patient's facial structure—including shape and size of the nose, lateral facial profile, presence of facial hair and presence of any anatomic irregularities—are critical to choosing a suitable interface (Borel et al., 2013).

Generally, a nasal interface is the first choice due to advantages with respect to comfort, number of available options and leak and other complications relative to the full-face mask design. Oral leak (escape of air flow out of the mouth) is a common issue, which can be addressed with the addition of a chinstrap for some patients, but may require use of a full-face mask that also covers the mouth.

EFFECTIVENESS

CPAP is a highly efficacious therapy in that application of PAP at appropriate levels results in resolution of the hallmark of OSA, namely upper airway obstruction in the vast majority of patients. On the other hand, effectiveness—defined as the ability of CPAP to relieve the symptoms and comorbidity of OSA in clinical practice—is dependent on additional factors, most importantly adherence to daily use over the entire sleep time. There are varying levels of confidence in the effectiveness of CPAP to ameliorate the specific patient outcomes associated with OSA based on the findings and robustness of the studies that have been performed. For example, there is a high level of confidence that CPAP can improve sleepiness and reduce blood pressure. On the other hand, the evidence regarding long-term outcomes (e.g., incident cardiovascular events and mortality) is less clear.

There is strong evidence supporting the effectiveness of CPAP in improving sleepiness related to OSA (Douglas, 1998; Douglas and Engleman, 1998; Engleman et al., 1998; Ballester et al., 1999; Jenkinson et al., 1999; Patel et al., 2003; Engleman and Douglas, 2004; Kushida et al., 2012). There are at least 36 randomized controlled trials (RCTs) that have compared the effect of CPAP in reducing subjective sleepiness, as measured by a commonly used measure, the Epworth Sleepiness Scale (ESS; Johns, 1991) to placebo.

The ESS is an instrument that queries how likely someone is to fall asleep (on a spectrum of "would never doze" to "high chance of dozing") in eight inactive situations (Fig. 4). Scores can range from 0 (never doze in all scenarios) to 24 (high chance of dozing in all scenarios) with levels greater than 10 interpreted as consistent with excessive sleepiness. With CPAP treatment for OSA, the mean reduction of ESS across studies is approximately 2.4 units, with reduction seen related to whether the study was performed in subject with elevated ESS at baseline (mean reduction = 2.8) or non-sleepy subjects (mean reduction = 1.1; Engleman et al., 1997, 1998, 1999; Ballester et al., 1999; Jenkinson et al., 1999; Hack et al., 2000; Faccenda et al., 2001; McArdle and Douglas, 2001; Monasterio et al., 2001; Montserrat et al., 2001; Barnes et al., 2002, 2004; Becker et al., 2003; Woodson et al., 2003; Hui et al., 2006; Coughlin et al., 2007; Lam et al., 2007; West et al., 2007; Kohler et al., 2008; Siccoli et al., 2008; Duran-Cantolla et al., 2010; Phillips et al., 2011; Huang et al., 2012; Kushida et al., 2012; Weaver et al., 2012; Martinez-Garcia et al., 2013; McMillan et al., 2014; Dalmases et al., 2015; McEvoy et al., 2016). Similarly, there are a number of RCTs (at least eight) that demonstrate an average reduction of 2.5 mmHg in mean blood pressure with higher magnitude of decrease seen in studies that included patients with more severe OSA and hypertension (mean reduction 10 mmHg; Becker et al., 2003).

There is a contradiction between the observational studies and the RCTs that have evaluated the impact of CPAP on incident cardiovascular events and/or mortality. Non-RCT data across thirteen studies showed a 67% average reduction in odds of incident events in patients using CPAP (Marti et al., 2002; Kanagala et al., 2003; Milleron et al., 2004; Doherty et al., 2005; Marin et al., 2005; Buchner et al., 2007; Cassar et al., 2007; Wang et al., 2007; Kasai et al., 2008; Abe et al., 2010; Campos-Rodriguez et al., 2014; Holmqvist et al., 2015). In contrast, RCT data from four studies suggested no reduction in incident cardiovascular events (Barbe et al., 2012; Parra et al., 2015; McEvoy et al., 2016; Peker et al., 2016). Details from specific RCT and non-RCT studies potentially reveal why findings have been discrepant.

Marin and colleagues (2005) published observational data comparing fatal and non-fatal cardiovascular events (myocardial infarction, stroke, coronary revascularization procedure) over a ten-year follow-up period in male patients with OSA seen in a Spanish sleep clinic. The study found that CPAP use was associated with a lower risk of both types of events (Fig. 5). The cumulative incidence on non-fatal events (> 30% *versus* < 10%) and fatal events (> 15% *versus* < 5%) was higher significantly in untreated severe OSA patients than in CPAP users. Patients with

Epworth Sleepiness Scale	Chance of dozing (0-3)			
Situation	0	1	2	3
Sitting and reading				
Watching television				
Sitting inactive in a public place - for example, a theater or meeting				
As a passenger in a car for an hour without a break				
Lying down to rest in the afternoon				
Sitting and talking to someone				
Sitting quietly after lunch (when you've had no alcohol)				
In a car, while stopped in traffic				

Figure 4. The Epworth Sleepiness Scale (ESS; Johns, 1991) is a validated measure of sleepiness, assessing the likelihood of dozing in various sedentary situations. Scores range from 0 to 24; a score of 11 or higher is considered indicative of clinically significant sleepiness. Reprinted with permission from Mapi Research Trust, Lyon, France: https://eprovide.mapi-trust.org.

treated severe OSA had higher risk of both cardiovascular death and non-fatal cardiovascular events after adjustment for several cardiovascular risk factors.

The major limitation of this study is potential bias introduced by non-random allocation of patients to CPAP *versus* untreated severe OSA groups. Those in the untreated group consisted predominantly of patients unable to achieve adequate adherence (> four hours/day needed to maintain CPAP prescription) or those who refused CPAP therapy.

The Sleep Apnea Cardiovascular Endpoints (SAVE) Study (McEvoy et al., 2016) is a multi-center RCT that was performed to determine the effects of CPAP on cardiovascular disease in high-risk patients with moderate to severe OSA. In this study, 2,717 subjects with prior coronary artery disease or cerebrovascular disease, but without severe daytime sleepiness (defined as ESS > 15) or severe sleep-related hypoxemia were

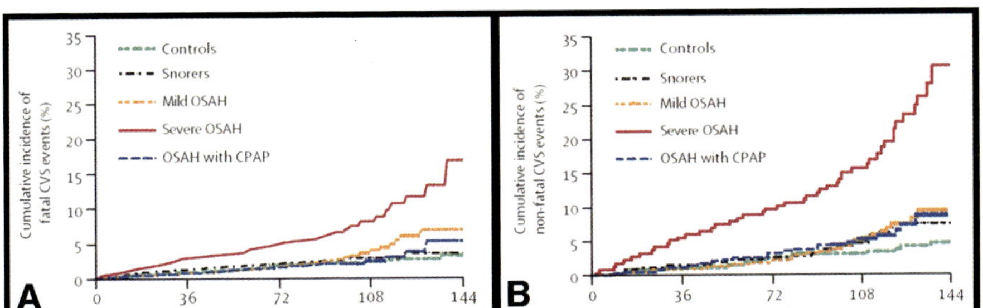

Figure 5. Cumulative percentages of men with new fatal (*A*) and non-fatal (*B*) cardiovascular events. From Marin et al., 2005; reprinted with permission from Elsevier.

assigned randomly to CPAP *versus* usual care and followed for a mean period of 3.7 years to measure the occurrence of the composite primary outcome: cardiovascular death, myocardial infarction (MI), stroke or hospitalization for heart failure, acute coronary syndrome or transient ischemic attack (TIA).

The primary outcome occurred in 17% of subjects in the CPAP group and 15% of subjects in the usual-care group; no significant effect on any individual or composite cardiovascular endpoint was associated with randomization to the CPAP group. CPAP significantly reduced snoring and daytime sleepiness and improved health-related quality of life and mood. A potentially important factor which reduced the effectiveness of CPAP was the relatively poor adherence among users (mean use of 3.3 hours/night and only 42% with average use > four hours/night). When adherent CPAP users (> = four hours/night) were compared controls (matched by propensity score), they had a significantly reduced risk (44%) of stroke.

Comparing the details of the study by Marin and coworkers (2005) to the SAVE trial (McEvoy et al., 2016) highlights two potential reasons for the discrepancy between findings from observational and RCT studies. First, achieving adequate CPAP adherence (or comparable adherence to that seen in clinical practice among highly adherent patients) is challenging in large RCTs. Second, patients with very severe OSA typically are excluded from RCTs, given the potential harm of allocating such severely affected patients to a non-treatment arm. Results from the SAVE trial (McEvoy et al., 2016) also suggest that the level of CPAP adherence needed to improve cardiovascular outcomes may be higher than that needed to improve outcomes (e.g., sleepiness, mood and quality of life).

ADHERENCE

The preceding discussion highlights the key role of CPAP adherence in achieving optimal outcomes. The extent to which an "optimal" level of adherence is achievable in clinical practice remains challenging. In a study by Weaver and colleagues (2007), 149 uncomplicated patients, age 21 to 60 years, with moderate to severe OSA were managed across seven academic centers according to usual practice. The distribution of average hours of CPAP use at three months across patients is shown in Figure 6. Approximately 45% of patients had very high use (more than six hours/day) and 66% had more than four hours of use/day. Thus, over the short term, a significant proportion of uncomplicated OSA patients under usual care achieve levels of adherence that provide at least symptomatic benefits, though a significant minority do not. Over longer periods, it is likely that the level of adherence declines, although in at least one study, 68% of patients had continued CPAP use after five years (McArdle et al., 1999).

Importantly, there appears to be a dose-response relationship between hours of CPAP use and functional outcomes (e.g., sleepiness). Cumulative benefit from CPAP has been shown to plateau after four (ESS score), six (sleep latencies on Multiple Sleep Latency Testing; MSLT) and 7.5 (Functional Outcomes of Sleep Questionnaire) hours of nightly use (Weaver et al., 2007; Fig. 7). Although use of CPAP for the entire sleep period is optimal, some use of CPAP on a nightly basis provides benefit, especially in terms of subjective sleepiness (Weaver et al., 2007; Antic et al., 2011).

During the initiation of CPAP therapy, a number of potential problems can interfere with use and present barriers to achieving optimal adherence (e.g., nasal congestion, mask leak, mask discomfort, oral air leak and claustrophobia) for which there are remedies (Table 1). Among

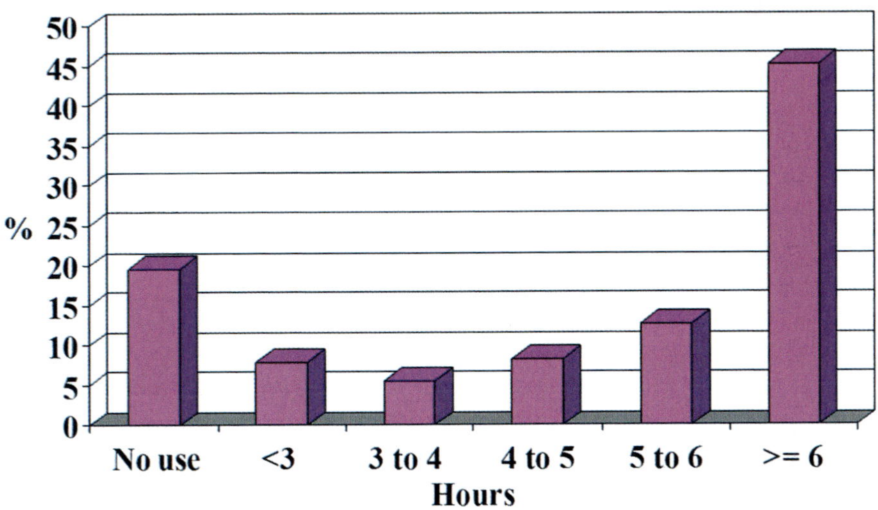

Figure 6. Average nightly CPAP use after three months. Adapted from Weaver et al., 2007; reprinted with permission from Oxford University Press.

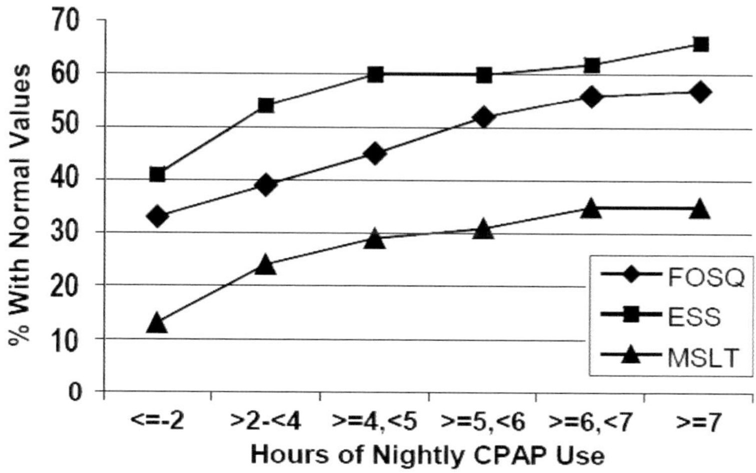

Figure 7. Association between nightly hours of CPAP use and percent of patients with normal values on the Functional Outcomes of Sleep Questionnaire (FOSQ), Epworth Sleepiness Scale (ESS) and Multiple Sleep Latency Test (MSLT). From Weaver et al., 2007; reprinted with permission from Oxford University Press.

the many advancements to CPAP technology mentioned earlier (APAP, bilevel PAP and EPR), none have been demonstrated consistently and rigorously in RCTs to improve adherence and outcomes (Rauscher et al., 1993; Engleman et al., 1994; Reeves-Hoche et al., 1994; Douglas and Engleman, 1998).

Surprisingly, even the use of heated humidification, which is felt by many clinicians to be an essential component of CPAP therapy, has not been demonstrated to improve adherence consistently (Martins De Araujo et al., 2000; Rakotonanahary et al., 2001; Ryan et al., 2009), though studies show significant reductions in the side effect of dry mouth and large magnitude trends toward improvement in dry nose and nasal congestion (Massie et al., 1999; Nilius et al., 2016).

Acknowledging the importance of adherence, several interventions aimed at increasing CPAP use and improving the user experience have been studied. Of these, the most consistently effective intervention is education and supportive care. A recent large Cochrane review (Wozniak et al., 2014) examined 30 studies including a total of 2,047 participants, most of whom were new users of CPAP, who received individual or group education about OSA and CPAP, as well as periodic reminders and encouragement to use their CPAP. As a whole, such programs tend to increase nightly CPAP usage by 35 to 50 minutes. Intensive programs

Table 1. Problems associated with CPAP compliance and possible solutions.

Problems	Solutions
Nasal congestion	Heated humidity, nasal steroids, surgical evaluation, nasal-oral mask
Mask leak	Proper fitting and alternative interfaces
Mask discomfort	Proper fitting and alternative interfaces
Oral air leak	Heated humidity, treat nasal congestion, chinstrap, nasal-oral mask
Claustrophobia	Desensitization, alternative interfaces
Difficulty exhaling	Ramp pressure, EPR, auto-titrating machine, bi-level positive airway pressure machine
Inconvenience	Education and CPAP trial

of targeted motivational interviewing tend to increase CPAP usage by 1.5 hours/night on average (Wozniak et al., 2014). Similarly, telemonitoring may be integrated into behavioral and educational interventions, particularly now that new PAP units typically are manufactured with built-in modems capable of transmitting data on usage and therapy.

The use of telemonitoring has been associated with improved CPAP usage compared to scheduled non-monitored clinic follow-up (by 86 minutes/night on average; Fox et al., 2012) and the use of automated reminders based on telemonitoring data has been shown to be equivalent to regular face-to-face evaluations in achieving adherence while significantly reducing labor requirements (Munafo et al., 2016). The use of hypnotic agents generally has not been found to be beneficial in improving CPAP compliance (Bradshaw et al., 2006; Park et al., 2013), although some studies have shown a benefit of short-term use of eszopiclone (a drug used to treat insomnia) in improving CPAP adherence over the first six months of therapy (Lettieri et al., 2008, 2009).

Nonetheless, a significant proportion of patients with OSA do not achieve adequate levels of adherence to CPAP under usual care. For these patients who do not tolerate CPAP, alternative therapies should be considered. Mandibular advancement devices can be offered to patients with mild or moderate OSA and to patients with more severe OSA who do not tolerate CPAP. This therapy tends to result in more modest improvements in Apnea-Hypopnea Index (AHI), oxyhemoglobin desaturations and hypertension compared to CPAP (Iftikhar et al., 2013; Phillips et al., 2013), but improvements in daytime sleepiness may be similar with these therapeutic modalities (Sutherland et al., 2014), potentially related to higher compliance with oral appliance use (Doff et al., 2013). A variety of surgical options to address correction of anatomic issues contributing to OSA is available and may be appropriate in carefully selected patients.

CONCLUSIONS

CPAP, which first was used to treat OSA in 1980, has undergone refinement over time and is considered a first-line therapy. CPAP not only is an efficacious treatment, but it also is tolerated well by many patients. Achieving adequate CPAP adherence is a crucial component of effectiveness, particularly for prevention of long-term cardiovascular morbidity. A significant proportion of patients with OSA do not achieve adequate levels of adherence under usual care. Promising approaches to improve adherence include educational and behavioral programs, as well as telemonitoring. It is important to realize that despite its merits, CPAP will not be tolerated or preferred by some patients. Fortunately, there are alternative therapies available that can help some of these patients.

REFERENCES

Abe H, Takahashi M, Yaegashi H, Eda S, Tsunemoto H, Kamikozawa M, Koyama J, Yamazaki K, Ikeda U. Efficacy of continuous positive airway pressure on arrhythmias in obstructive sleep apnea patients. Heart Vessels 2010;25(1):63-69.

Aloia MS, Stanchina M, Arnedt JT, Malhotra A, Millman RP. Treatment adherence and outcomes in flexible *vs* standard continuous positive airway pressure therapy. Chest 2005;127(6):2085-2093.

Antic NA, Catcheside P, Buchan C, Hensley M, Naughton MT, Rowland S, Williamson B, Windler S, McEvoy RD. The effect of CPAP in normalizing daytime sleepiness, quality of life, and neurocognitive function in patients with moderate to severe OSA. Sleep 2011;34(1):111-119.

Ayas NT, Patel SR, Malhotra A, Schulzer M, Malhotra M, Jung D, Fleetham J, White DP. Auto-titrating *versus* standard continuous positive airway pressure for the treatment of obstructive sleep apnea: Results of a meta-analysis. Sleep 2004;27(2):249-253.

Bakker JP, Marshall NS. Flexible pressure delivery modification of continuous positive airway pressure for obstructive sleep apnea does not improve compliance with therapy: Systematic review and meta-analysis. Chest 2011;139(6):1322-1330.

Ballester E, Badia JR, Hernández L, Carrasco E, de Pablo J, Fornas C, Rodriguez-Roisin R, Montserrat JM. Evidence of the effectiveness of continuous positive airway pressure in the treatment of sleep apnea/hypopnea syndrome. Am J Respir Crit Care Med 1999;159(5 Pt 1):495-501.

Barbé F, Durán-Cantolla J, Sánchez-de-la-Torre M, Martínez-Alonso M, Carmona C, Barceló A, Chiner E, Masa JF, Gonzalez M, Marín JM, Garcia-Rio F, Diaz de Atauri J, Terán J, Mayos M, de la Peña M, Monasterio C, del Campo F, Montserrat JM; Spanish Sleep and Breathing Network. Effect of continuous positive airway pressure on the incidence of hypertension and cardiovascular events in nonsleepy patients with obstructive sleep apnea: A randomized controlled trial. JAMA 2012;307(20):2161-2168.

Barnes M, Houston D, Worsnop CJ, Neill AM, Mykytyn IJ, Kay A, Trinder J, Saunders NA, McEvoy RD, Pierce RJ. A randomized controlled trial of continuous positive airway pressure in mild obstructive sleep apnea. Am J Respir Crit Care Med 2002;165(6):773-780.

Barnes M, McEvoy RD, Banks S, Tarquinio N, Murray CG, Vowles N, Pierce RJ. Efficacy of positive airway pressure and oral appliance in mild to moderate obstructive sleep apnea. Am J Respir Crit Care Med 2004; 170(6):656-664.

Becker HF, Jerrentrup A, Ploch T, Grote L, Penzel T, Sullivan CE, Peter JH. Effect of nasal continuous positive airway pressure treatment on blood pressure in patients with obstructive sleep apnea. Circulation 2003;107(1):68-73.

Borel JC, Tamisier R, Dias-Domingos S, Sapene M, Martin F, Stach B, Grillet Y, Muir JF, Levy P, Series F, Pepin JL; Scientific Council of The Sleep Registry of the French Federation of Pneumology (OSFP). Type of mask may impact on continuous positive airway pressure adherence in apneic patients. PLoS One 2013;8(5):e64382.

Bradshaw DA, Ruff GA, Murphy DP. An oral hypnotic medication does not improve continuous positive airway pressure compliance in men with obstructive sleep apnea. Chest 2006;130(5):1369-1376.

Buchner NJ, Sanner BM, Borgel J, Rump LC. Continuous positive airway pressure treatment of mild to moderate obstructive sleep apnea reduces cardiovascular risk. Am J Respir Crit Care Med 2007;176(12): 1274-1280.

Campos-Rodriguez F, Martinez-Garcia MA, Reyes-Nuñez N, Caballero-Martinez I, Catalan-Serra P, Almeida-Gonzalez CV. Role of sleep apnea and continuous positive airway pressure therapy in the incidence of stroke or coronary heart disease in women. Am J Respir Crit Care Med 2014;189(12):1544-1550.

Carlucci A, Ceriana P, Mancini M, Cirio S, Pierucci P, D'Artavilla Lupo N, Gadaleta F, Morrone E, Fanfulla F. Efficacy of bilevel-auto treatment in patients with obstructive sleep apnea not responsive to or intolerant of continuous positive airway pressure ventilation. J Clin Sleep Med 2015;11(9):981-985.

Cassar A, Morgenthaler TI, Lennon RJ, Rihal CS, Lerman A. Treatment of obstructive sleep apnea is associated with decreased cardiac death after percutaneous coronary intervention. J Am Coll Cardiol 2007; 50(14):1310-1314.

Coughlin SR, Mawdsley L, Mugarza JA, Wilding JP, Calverley PM. Cardiovascular and metabolic effects of CPAP in obese males with OSA. Eur Respir J 2007;29(4):720-727.

Cowie MR, Woehrle H, Wegscheider K, Angermann C, d'Ortho MP, Erdmann E, Levy P, Simonds A, Somers VK, Zannad F, Teschler H. Rationale and design of the SERVE-HF study: Treatment of sleep-disordered breathing with predominant central sleep apnoea with adaptive servo-ventilation in patients with chronic heart failure. Eur J Heart Fail 2013;15(8):937-943.

Dalmases M, Solé-Padullés C, Torres M, Embid C, Nuñez MD, Martínez-Garcia MÁ, Farré R, Bargalló N, Bartrés-Faz D, Montserrat JM. Effect of CPAP on cognition, brain function, and structure among elderly patients with OSA: A randomized pilot study. Chest 2015;148(5):1214-1223.

Doff MH, Hoeksema A, Wijkstra PJ, van der Hoeven JH, Huddleston Slater JJ, de Bont LG, Stegenga B. Oral appliance *versus* continuous positive airway pressure in obstructive sleep apnea syndrome: A 2-year follow-up. Sleep 2013;36(9):1289-1296.

Doherty LS, Kiely JL, Swan V, McNicholas WT. Long-term effects of nasal continuous positive airway pressure therapy on cardiovascular outcomes in sleep apnea syndrome. Chest 2005;127(6):2076-2084.

Douglas NJ. Systematic review of the efficacy of nasal CPAP. Thorax 1998;53(5):414-415.

Douglas NJ, Engleman HM. CPAP therapy: Outcomes and patient use. Thorax 1998;53(Suppl 3):S47-S48.

Durán-Cantolla J, Aizpuru F, Montserrat JM, Ballester E, Terán-Santos J, Aguirregomoscorta JI, Gonzalez M, Lloberes P, Masa JF, De La Peña M, Carrizo S, Mayos M, Barbé F; Spanish Sleep and Breathing Group. Continuous positive airway pressure as treatment for systemic hypertension in people with obstructive sleep apnoea: Randomised controlled trial. BMJ 2010;341:c5991.

Engleman HM, Douglas NJ. Sleep: 4. Sleepiness, cognitive function, and quality of life in obstructive sleep apnoea/hypopnoea syndrome. Thorax 2004;59(7):618-622.

Engleman HM, Kingshott RN, Wraith PK, Mackay TW, Deary IJ, Douglas NJ. Randomized placebo-controlled crossover trial of continuous positive airway pressure for mild sleep apnea/hypopnea syndrome. Am J Respir Crit Care Med 1999;159(2):461-467.

Engleman HM, Martin SE, Deary IJ, Douglas NJ. Effect of CPAP therapy on daytime function in patients with mild sleep apnoea/hypopnoea syndrome. Thorax 1997;52(2):114-119.

Engleman HM, Martin SE, Douglas NJ. Compliance with CPAP therapy in patients with the sleep apnoea/hypopnoea syndrome. Thorax 1994;49(3):263-266.

Engleman HM, Martin SE, Kingshott RN, Mackay TW, Deary IJ, Douglas NJ. Randomised placebo controlled trial of daytime function after continuous positive airway pressure (CPAP) therapy for the sleep apnoea/hypopnoea syndrome. Thorax 1998;53(5):341-345.

Epstein LJ, Kristo D, Strollo PJ Jr, Friedman N, Malhotra A, Patil SP, Ramar K, Rogers R, Schwab RJ, Weaver EM, Weinstein MD; Adult Obstructive Sleep Apnea Task Force of the American Academy of Sleep Medicine. Clinical guideline for the evaluation, management and long-term care of obstructive sleep apnea in adults. J Clin Sleep Med 2009;5(3):263-276.

Faccenda JF, Mackay TW, Boon NA, Douglas NJ. Randomized placebo-controlled trial of continuous positive airway pressure on blood pressure in the sleep apnea-hypopnea syndrome. Am J Respir Crit Care Med 2001;163(2):344-348.

Fox N, Hirsch-Allen AJ, Goodfellow E, Wenner J, Fleetham J, Ryan CF, Kwiatkowska M, Ayas NT. The impact of a telemedicine monitoring system on positive airway pressure adherence in patients with obstructive sleep apnea: A randomized controlled trial. Sleep 2012;35(4):477-481.

Gay P, Weaver T, Loube D, Iber C; Positive Airway Pressure Task Force; Standards of Practice Committee; American Academy of Sleep Medicine. Evaluation of positive airway pressure treatment for sleep related breathing disorders in adults. Sleep 2006;29(3):381-401.

Hack M, Davies RJ, Mullins R, Choi SJ, Ramdassingh-Dow S, Jenkinson C, Stradling JR. Randomised prospective parallel trial of therapeutic *versus* subtherapeutic nasal continuous positive airway pressure on simulated steering performance in patients with obstructive sleep apnoea. Thorax 2000;55(3):224-231.

Holmqvist F, Guan N, Zhu Z, Kowey PR, Allen LA, Fonarow GC, Hylek EM, Mahaffey KW, Freeman JV, Chang P, Holmes DN, Peterson ED, Piccini JP, Gersh BJ; ORBIT-AF Investigators. Impact of obstructive sleep apnea and continuous positive airway pressure therapy on outcomes in patients with atrial fibrillation: Results from the Outcomes Registry for Better Informed Treatment of Atrial Fibrillation (ORBIT-AF). Am Heart J 2015;169(5):647-654.e2.

Huang HC, Hillman DR, McArdle N. Control of OSA during automatic positive airway pressure titration in a clinical case series: Predictors and accuracy of device download data. Sleep 2012;35(9):1277A-1283A.

Hui DS, To KW, Ko FW, Fok JP, Chan MC, Ngai JC, Tung AH, Ho CW, Tong MW, Szeto CC, Yu CM. Nasal CPAP reduces systemic blood pressure in patients with obstructive sleep apnoea and mild sleepiness. Thorax 2006;61(12):1083-1090.

Iftikhar IH, Hays ER, Iverson MA, Magalang UJ, Maas AK. Effect of oral appliances on blood pressure in obstructive sleep apnea: A systematic review and meta-analysis. J Clin Sleep Med 2013;9(2):165-174.

Ip S, D'Ambrosio C, Patel K, Obadan N, Kitsios GD, Chung M, Balk EM. Auto-titrating *versus* fixed continuous positive airway pressure for the treatment of obstructive sleep apnea: A systematic review with meta-analyses. Syst Rev 2012;1:20.

Javaheri S, Smith J, Chung E. The prevalence and natural history of complex sleep apnea. J Clin Sleep Med 2009;5(3):205-211.

Jenkinson C, Davies RJ, Mullins R, Stradling JR. Comparison of therapeutic and subtherapeutic nasal continuous positive airway pressure for obstructive sleep apnoea: A randomised prospective parallel trial. Lancet 1999;353(9170):2100-2105.

Johns MW. A new method for measuring daytime sleepiness: The Epworth sleepiness scale. Sleep 1991;14(6):540-545.

Johnson KG, Johnson DC. Treatment of sleep-disordered breathing with positive airway pressure devices: Technology update. Med Devices (Auckl) 2015;8:425-437.

Kanagala R, Murali NS, Friedman PA, Ammash NM, Gersh BJ, Ballman KV, Shamsuzzaman AS, Somers VK. Obstructive sleep apnea and the recurrence of atrial fibrillation. Circulation 2003;107(20):2589-2594.

Kaneko Y, Floras JS, Usui K, Plante J, Tkacova R, Kubo T, Ando S, Bradley TD. Cardiovascular effects of continuous positive airway pressure in patients with heart failure and obstructive sleep apnea. N Engl J Med 2003;348(13):1233-1241.

Kasai T, Narui K, Dohi T, Yanagisawa N, Ishiwata S, Ohno M, Yamaguchi T, Momomura S. Prognosis of patients with heart failure and obstructive sleep apnea treated with continuous positive airway pressure. Chest 2008;133(3):690-696.

Kohler M, Peppernell JC, Casadei B, Craig S, Crosthwaite N, Stradling JR, Davies RJ. CPAP and measures of cardiovascular risk in males with OSAS. Eur Respir J 2008;32(6):1488-1496.

Kushida CA, Littner MR, Hirshkowitz M, Morgenthaler TI, Alessi CA, Bailey D, Boehlecke B, Brown TM, Coleman J Jr, Friedman L, Kapen S, Kapur VK, Kramer M, Lee-Chiong T, Owens J, Pancer JP, Swick TJ, Wise MS; American Academy of Sleep Medicine. Practice parameters for the use of continuous and bilevel positive airway pressure devices to treat adult patients with sleep-related breathing disorders. Sleep 2006; 29(3):375-380.

Kushida CA, Nichols DA, Holmes TH, Quan SF, Walsh JK, Gottlieb DJ, Simon RD Jr, Guilleminault C, White DP, Goodwin JL, Schweitzer PK, Leary EB, Hyde PR, Hirshkowitz M, Green S, McEvoy LK, Chan C, Gevins A, Kay GG, Bloch DA, Crabree T, Dement WC. Effects of continuous positive airway pressure on neurocognitive function in obstructive sleep apnea patients: The Apnea Positive Pressure Long-term Efficacy Study (APPLES). Sleep 2012;35(12):1593-1602.

Lam B, Sam K, Mok WY, Cheung MT, Fong DY, Lam JC, Lam DC, Yam LY, Ip MS. Randomised study of three non-surgical treatments in mild to moderate obstructive sleep apnoea. Thorax 2007;62(4):354-359.

Lettieri CJ, Quast TN, Eliasson AH, Andrada T. Eszopiclone improves overnight polysomnography and continuous positive airway pressure titration: A prospective, randomized, placebo-controlled trial. Sleep 2008;31(9):1310-1316.

Lettieri CJ, Shah AA, Holley AB, Kelly WF, Chang AS, Roop SA; CPAP Promotion and Prognosis: The Army Sleep Apnea Program Trial. Effects of a short course of eszopiclone on continuous positive airway pressure adherence: A randomized trial. Ann Intern Med 2009;151(10):696-702.

Loube DI, Gay PC, Strohl KP, Pack AI, White DP, Collop NA. Indications for positive airway pressure treatment of adult obstructive sleep apnea patients: A consensus statement. Chest 1999;115(3):863-866.

Mansfield DR, Gollogly NC, Kaye DM, Richardson M, Bergin P, Naughton MT. Controlled trial of continuous positive airway pressure in obstructive sleep apnea and heart failure. Am J Respir Crit Care Med 2004;169(3):361-366.

Marin JM, Carrizo SJ, Vicente E, Agusti AG. Long-term cardiovascular outcomes in men with obstructive sleep apnoea-hypopnoea with or without treatment with continuous positive airway pressure: An observational study. Lancet 2005;365(9464):1046-1053.

Marti S, Sampol G, Muñoz X, Torres F, Roca A, Lloberes P, Sagalés T, Quesada P, Morell F. Mortality in severe sleep apnoea/hypopnoea syndrome patients: Impact of treatment. Eur Respir J 2002;20(6): 1511-1518.

Martínez-García MA, Capote F, Campos-Rodríguez F, Lloberes P, Díaz de Atauri MJ, Somoza M, Masa JF, González M, Sacristán L, Barbé F, Durán-Cantolla J, Aizpuru F, Manas E, Barreiro B, Mosteiro M, Cebrián JJ, de la Peña M, García-Río F, Maimó A, Zapater J, Hernández C, Grau SanMarti N, Montserrat JM; Spanish Sleep Network. Effect of CPAP on blood pressure in patients with obstructive sleep apnea and resistant hypertension: The HIPARCO randomized clinical trial. JAMA 2013; 310(22):2407-2415.

Martins De Araújo MT, Vieira SB, Vasquez EC, Fleury B. Heated humidification or face mask to prevent upper airway dryness during continuous positive airway pressure therapy. Chest 2000;117(1):142-147.

Massie CA, Hart RW, Peralez K, Richards GN. Effects of humidification on nasal symptoms and compliance in sleep apnea patients using continuous positive airway pressure. Chest 1999;116(2):403-408.

McArdle N, Devereux G, Heidarnejad H, Engleman HM, Mackay TW, Douglas NJ. Long-term use of CPAP therapy for sleep apnea/hypopnea syndrome. Am J Respir Crit Care Med 1999;159(4 Pt 1):1108-1114.

McArdle N, Douglas NJ. Effect of continuous positive airway pressure on sleep architecture in the sleep apnea-hypopnea syndrome: A randomized controlled trial. Am J Respir Crit Care Med 2001;164(8 Pt 1):1459-1463.

McEvoy RD, Antic NA, Heeley E, Luo Y, Ou Q, Zhang X, Mediano O, Chen R, Drager LF, Liu Z, Chen G, Du B, McArdle N, Mukherjee S, Tripathi M, Billot L, Li Q, Lorenzi-Filho G, Barbe F, Redline S, Wang J, Arima H, Neal B, White DP, Grunstein RR, Zhong N, Anderson CS; SAVE Investigators and Coordinators. CPAP for prevention of cardiovascular events in obstructive sleep apnea. N Engl J Med 2016;375(10):919-931.

McMillan A, Bratton DJ, Faria R, Laskawiec-Szkonter M, Griffin S, Davies RJ, Nunn AJ, Stradling JR, Riha RL, Morrell MJ; PREDICT Investigators. Continuous positive airway pressure in older people with obstructive sleep apnoea syndrome (PREDICT): A 12-month, multicentre, randomised trial. Lancet Respir Med 2014;2(10):804-812.

Milleron O, Pilliere R, Foucher A, de Roquefeuil F, Aegerter P, Jondeau G, Raffestin BG, Dubourg O. Benefits of obstructive sleep apnoea treatment in coronary artery disease: A long-term follow-up study. Eur Heart J 2004;25(9):728-734.

Monasterio C, Vidal S, Duran J, Ferrer M, Carmona C, Barbé F, Mayos M, Gonzalez-Mangado N, Juncadella M, Navarro A, Barreira R, Capote F, Mayoralas LR, Peces-Barba G, Alonso J, Montserrat JM. Effectiveness of continuous positive airway pressure in mild sleep apnea-hypopnea syndrome. Am J Respir Crit Care Med 2001;164(6):939-943.

Montserrat JM, Ferrer M, Hernandez L, Farré R, Vilagut G, Navajas D, Badia JR, Carrasco E, De Pablo J, Ballester E. Effectiveness of CPAP treatment in daytime function in sleep apnea syndrome: A randomized controlled study with an optimized placebo. Am J Respir Crit Care Med 2001;164(4):608-613.

Morgenthaler TI, Aurora RN, Brown T, Zak R, Alessi C, Boehlecke B, Chesson AL Jr, Friedman L, Kapur V, Maganti R, Owens J, Pancer J, Swick TJ; Standards of Practice Committee of the AASM (American Academy of Sleep Medicine). Practice parameters for the use of autotitrating continuous positive airway pressure devices for titrating pressures and treating adult patients with obstructive sleep apnea syndrome: An update for 2007. An American Academy of Sleep Medicine report. Sleep 2008;31(1):141-147.

Munafo D, Hevener W, Crocker M, Willes L, Sridasome S, Muhsin M. A telehealth program for CPAP adherence reduces labor and yields similar adherence and efficacy when compared to standard of care. Sleep Breath 2016;20(2):777-785.

Nilius G, Franke KJ, Domanski U, Schroeder M, Ruhle KH. Effect of APAP and heated humidification with a heated breathing tube on adherence, quality of life, and nasopharyngeal complaints. Sleep Breath 2016;20(1):43-49.

Park JG, Olson EJ, Morgenthaler TI. Impact of zaleplon on continuous positive airway pressure therapy compliance. J Clin Sleep Med 2013; 9(5):439-444.

Parra O, Sánchez-Armengol Á, Capote F, Bonnin M, Arboix A, Campos-Rodríguez F, Pérez-Ronchel J, Durán-Cantolla J, Martínez-Null C, de la Peña M, Jiménez MC, Masa F, Casadon I, Alonso ML, Macarrón JL. Efficacy of continuous positive airway pressure treatment on 5-year survival in patients with ischaemic stroke and obstructive sleep apnea: A randomized controlled trial. J Sleep Res 2015;24(1):47-53.

Patel SR, White DP, Malhotra A, Stanchina ML, Ayas NT. Continuous positive airway pressure therapy for treating sleepiness in a diverse population with obstructive sleep apnea: Results of a meta-analysis. Arch Intern Med 2003;163(5):565-571.

Peker Y, Glantz H, Eulenburg C, Wegscheider K, Herlitz J, Thunström E. Effect of positive airway pressure on cardiovascular outcomes in coronary artery disease patients with nonsleepy obstructive sleep apnea: The RICCADSA randomized controlled trial. Am J Respir Crit Care Med 2016;194(5):613-620.

Phillips CL, Grunstein RR, Darendeliler MA, Mihailidou AS, Srinivasan VK, Yee BJ, Marks GB, Cistulli PA. Health outcomes of continuous positive airway pressure *versus* oral appliance treatment for obstructive sleep apnea: A randomized controlled trial. Am J Respir Crit Care Med 2013; 187(8):879-887.

Phillips CL, Yee BJ, Marshall NS, Liu PY, Sullivan DR, Grunstein RR. Continuous positive airway pressure reduces postprandial lipidemia in obstructive sleep apnea: A randomized, placebo-controlled crossover trial. Am J Respir Crit Care Med 2011;184(3):355-361.

Powell ED, Gay PC, Ojile JM, Litinski M, Malhotra A. A pilot study assessing adherence to auto-bilevel following a poor initial encounter with CPAP. J Clin Sleep Med 2012;8(1):43-47.

Rakotonanahary D, Pelletier-Fleury N, Gagnadoux F, Fleury B. Predictive factors for the need for additional humidification during nasal continuous positive airway pressure therapy. Chest 2001;119(2):460-465.

Rauscher H, Formanek D, Popp W, Zwick H. Self-reported *vs* measured compliance with nasal CPAP for obstructive sleep apnea. Chest 1993; 103(6):1675-1680.

Reeves-Hoché MK, Hudgel DW, Meck R, Witteman R, Ross A, Zwilich CW. Continuous *versus* bilevel positive airway pressure for obstructive sleep apnea. Am J Respir Crit Care Med 1995;151(2 Pt 1):443-449.

Reeves-Hoche MK, Meck R, Zwillich CW. Nasal CPAP: An objective evaluation of patient compliance. Am J Respir Crit Care Med 1994; 149(1):149-154.

Roux FJ, Hilbert J. Continuous positive airway pressure: New generations. Clin Chest Med 2003;24(2):315-342.

Ryan S, Doherty LS, Nolan GM, McNicholas WT. Effects of heated humidification and topical steroids on compliance, nasal symptoms, and quality of life in patients with obstructive sleep apnea syndrome using nasal continuous positive airway pressure. J Clin Sleep Med 2009;5(5):422-427.

Sanders MH, Kern N. Obstructive sleep apnea treated by independently adjusted inspiratory and expiratory positive airway pressures *via* nasal mask: Physiologic and clinical implications. Chest 1990;98(2):317-324.

Sanders MH, Kern NB, Costantino JP, Stiller RA, Studnicki K, Coates J, Orris S, Schimerman S. Adequacy of prescribing positive airway pressure therapy by mask for sleep apnea on the basis of a partial-night trial. Am Rev Respir Dis 1993;147(5):1169-1174.

Siccoli MM, Pepperell JC, Kohler M, Craig SE, Davies RJ, Stradling JR. Effects of continuous positive airway pressure on quality of life in patients with moderate to severe obstructive sleep apnea: Data from a randomized controlled trial. Sleep 2008;31(11):1551-1558.

Strohl KP, Redline S. Nasal CPAP therapy, upper airway muscle activation, and obstructive sleep apnea. Am Rev Respir Dis 1986;134(3):555-558.

Sullivan CE, Issa FG, Berthon-Jones M, Eves L. Reversal of obstructive sleep apnoea by continuous positive airway pressure applied through the nares. Lancet 1981;1(8225):862-865.

Sun H, Shi J, Li M, Chen X. Impact of continuous positive airway pressure treatment on left ventricular ejection fraction in patients with obstructive sleep apnea: A meta-analysis of randomized controlled trials. PLoS One 2013;8(5):e62298.

Sutherland K, Vanderveken OM, Tsuda H, Marklund M, Gagnadoux F, Kushida CA, Cistulli PA. Oral appliance treatment for obstructive sleep apnea: An update. J Clin Sleep Med 2014;10(2):215-227.

Wang H, Parker JD, Newton GE, Floras JS, Mak S, Chiu KL, Ruttanaumpawan P, Tomlinson G, Bradley TD. Influence of obstructive sleep apnea on mortality in patients with heart failure. J Am Coll Cardiol 2007;49(15):1625-1631.

Weaver TE, Maislin G, Dinges DF, Bloxham T, George CF, Greenberg H, Kader G, Mahowald M, Younger J, Pack AI. Relationship between hours of CPAP use and achieving normal levels of sleepiness and daily functioning. Sleep 2007;30(6):711-719.

Weaver TE, Mancini C, Maislin G, Cater J, Staley B, Landis JR, Ferguson KA, George CF, Schulman DA, Greenberg H, Rapoport DM, Walsleben JA, Lee-Chiong T, Gurubhagavatula I, Kuna ST. Continuous positive airway pressure treatment of sleepy patients with milder obstructive sleep apnea: Results of the CPAP Apnea Trial North American Program (CATNAP) randomized clinical trial. Am J Respir Crit Care Med 2012; 186(7):677-683.

West SD, Nicoll DJ, Wallace TM, Matthews DR, Stradling JR. Effect of CPAP on insulin resistance and HbA1c in men with obstructive sleep apnoea and type 2 diabetes. Thorax 2007;62(11):969-974.

Woodson BT, Steward DL, Weaver EM, Javaheri S. A randomized trial of temperature-controlled radiofrequency, continuous positive airway pressure, and placebo for obstructive sleep apnea syndrome. Otolaryngol Head Neck Surg 2003;128(6):848-861.

Wozniak DR, Lasserson TJ, Smith I. Educational, supportive and behavioural interventions to improve usage of continuous positive airway pressure machines in adults with obstructive sleep apnoea. Cochrane Database Syst Rev 2014(1):CD007736.

PEDIATRIC SLEEP-DISORDERED BREATHING: BASIC CONCEPTS PERTINENT TO ORTHODONTISTS

Joseane Pizzatto and Carlos Flores-Mir

ABSTRACT

Pediatric sleep-disordered breathing (PSDB) refers to a wide spectrum of sleep-related breathing abnormalities that are associated with increased upper airway resistance, arousal and gas exchange anomalies during sleep. Based on an increased level of complexity, PSDB as an overall term includes snoring, upper airway resistance syndrome and obstructive sleep apnea-hypopnea syndrome.

Pediatric obstructive sleep apnea (POSA), the most advanced and threatening form of PSDB, can be a cause of significant morbidity among children and adolescents. Some common signs and symptoms include snoring, nocturnal enuresis, disturbed non-refreshing sleep, increased numbers of arousals, sleep awakenings, excessive daytime sleepiness, irritability, inattention, hyperactivity and confrontational behavior.

A growing number of risk factors (e.g., hypertrophic adenoids and/or tonsils, asthma, obesity, socio-economic status, pre-term birth and craniofacial anomalies) have been linked to PSDB. Craniofacial characteristics associated with PSDB include retrognathic mandible, large facial height, narrow maxillary arch and posterior crossbite.

Among conventional treatment alternatives, adenotonsillectomy and, in more severe cases, continuous positive airway pressure (CPAP), have been proposed. Most importantly, the impact of orthodontic treatment modalities that have been linked to notable reduction of some of the symptoms of POSA will be emphasized.

In summary, this chapter will discuss the epidemiology, outcome, diagnosis and treatment of PSDB succinctly. Key pertinent manuscripts concerning PSDB will be included. This PSDB overview is intended to enhance awareness by orthodontists regarding their potential role in the multi-disciplinary management of PSDB. Timely multi-disciplinary management can prevent significant health consequences in addition to craniofacial growth and development alterations.

KEY WORDS: children, adenoids, sleep-disordered breathing, obstructive sleep apnea, polysomnography

---

INTRODUCTION

For more than three decades, pediatric sleep medicine has made significant strides in terms of enhanced awareness of pediatric sleep-disordered breathing (PSDB) and its health consequences (Aurora et al., 2011). However, obstructive sleep apnea (OSA), the most advanced and dangerous form of sleep-disordered breathing (SDB), still mainly is underdiagnosed (Chervin et al., 2002) and/or undertreated (Gozal, 2012). In general terms, OSA can be a cause of significant morbidity and its multifactorial nature demands multi-disciplinary management (Zettergren-Wijk et al., 2006; Gozal, 2012).

BASIC CONCEPTS

Sleep-disordered Breathing

PSDB reflects a continuum of signs and symptoms that vary on severity based on the degree of upper airway narrowing, arousal and gas exchange abnormality during sleep (Ali et al., 1993). These disturbances disrupt pulmonary ventilation, oxygenation and/or sleep quality (Carroll, 2003). SDB spectrum ranges from snoring to upper airway resistance syndrome (UARS) and from it, to OSA (Ali et al., 1993). Potential causes include upper and/or lower airway obstruction and/or abnormal breathing control by the central nervous system (Ross, 2013).

Obstructive Sleep Apnea

As mentioned before, OSA is the most severe form of SDB characterized by recurrent episodes of partial or complete airway obstruction during sleep with associated abnormalities in gas exchange, sleep disruption or both (Guilleminault et al., 1976). The obstruction occurs because reduced muscle tone and gravity in the supine position during sleep both can cause a narrow upper airway to collapse in susceptible individuals (Pliska et al., 2012).

Children and adults can be affected; however, the two groups manifest the disease with different etiology, pathophysiology and prevalence (Ross, 2013). A frequent risk factor for OSA in adults is obesity and the first-line treatment is continuous positive airway pressure (CPAP). In children, however, adenotonsillar hypertrophy is the most often associated factor and, therefore, the first option therapy is adenotonsillectomy (Ross, 2013).

Pediatric OSA (POSA) fits into a model of a chronic disease with several important characteristics:

1. Individual variation in adverse clinical outcome, which may be related to variable degree of hypoxia and arousals;
2. Variability in disease severity, with potential deterioration during acute upper respiratory infection;
3. Potential differences in individual susceptibility; and
4. Uncertainty in measurement of disease manifestation due to inherent differences in baseline neurocognitive and cardiovascular function across individuals (Wong, 2011).

OSA SIGNS AND SYMPTOMS

The subjective symptoms and clinical sequelae of OSA vary according to the severity of the disease and patient-specific factors (e.g., neuromotor tone; Carroll, 2003). Sleep-related OSA symptoms include snoring (Guilleminault et al., 1981; Guilleminault and Stoohs, 1990; Carroll, 2003), reporting of disturbed unrefreshing sleep (Guilleminault et al., 1981; Carroll, 2003), increased breathing effort (Guilleminault and Stoohs, 1990; Carroll, 2003), nocturnal enuresis (Guilleminault et al., 1981; Weider and Hauri, 1985; Guilleminault and Stoohs, 1990; Brooks and Topol, 2003), episodic hypoxemia (Carroll, 2003), CO_2 retention (Carroll, 2003) and increased number of arousals and awakenings from sleep (Carroll, 2003).

Daytime symptoms include excessive irritability, excessive daytime sleepiness (Guilleminault et al., 1981; Chervin et al., 2002; Marcus et al., 2012), fatigue (Carroll, 2003), obesity (Guilleminault et al., 1981; Guilleminault and Stoohs, 1990; Arens and Muzumdar, 2010), daytime tiredness and poor school performance (Guilleminault et al., 1982; Guilleminault and Stoohs, 1990), inattention (Gozal, 1998; Chervin et al., 2002), hyperactivity (Guilleminault et al., 1981; Guilleminault and Stoohs, 1990; Chervin and Archbold, 2001; Chervin et al., 2002), confrontational behavior (Carroll, 2003) and other behavioral disturbances (Gozal, 1998).

PREVALENCE

The prevalence of POSA is reported as ranging between 1% and 4% (Carroll, 2003; Lumeng and Chervin, 2008; Bixler et al., 2009). OSA seems to be more frequent in African-American children (Redline et al., 1999; Rosen et al., 2003; Calhoun et al., 2010), obese children (Redline et al., 1999), children with respiratory disease (Redline et al., 1999; Kheirandish-Gozal et al., 2011; Ross et al., 2012; Ross, 2013) and children with a family history of OSA (Redline et al., 1999). Whether sex is a predisposing factor for OSA in children is unclear as some researchers reported no significant association (Ali et al., 1993; Redline et al., 1999; Calhoun et al., 2010).

Concerns also have been raised in regard to high SDB prevalence in populations with severe asthma (55%; Ross, 2013) or poorly controlled asthma (63%; Kheirandish-Gozal et al., 2011; Ross et al., 2012).

The American Academy of Sleep Medicine recommends diagnostic polysomnography (PSG) in children with chronic respiratory disorders only if there is clinical suspicion of SDB (Aurora et al., 2011). Some have argued that high prevalence is enough to warrant clinical suspicion and, therefore, referral for testing in these populations (Ross, 2013).

Even though habitual snoring is the most common reason for suspecting SDB in clinical practice, it has been described that only 49% of children with OSA reported habitual snoring, suggesting that clinicians should think carefully about referring children with severe or poorly controlled asthma for OSA even in the absence of habitual snoring. Better controlled studies are needed to confirm these findings and to improve our understanding of how asthma and SDB interact (Ross, 2013).

Furthermore, the effects of adenotonsillectomy on asthma outcomes were remarkable, although uncontrolled. Preliminary data suggest that treatment of SDB may result in significant improvements in asthma outcomes (Kheirandish-Gozal et al., 2011).

PATHOPHYSIOLOGY OF POSA

Sleep-related upper airway obstruction during childhood is a dynamic airway collapse related to muscle tone, structure and motor control (Carroll, 2003). The most common cause of POSA is adenotonsillar hypertrophy (Guilleminault and Stoohs, 1990; Ross, 2013). Other risk factors include exposure to environmental tobacco smoke (Ali et al., 1993; Calhoun et al., 2010), asthma (Valery et al., 2004), low socio-economic status (Ali et al., 1993; Oliveti et al., 1996; Calhoun et al., 2010), neuromuscular disorders (Marcus et al., 2012), obesity (Robertson, 1996; Kheirandish-Gozal et al., 2011; Ross et al., 2012), perinatal complications (Calhoun et al., 2010) and pre-term birth (Rosen et al., 2003).

There also are several additional craniofacial risk factors such as Class II malocclusion tendency (Flores-Mir et al., 2013), vertical facial growth direction (Flores-Mir et al., 2013), retrusive chin (Guilleminault and Stoohs, 1990; Flores-Mir et al., 2013), steep mandibular plane (Guilleminault and Stoohs, 1990; Flores-Mir et al., 2013), bruxism (Sjöholm et al., 2000; Archbold et al., 2002; Ng et al., 2002; DiFrancesco et al., 2004; Lam et al., 2011) and/or constricted and arched palate (Guilleminault and Stoohs, 1990). A recent systematic review (De Luca Canto et al., 2014*a*) did not find sufficient scientific evidence either to support or refute the association between bruxism and SDB.

POSA CONSEQUENCES

If POSA is left untreated, it can be a cause of significant morbidity in children that could lead to growth failure, cardiovascular effects including *cor pulmonale*, ventricular dysfunction, systemic hypertension and neurocognitive/behavioral abnormalities (Goldstein et al., 2012). Behavioral abnormalities include hyperactivity and attention deficit (Chervin et al., 2002), aggression (Chervin et al., 2003) and poor academic performance (Gozal, 1998). With timely diagnosis and treatment, most sequela can be avoided or reversed (Zettergren-Wijk et al., 2006; Wysocki et al., 2009).

ADENOID AND TONSILS

The adenoids and tonsils (Fig. 1) are part of a conglomerate of lymphatic tissue located around the nasopharyngeal airway. Adenoid and/or tonsil hypertrophy is a common cause of impaired nasal airflow and nasopharyngeal obstruction (Cho et al., 1999; Zettergren-Wijk et al., 2006; Abreu et al., 2008; Chandra et al., 2009). This condition is associated with the fact that the facial bones grow more slowly than the lymphoid

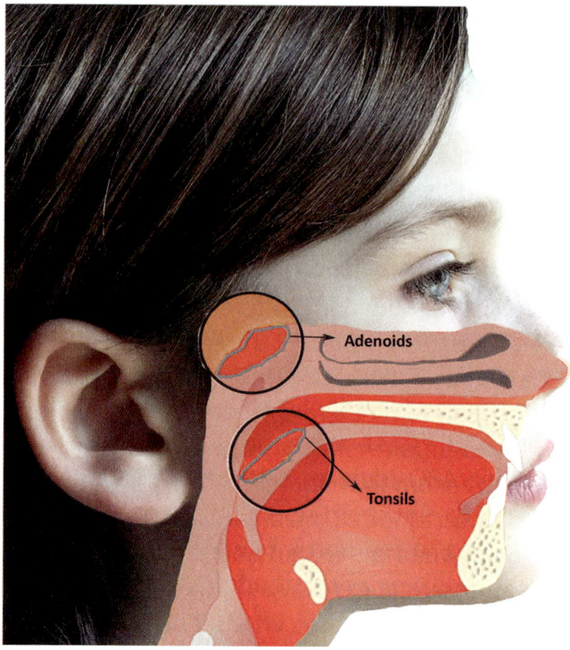

Figure 1. Adenoids and tonsils in normal relationship with nasopharynx.

tissue during childhood. Lymphoid tissues proliferate far beyond their adult size in late childhood and then undergo involution to reach adult size during puberty (Proffit, 2013). The estimated frequency of adenoid hypertrophy among children, six months through 14 years of age, range between 19.5 and 57.7% (Aydin et al., 2008; Bitar et al., 2009).

Nasal obstruction also is a potential cause of SDB and can be evaluated both objectively and subjectively. It is important to remember that patients may have a combination of multiple factors contributing to the symptoms of nasal obstruction (Chandra et al., 2009).

Constriction of the posterior upper airway caused by repeated adenoidal infection and inflammation or genetic factors has been a suggested reason of altered craniofacial development (Linder-Aronson and Henrikson, 1973; Behlfelt et al., 1990; Woodside et al., 1991; Katyal et al., 2013*b*). These characteristics have been associated with an "adenoid face" morphology that includes large facial height, narrow maxillary arch (Kim and Guilleminault, 2011; Katyal et al., 2013*a*), retrognathic mandible

(Gunn et al., 2000; Kim and Guilleminault, 2011) and posterior crossbite (Linder-Aronson and Henrikson, 1973; Huynh et al., 2011; Flores-Mir et al., 2013; Korayem et al., 2013). Additionally, anterior and lower anterior facial height (Wysocki et al., 2009; Flores-Mir et al., 2013) and mandibular plane angle (Wysocki et al., 2009; Huynh et al., 2011; Flores-Mir et al., 2013) have been linked to this clinical profile. Many of these same facial dimensions often are found in adults with OSA (Miles et al., 1996).

General dentists and orthodontists, therefore, should be familiar with nasopharyngeal obstruction management, not just because of its influence on craniofacial growth and malocclusion development, but also because of its potential relationship with SDB. If screened properly, dentists and orthodontists may be able to provide a timely referral that could have a significant impact in the management of this medical condition.

Pharyngoscopy is the reference standard test used by otorhinolaryngologists for a definitive diagnosis of nasal and nasopharyngeal obstruction (Kubba and Bingham, 2001; Wood, 2008; Yilmaz et al., 2008; Ysunza et al., 2008; Chandra et al., 2009; Aziz et al., 2014). Alternative tools for screening patients for nasal obstruction include clinical history (Bitar et al., 2006; Ciprandi et al., 2010), lateral cephalometry (Major et al., 2006; Wood, 2008; Fig. 2), peak nasal inspiratory flow (Chandra et al., 2009), fluoroscopy (Ysunza et al., 2008), rhinomanometry (Wood, 2008; André et al., 2009; Chandra et al., 2009), acoustic rhinometry (Cho et al., 1999; Kubba and Bingham, 2001; Abreu et al., 2008; André et al., 2009; Chandra et al., 2009; Aziz et al., 2014; Isaac et al., 2015), fiberoptic examination (Wang et al., 1992), computed tomography (CT; Osorio et al., 2008; Wood, 2008; Major et al., 2014*b*) and magnetic resonance imaging (Suto et al., 1996; Kao et al., 2008).

Because pharyngoscopy is outside the scope of dental practice, the challenge facing dentists is deciding which alternative diagnostic modality will provide them with the best indication of potential upper airway obstructive problem (Major et al., 2014*a*). Medical histories combined with a lateral cephalogram traditionally have been the best available tools for dentists justifying an adenoid obstruction diagnosis (Major et al., 2014*b*). However, a recent systematic review (Major et al., 2014*a*) suggested that lateral cephalogram is a tool more reliable when evaluating airway patency because it has a tendency to overestimate adenoid size.

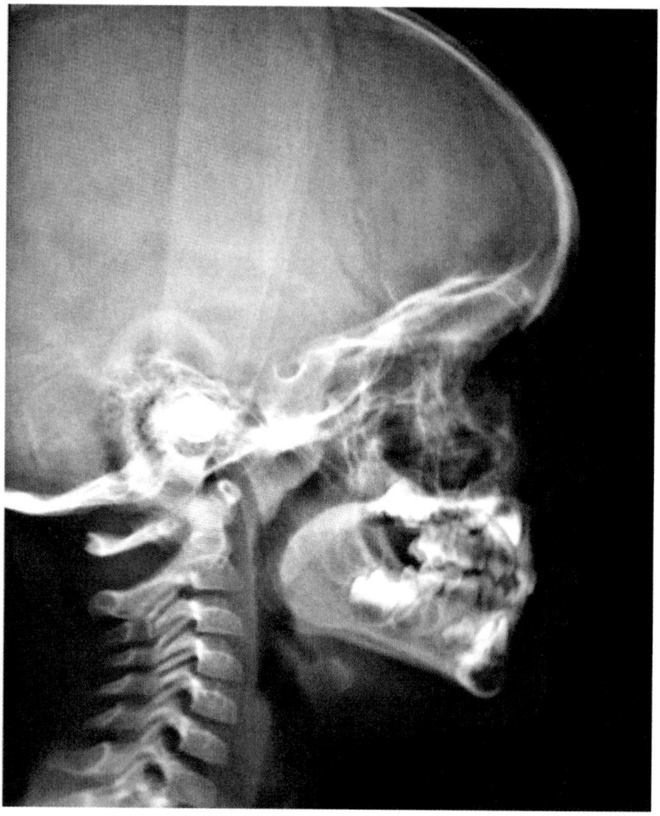

Figure 2. Lateral cephalogram traditionally used for airway evaluation.

In addition, this publication considered that clinical examination had only a fair diagnostic value.

Besides lateral cephalograms, the same systematic review determined the accuracy of other alternative upper-airway diagnostic tools compared with pharyngoscopy for diagnosing adenoid hypertrophy and concluded that video fluoroscopy and conventional CT had excellent accuracy for diagnosing adenoid hypertrophy, although both procedures require a significant amount of radiation (Major et al., 2014*a*). A clinical examination can be used to identify healthy patients, but it cannot be used to differentiate adenoid hypertrophy from any other cause of nasopharyngeal obstruction (Major et al., 2014*a*).

An emerging option in the dental field is cone-bean computed tomography (CBCT). This tool has shown to be able to identify adenoid size accurately with 88% sensitivity and 93% specificity (Major et al., 2014*b*). CBCT, along with a detailed clinical history, could be adequate for screening patients who would benefit from further otorhinolaryngology and/or sleep medicine follow-up. The caveat lies in the availability of CBCT imaging for those patients.

Screening for adenoid hypertrophy as the sole justification for CBCT imaging cannot be supported as of yet. The use of CBCT should remain a secondary assessment to imaging previously acquired for orthodontic reasons. As low as reasonably achievable (ALARA) principles of radiation hygiene should be respected, and dentists and orthodontists need to invest in CBCT continuing education to understand better CBCT imaging handling and interpretation (Major et al., 2014*b*).

Therefore, we still need to develop a reliable, safe, clinically practical and readily accessible tool for screening patients who may have adenoid hypertrophy (Major et al., 2014*a*). Future research should be focused on improving dentists' roles as early detectors of airway disturbances, whether by means of validating new methods (e.g., CBCT) or improving algorithms of existent diagnostic tools (Major et al., 2014*a*).

POSA DIAGNOSIS MODALITY

A detailed clinical history should include discussion of sleep-related symptoms and daytime consequences suggestive of SDB. The physical exam should include a detailed assessment of the upper airway, including craniofacial morphology, tonsillar size and nasal patency. Questionnaires, imaging and other diagnostic tools may be considered to support the diagnosis (Witmans and Young, 2011), but should be used only in the appropriate clinical context.

POSA currently is underdiagnosed (Gozal, 2012; Pliska et al., 2012) and undertreated (Gozal, 2012) due to a shortage of pediatric sleep laboratories and the high cost associated with such diagnostic methods. Waiting times between referral for evaluation and diagnosis commonly may take five to six months worldwide (Gozal and Kheirandish-Gozal, 2010; Gozal, 2012).

In adults, OSA is diagnosed after clinical signs and symptoms suggest the disorder is present; polysomnography (PSG) demonstrates recurrent complete obstruction (apneas) and/or partial obstructions associated with oxygen desaturation of at least 3% or an arousal from sleep (hypopneas). There must be at least five of these events per hour on average, resulting in an Apnea-Hypopnea Index (AHI) of 5 or more. In the absence of clinical symptoms, an AHI of 15 or greater is diagnostic of OSA.

Upper Airway Resistance Syndrome (UARS) is diagnosed when the snoring and resistance through the airway is significant enough to disrupt the quality of sleep and cause daytime impairment, but the PSG findings do not meet criteria for OSA. In UARS, PSG shows repetitive inspiratory flow limitation events that are associated with arousals but do not meet hypopnea criteria and/or there are fewer than five apneas/hypopneas per hour (Marcus et al., 2012).

Although the American Academy of Pediatrics' Clinical Practice Guideline defines childhood OSA similarly to the diagnostic criteria in adults described above, without the requirement for five respiratory events/hour determined by PSG (Marcus et al., 2012), there is no clear consensus on the required AHI index to diagnose POSA (Redline et al., 1999). Because of the cost and lack of availability of pediatric sleep laboratories, many pediatric patients are treated for OSA without having a PSG-confirmed diagnosis (Ross, 2013). Approximately 10% of children with chronic snoring referred for adenotonsillectomy undergo PSG to confirm the diagnosis (Mitchell et al., 2006; Fig. 3).

SDB is of significant relevance to practicing dentists and orthodontists, as it has been associated with a variety of oral and craniofacial problems, such as a retrusive chin (Flores-Mir et al., 2013), steep mandibular plane, Class II malocclusion (Flores-Mir et al., 2013), vertical growth direction (Flores-Mir et al., 2013) and bruxism (Sjöholm et al., 2000; Archbold et al., 2002; Ng et al., 2002; Lam et al., 2011).

Ideally, when children with a craniofacial morphology consistent with POSA are identified, it is necessary to investigate further into their medical histories of snoring, obesity, gasping for breath while sleeping, inability to breathe through the nose, significant environmental allergies and asthma. A full otorhinolaryngology assessment may be necessary for some patients.

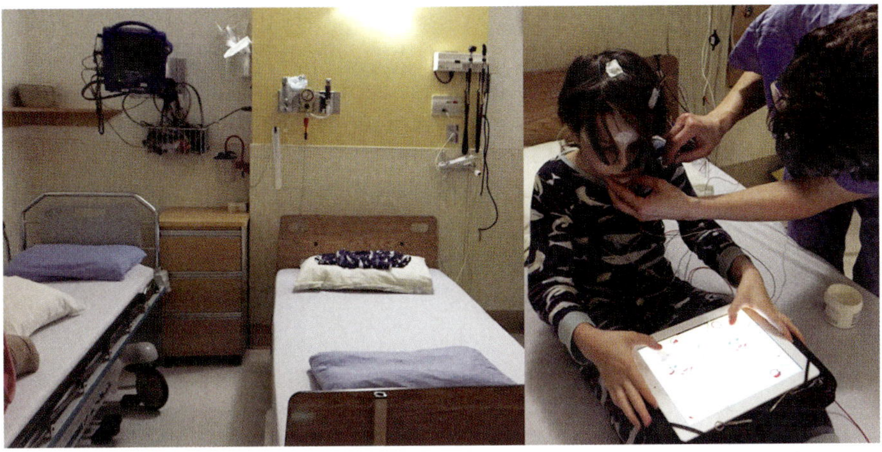

Figure 3. Pediatric sleep lab room for overnight sleep laboratory-based PSG.

Polysomnography (PSG)

An overnight sleep laboratory-based PSG is the benchmark for the diagnosis of SDB in children (Chervin et al., 2000; Gozal and Kheirandish-Gozal, 2010; Goldstein et al., 2012).

It is difficult, however, to define OSA exclusively using PSG-based criteria because there is no definitive set of PSG criteria that reliably will discriminate those patients who have obstructive sleep apnea syndrome (OSAS) and require treatment (Gozal and Kheirandish-Gozal, 2010). The AHI—the diagnostic PSG measure of OSA severity—is associated with two major limitations: 1) the clinically valid cut-off for normal AHI is unclear in children; and 2) no consensus has been achieved as to whether children with AHI values between the normal cut-off (< 1/hour total sleep time [TST]) and 5/hour TST should undergo adenotonsillectomy (Wong, 2011).

In addition, PSG measures are poor predictors of OSA-associated morbidities (Kheirandish-Gozal, 2010). Patients with similar OSA severity may present with greatly different clinical phenotypes. Children who are very symptomatic may have a "normal PSG" in the presence of habitual snoring. Conversely, asymptomatic snoring children may have concurrent and severe respiratory disturbance seen during their PSG (Gozal and Kheirandish-Gozal, 2010).

Supplemental Methods

Other clinical tools that have been evaluated include:

1. Medical history and physical examination (Goldstein et al., 2012; Spruyt and Gozal, 2012);
2. Audiotaping (Lamm et al., 1999);
3. Videotaping (Sivan et al., 1996);
4. Pulse oximetry (Brouillette et al., 2000);
5. Abbreviated polysomnography (aPSG; Finn and McNally, 2011);
6. Questionnaires (Chervin et al., 2000; Spruyt and Gozal, 2012); or
7. Multi-channel recordings.

These methods all have been assessed, albeit with variable success. Moreover, among these, special interest recently has centered on the identification of biomarkers (Kheirandish-Gozal, 2010; Wong, 2011; Gozal, 2012; De Luca Canto et al., 2015*a,b*).

Sleep Questionnaires and/or Clinical Signs/Symptoms. Only a few published questionnaires have been designed to assess for SDB-associated symptoms as they occur in children (Chervin et al., 2000; Spruyt and Gozal, 2012). For example, although excessive daytime sleepiness can affect adults or children with SDB, inattention and hyperactivity—often sufficient to result in a diagnosis of attention-deficit/hyperactivity disorder (ADHD)—may be more specific to children with SDB (Guilleminault et al., 1981, 1982; Stradling et al., 1990; Ali et al., 1996).

The Pediatric Sleep Questionnaire (PSQ; Fig. 4) is a validated questionnaire that includes three prominent symptom domains: snoring; excessive daytime sleepiness; and inattentive/hyperactive behavior. The PSQ was developed to explore potential utility in clinical research rather than to elaborate an instrument that reduces the need for PSG in clinical practice (Chervin et al., 2000).

Compared with the clinical history, physical examination or both, the PSQ has the best screening accuracy to identify possible SDB in children (Chervin et al., 2000; De Luca Canto et al., 2014*b*). It does not achieve diagnostic values high enough to replace the current reference

Pediatric Sleep Questionnaire

(Screening)

Name of the child:______________________________ Date of birth:__________

Person completing this form: ____

Date that you are completing the questionnaire: ____________

Instructions: Please answer the questions about how your child **IN THE PAST MONTH**. Circle the correct response or *print* your answers in the space provided. "Y" means "yes," "N" means "no," and "DK" means "don't know." For this questionnaire, the word "usually" means "more than half the time" or "on more than half the nights."

Please answer the following questions as they pertain to your child in the past month.

		YES	**NO**	**Don't Know**
1.	**While sleeping, does your child:**			
	Snore more than half the time?	Y	N	DK
	Always snore?	Y	N	DK
	Snore loudly?	Y	N	DK
	Have "heavy" or loud breathing?	Y	N	DK
	Have trouble breathing, or struggle to breath?	Y	N	DK
2.	**Have you ever seen your child stop breathing during the night?**	Y	N	DK
3.	**Does your child:**			
	Tend to breathe through the mouth during the day?	Y	N	DK
	Have a dry mouth on waking up in the morning?	Y	N	DK
	Occasionally wet the bed?	Y	N	DK
4.	**Does your child:**			
	Wake up feeling unrefreshed in the morning?	Y	N	DK
	Have a problem with sleepiness during the day?	Y	N	DK
5.	**Has a teacher or other supervisor commented that your child appears sleepy during the day?**	Y	N	DK
6.	**Is it hard to wake your child up in the morning?**	Y	N	DK
7.	**Does your child wake up with headaches in the morning?**	Y	N	DK
8.	**Did your child stop growing at a normal rate at any time since birth?**	Y	N	DK
9.	**Is your child overweight?**	Y	N	DK
10.	**This child often:**			
	Does not seem to listen when spoken to directly........	Y	N	DK
	Has difficulty organizing tasks and activities........	Y	N	DK
	Is easily distracted by extraneous stimuli	Y	N	DK
	Fidgets with hands or feet, or squirms in seat	Y	N	DK
	Is "on the go" or often acts as if "driven by a motor"	Y	N	DK
	Interrupts or intrudes on others (eg butts into conversations or games)	Y	N	DK

Figure 4. Pediatric Sleep Questionnaire (PSQ). ©2007 The Regents of The University of Michigan. From Chervin et al., 2000. Reprinted with permission of The University of Michigan.

standard, PSG; however, dentists could use the PSQ as a screening tool to identify pediatric patients at risk for SDB, which should improve the referral process to pediatric sleep specialists (De Luca Canto et al., 2014*b*).

Biomarkers. A biomarker is a "biological molecule found in blood, other body fluids, or tissues that is a sign of a normal or abnormal process, or a condition or disease." Gene expression (arrangement) has revealed significant and reproducible changes in a restricted number of genes that could enable discriminatory ability in the recognition of OSA. Similarly, a number of serum and urinary proteins have been identified that display favorable significant receiver-operator properties toward the diagnosis of OSA (Gozal, 2012).

Provided that acceptable sensitivity and specificity are achieved, a unique set of disease biomarkers would enable greatly simplified, user-friendly and context-relevant approaches to the diagnosis of OSA in the future (Wong, 2011). A biomarker, therefore, may be used to determine whether a disease is present or absent, or to delineate how well the patient responds to a treatment for a disease or condition. Of note, a biomarker also may be termed as "molecular marker and signature molecule." Research into the deep urinary proteome in children could lead the way for development of diagnostic kits using combinatorial biomarker approaches of utility in the clinical evaluation of the habitually snoring child (Gozal, 2012). These strategies presently are not part of routine clinical care.

The phenotypic variance in the clinical morbidity of OSA has prompted exploration of biomarkers that would enable the identification of the more "vulnerable" patients, who more likely would benefit from timely and targeted therapeutic interventions (Kheirandish-Gozal, 2010).

Over the last fourteen years, a substantial number of studies have been undertaken to identify an ideal biomarker for OSA. Although there still is no available simple and useful disease marker for OSA, considerable progress has been accomplished (Wong, 2011). So far, the most promising biomarker identified is centered on the identification of cardiovascular disease in OSA patients (De Luca Canto et al., 2015*a*).

In a recent systematic review (De Luca Canto et al., 2015*a*), blood-based biomarkers accounted for the majority of studies and most of the explored approaches did not identify definitive biomarkers of OSA morbidity. IL-6 and hsCRP appear to exhibit a favorable profile as biomarkers aiming to discriminate OSA patients with and without morbidity in adults, as well as MRP 8/14 in children. Additionally, urinary neurotransmitters potentially can be a good tool for screening cognitive function in children with OSA.

The ideal biomarker should reflect the severity of the clinical morbid manifestation and provide an indicator of cumulative adverse effects over time, as well as:

1. Show a value with minimum overlap between normal and diseased;
2. Be present in all patients with the end-organ dysfunction of interest;
3. Be detected before patients develop several clinical manifestations;
4. Reflect the severity of the disease over the cumulative period of the past; and
5. Demonstrate reversibility following proper treatment (Wong, 2011).

The study of De Luca Canto and associates (2015*b*) indicates that the combination of Kallikrein-1, uromodulin, urocortin-3 and orosomucoid-1 have enough accuracy to be used as an OSA diagnostic test in children. Plasma IL-6 and -10 levels potentially could prove to be good biomarkers in identifying adult individuals with and without OSA.

Thus, there is a critical need for future studies to assess and provide a cut-off value for each of the biomarkers being evaluated. This also will enable comparisons across studies, as well as across age groups (e.g., adults *versus* children; De Luca Canto et al., 2015*a*).

TREATMENT

Adenotonsillectomy

Treatment of POSA remains a challenge. The most common treatment approach for SDB for children is adenotonsillectomy (Guilleminault et al., 2007; Marcus et al., 2013); however, 47% of treated samples still had abnormal post-surgical sleep parameters (Guilleminault et al., 2007). This outcome likely was due to the multi-factorial nature of POSA because adenotonsillectomy does appear to reduce symptoms and improve quality of life (Gozal, 1998; Tauman et al., 2006), compared to watchful waiting (Marcus et al., 2013).

Adenotonsillectomy allows better air exchange by the elimination of soft tissues occupancy space, but the surgery itself does not resolve other anatomic problems that may be present in children with OSA. Therefore, it is crucial to address completely the anatomic impairment and residual abnormal breathing that persist after surgery. In this regard, orthodontic approaches have been shown to be useful in children with persistent SDB post-adenotonsillectomy (Tauman et al., 2006).

Continuous Positive Airway Pressure (CPAP)

Although CPAP usually is offered as first-line therapy to adults (Epstein et al., 2009), it has not been advised in children that do not have severe signs and symptoms due to its craniofacial side effects after prolonged use. Poor compliance is another limiting factor (Nazarali et al., 2015). CPAP, therefore, has been recommended in children when adenotonsillectomy is not indicated or is only partial effective (Katz and D'Ambrosio, 2008).

Orthodontics Appliances

There are a number of orthodontic treatment modalities that have been suggested to reduce the symptoms of POSA and also improve the associated craniofacial abnormalities. The success of orthodontic appliances in improving symptoms of OSA has been attributed to enlarging the airway. Examples of orthodontic methods used for treating OSA include:

1. Mandibular advancement (MA; Marklund et al., 2012; Villa et al., 2012; Huynh et al., 2016);
2. Rapid maxillary expansion (RME; Pirelli et al., 2004; Villa et al., 2007; Huynh et al., 2016); and
3. Orthopedic maxillary protraction (Hiyama et al., 2002).

Mandibular Advancement Appliances (MAAs). MAAs are the most common class of oral appliance used for the treatment of OSA in adults (Chan et al., 2010; Marklund et al., 2012). These devices aim to increase the upper airway size and, therefore, reduce the risk of sleep apneas and snoring in that group of patients (Marklund et al., 2012). Such appliances may be indicated in the treatment of patients with mild-to-moderate OSA, including patients who are unwilling or unable to wear CPAP.

In that a smaller mandible can be a risk factor of POSA, correction of craniofacial imbalances redirecting mandibular growth into a more forward and downward directions has been suggested (Huynh et al., 2016). During orthodontic treatment, the purpose of the MAAs is to alter the neuromuscular forces on the craniofacial skeleton and dentition, promoting a combination of dentoalveolar changes and skeletal growth (Katz and D'Ambrosio, 2008).

MAAs can be useful to reduce AHI in POSA patients; however, exploration of its long-term efficacy is required (Huynh et al., 2016). Although studies reported improvement in AHI during MAA treatment, none of the respiratory variables returned to normal pediatric reference values. This may suggest that other etiologic factors, not just an anatomical problem located in the oropharynx, play a role in POSA (Nazarali et al., 2015). Furthermore, the performance of MAA on OSA health outcomes (e.g., quality of life, neurocognitive function and cardiovascular health) remain significantly underinvestigated (Huynh et al., 2016). As a consequence, even though treatment using a MAA appears promising, patients still require follow-up and long-term monitoring from their physician (Nazarali et al., 2015).

The craniofacial abnormalities often seen associated with POSA patients are suggestive of the pertinence for interceptive orthodontic treatment, while simultaneously managing some symptoms of POSA. It is important to note that any oral appliance that repositions the mandible forward immediately will enlarge the upper airway space. A PSG should be performed with the MAA at the maximal protrusion setting to determine if this treatment fully addresses the patient's OSA. Because the main goals of treatment with MAAs are to achieve permanent changes in skeletal and dental relationships longer than one year, continuous treatment and adequate compliance are key factors that influence the response to MAA (Nazarali et al., 2015).

If treatment with MAAs in fact does demonstrate long-term stability, the correction of the craniofacial and dentoalveolar morphology may improve symptoms of POSA while taking advantage of the adolescent growth spurt to improve the associated malocclusion. Additionally, if permanent change is demonstrated, children may not need to wear the MAA permanently thereafter, as skeletal growth may have resolved one of the main contributing factors of the patient's POSA (Nazarali et al.,

2015). These patients should have ongoing clinical follow-up with consideration for a repeat PSG to confirm that OSA has resolved fully.

Rapid Maxillary Expansion (RME). RME is a well-known orthodontic procedure that has been used for many years in children. The bone distraction happening at the palatal suture level during RME likely causes an actual widening of the maxilla with increases to both the cross-section and the volumetric space of the nasal cavity. In addition, a downward and forward movement of the maxillary complex with an improvement in nasal airflow has been suggested (Pirelli et al., 2004). Radiographs of the region indicate that the RME position changes nasal and palatal bones (Pirelli et al., 2004).

Many patients with OSA exhibited not only craniofacial abnormalities, but also an affected respiratory dynamic space (Fransson et al., 2002). Nasal septal deviation (NSD) is a problem that often is congenital and perhaps determined genetically (Pirelli et al., 2004). NSD may cause abnormal nasal resistance (Hiyama et al., 2002; Villa et al., 2007) with consequent inadequate nasal airway flow (Aziz et al., 2015) and can lead to maxillary deficiency early in life (Pirelli et al., 2004). RME may be a useful approach in addressing NSD in early childhood, but its clinical impact on NSD remains questionable in adolescence (Aziz et al., 2015).

Furthermore, RME and associated orthodontic movements might modify the resting posture of the tongue (Huynh et al., 2016), causing indirect improvement of the oropharyngeal space (Harvold et al., 1981; Principato, 1991; Pirelli et al., 1995). It also has been suggested that RME is a valid treatment for OSA in children without enlarged tonsils and adenoids (Pirelli et al., 2004).

It has been shown that both RME (Villa et al., 2007; Katyal et al., 2013*a*; Huynh et al., 2016) and MAA (Huynh et al., 2016; Nazarali et al., 2015) reduced OSA symptoms immediately after using those appliances. Long-term stability of those changes remains questionable (Villa et al., 2007; Katyal et al., 2013*a*; Huynh et al., 2016). In addition, research combining mandibular advancement and RME to manage OSA is desirable (Huynh et al., 2016).

Maxillary Protraction Appliance (MPA). The facilitation of maxillary development in growing patients during MPA treatment might contribute to an increase in the nasopharyngeal dimension and, therefore, may improve the respiratory function of patients with maxillary hypoplasia (Ozbek et al., 1998; Hiyama et al., 2002).

Although the exact underlying mechanism of the increase in upper airway dimensions by maxillary protraction is uncertain, potential explanations include:

1. An increase in nasopharyngeal volume, possibly induced by maxillary forward growth (Hiyama et al., 2002);
2. A change in tongue posture could induce the soft palate to a more anterior position, which might result in an increase in nasopharyngeal dimensions (Ozbek et al., 1998); and
3. A clockwise rotation of the mandible also might influence the tongue posture (Hiyama et al., 2002).

CONCLUSIONS

PSDB is a multi-disciplinary, complex and dynamic disease. Professionals involved with the evaluation and management of PSDB face an ever-evolving landscape. Orthodontists need to recognize symptoms, signs, risk factors and comorbidities related to SDB. Multi-disciplinary early diagnosis and evaluation are important and demands close coordination between general dentists, pediatric dentists, orthodontists, pediatricians, otolaryngologists and sleep medicine specialists.

The craniofacial growth and development expertise of orthodontists makes them important in early detection of altered growth patterns that could be linked to airway dysfunction. Orthopedic facial management may become a meaningful management alternative for specific forms of PSDB through MAA, RME and MPA. The long-term impact of these approaches remains unknown. Finally, use of these approaches when the malocclusion does not justify them orthodontically is an unexplored area that merits further study.

REFERENCES

Abreu RR, Rocha RL, Lamounier JA, Guerra AF. Etiology, clinical manifestations and concurrent findings in mouth-breathing children. J Pediatr (Rio J) 2008;84(6):529-535. [In English and Portuguese.]

Ali NJ, Pitson D, Stradling JR. Sleep disordered breathing: Effects of adenotonsillectomy on behaviour and psychological functioning. Eur J Pediatr 1996;155(1):56-62.

Ali NJ, Pitson DJ, Stradling JR. Snoring, sleep disturbance, and behavior in 4-5 year olds. Arch Dis Child 1993;68(3):360-366.

André RF, Vuyk HD, Ahmed A, Graamans K, Nolst Trenité GJ. Correlation between subjective and objective evaluation of the nasal airway: A systematic review of the highest level of evidence. Clin Otolaryngol 2009;34(6):518-525.

Archbold KH, Pituch KJ, Panahi P, Chervin RD. Symptoms of sleep disturbances among children at two general pediatric clinics. J Pediatr 2002;140(1):97-102.

Arens R, Muzumdar H. Childhood obesity and obstructive sleep apnea syndrome. J Appl Physiol 2010;108(2):436-444.

Aurora RN, Zak RS, Karippot A, Lamm CI, Morganthaler TI, Auerbach SH, Bista SR, Casey KR, Chowdhuri S, Kristo DA, Ramar K; American Academy of Sleep Medicine. Practice parameters for the respiratory indications for polysomnography in children. Sleep 2011;34(3):379-388.

Aydin S, Sanli A, Celebi O, Tasdemir O, Paksoy M, Eken M, Hardal U, Ayduran E. Prevalence of adenoid hypertrophy and nocturnal enuresis in primary school children in Istanbul, Turkey. Int J Pediatr Otorhinolaryngol 2008;72(5):665-668.

Aziz T, Ansari K, Lagravère MO, Major MP, Flores-Mir C. Effect of nonsurgical maxillary expansion on the nasal septum deviation: A systematic review. Prog Orthod 2015;16:15.

Aziz T, Biron VL, Ansari K, Flores-Mir C. Measurement tools for the diagnosis of nasal septal deviation: A systematic review. J Otolaryngol Head Neck Surg 2014;43:11.

Behlfelt K, Linder-Aronson S, McWilliam J, Neander P, Laage-Hellman J. Cranio-facial morphology in children with and without enlarged tonsils. Eur J Orthod 1990;12(3):233-243.

Bitar MA, Birjawi G, Youssef M, Fuleihan N. How frequent is adenoid obstruction? Impact on the diagnostic approach. Pediatr Int 2009;51(4):478-483.

Bitar MA, Rahi A, Khalifeh M, Madanat LM. A suggested clinical score to predict the severity of adenoid obstruction in children. Eur Arch Otorhinolaryngol 2006;263(10):924-928.

Bixler EO, Vgontzas AN, Lin HM, Liao D, Calhoun S, Vela-Bueno A, Fedok F, Vlasic V, Graff G. Sleep disordered breathing in children in a general population sample: Prevalence and risk factors. Sleep 2009;32(6):731-736.

Brooks LJ, Topol HI. Enuresis in children with sleep apnea. J Pediatr 2003;142(5):515-518.

Brouillette RT, Morielli A, Leimanis A, Waters KA, Luciano R, Ducharme FM. Nocturnal pulse oximetry as an abbreviated testing modality for pediatric obstructive sleep apnea. Pediatrics 2000;105(2):405-412.

Calhoun SL, Vgontzas AN, Mayes SD, Tsaoussoglou M, Sauder K, Mahr F, Karippot A, Wisner K, Bixler EO. Prenatal and perinatal complications: Is it the link between race and SES and childhood sleep disordered breathing? J Clin Sleep Med 2010;6(3):264-269.

Carroll JL. Obstructive sleep-disordered breathing in children: New controversies, new directions. Clin Chest Med 2003;24(2):261-282.

Chan AS, Phillips CL, Cistulli PA. Obstructive sleep apnoea: An update. Intern Med J 2010;40(2):102-106.

Chandra RK, Patadia MO, Raviv J. Diagnosis of nasal airway obstruction. Otolaryngol Clin North Am 2009;42(2):207-225.

Chervin RD, Archbold KH. Hyperactivity and polysomnographic findings in children evaluated for sleep-disordered breathing. Sleep 2001;24(3): 313-320.

Chervin RD, Archbold KH, Dillon JE, Panahi P, Pituch KJ, Dahl RE, Guilleminault C. Inattention, hyperactivity and symptoms of sleep-disordered breathing. Pediatrics 2002;109(3):449-456.

Chervin RD, Dillon JE, Archbold KH, Ruzicka DL. Conduct problems and symptoms of sleep disorders in children. J Am Acad Child Adolesc Psychiatry 2003;42(2):201-208.

Chervin RD, Hedger K, Dillon JE, Pituch KJ. Pediatric sleep questionnaire (PSQ): Validity and reliability of scales for sleep-disordered breathing, snoring, sleepiness, and behavioral problems. Sleep Med 2000;1(1): 21-32.

Cho JH, Lee DH, Lee NS, Won YS, Yoon HR, Suh BD. Size assessment of adenoid and nasopharyngeal airway by acoustic rhinometry in children. J Laryngol Otol 1999;113(10):899-905.

Ciprandi G, Tosca MA, Signori A, Ameli F. Comparison between symptoms and endoscopy in children with nasal obstruction. Int J Pediatr Otorhinolaryngol 2010;74(12):1405-1408.

De Luca Canto G, Pachêco-Pereira C, Aydinoz S, Major PW, Flores-Mir C, Gozal D. Biomarkers associated with obstructive sleep apnea and morbidities: A scoping review. Sleep Med 2015*a*;16(3):347-357.

De Luca Canto G, Pachêco-Pereira C, Aydinoz S, Major PW, Flores-Mir C, Gozal D. Diagnostic capability of biological markers in assessment of obstructive sleep apnea: A systematic review and meta-analysis. J Clin Sleep Med 2015*b*;11(1):27-36.

De Luca Canto G, Singh V, Gozal D, Major PW, Flores-Mir C. Sleep bruxism and sleep disordered breathing: A systematic review. J Oral Facial Pain Headache 2014*a*;28(4):299-305.

De Luca Canto G, Singh V, Major MP, Witmans M, El-Hakim H, Major PW, Flores-Mir C. Diagnostic capability of questionnaires and clinical examinations to assess sleep-disordered breathing in children: A systematic review and meta-analysis. J Am Dent Assoc 2014*b*;145(2):165-178.

DiFrancesco RC, Junqueira PA, Trezza PM, de Faria ME, Frizzarini R, Zerati FE. Improvement of bruxism after T & A surgery. Int J Pediatr Otorhinolaryngol 2004;68(4):441-445.

Epstein LJ, Kristo D, Strollo PJ Jr, Friedman N, Malhotra A, Patil SP, Ramar K, Rogers R, Schwab RJ, Weaver EM, Weinstein MD; Adult Obstructive Sleep Apnea Task Force of the American Academy of Sleep Medicine. Clinical guideline for the evaluation, management and long-term care of obstructive sleep apnea in adults. J Clin Sleep Med 2009;5(3):263-276.

Finn D, McNally P. Estimated population prevalence of obstructive sleep apnoea in a community of German third graders. Eur Respir J 2011; 37(4):975.

Flores-Mir C, Korayem M, Heo G, Witmans M, Major MP, Major PW. Craniofacial morphological characteristics in children with obstructive sleep apnea syndrome: A systematic review and meta-analysis. J Am Dent Assoc 2013;144(3):269-277.

Fransson AM, Tegelberg A, Svenson BA, Lennartsson B, Isacsson G. Influence of mandibular protruding device on airway passages and dentofacial characteristics in obstructive sleep apnea and snoring. Am J Orthod Dentofacial Orthop 2002;122(4):371-379.

Goldstein NA, Stefanov DG, Graw-Panzer KD, Fahmy SA, Fishkin S, Jackson A, Sarhis JS, Weedon J. Validation of a clinical assessment score for pediatric sleep-disordered breathing. Laryngoscope 2012;122(9): 2096-2104.

Gozal D. Serum, urine and breath-related biomarkers in the diagnosis of obstructive sleep apnea in children: Is it for real? Curr Opin Pulm Med 2012;18(6):561-567.

Gozal D. Sleep-disordered breathing and school performance in children. Pediatrics 1998;102(3 Pt 1):616-620.

Gozal D, Kheirandish-Gozal L. New approaches to the diagnosis of sleep-disordered breathing in children. Sleep Med 2010;11(7):708-713.

Guilleminault C, Huang YS, Glamann C, Li K, Chan A. Adenotonsillectomy and obstructive sleep apnea in children: A prospective study. Otolaryngol Head Neck Surg 2007;136(2):169-175.

Guilleminault C, Korobkin R, Winkle R. A review of 50 children with obstructive sleep apnea syndrome. Lung 1981;159(5):275-287.

Guilleminault C, Stoohs R. Chronic snoring and obstructive sleep apnea syndrome in children. Lung 1990;168(Suppl):912-919.

Guilleminault C, Tilkian A, Dement WC. The sleep apnea syndromes. Annu Rev Med 1976;27:465-484.

Guilleminault C, Winkle R, Korobkin R, Simmons B. Children and nocturnal snoring: Evaluation of the effects of sleep related respiratory resistive load and daytime functioning. Eur J Pediatr 1982;139(3):165-171.

Gunn TR, Tonkin SL, Hadden W, Davis SL, Gunn AJ. Neonatal micrognathia is associated with small upper airways on radiographic measurement. Acta Paediatr 2000;89(1):82-87.

Harvold EP, Tomer BS, Vargervik K, Chierici G. Primate experiments on oral respiration. Am J Orthod 1981;79(4):359-372.

Hiyama S, Suda N, Ishii-Suzuki M, Tsuiki S, Ogawa M, Tsuiki S, Ogawa M, Suzuki S, Kuroda T. Effects of maxillary protraction on craniofacial structures and upper-airway dimension. Angle Orthod 2002;72(1):43-47.

Huynh NT, Desplats E, Almeida FR. Orthodontics treatments for managing obstructive sleep apnea syndrome in children: A systematic review and meta-analysis. Sleep Med Rev 2016;25:84-94.

Huynh NT, Morton PD, Rompré PH, Papadakis A, Remise C. Associations between sleep-disordered breathing symptoms and facial and dental morphometry, assessed with screening examinations. Am J Orthod Dentofacial Orthop 2011;140(6):762-770.

Isaac A, Major M, Witmans M, Alrajhi Y, Flores-Mir C, Major P, Alsufyani N, Korayem M, El-Hakim H. Correlations between acoustic rhinometry, subjective symptoms, and endoscopic findings in symptomatic children with nasal obstruction. JAMA Otolaryngol Head Neck Surg 2015; 141(6):550-555.

Kao DS, Soltysik DA, Hyde JS, Gosain AK. Magnetic resonance imaging as an aid in the dynamic assessment of the velopharyngeal mechanism in children. Plast Reconstr Surg 2008;122(2):572-577.

Katyal V, Pamula Y, Daynes CN, Martin J, Dreyer CW, Kennedy D, Sampson WJ. Craniofacial and upper airway morphology in pediatric sleep-disordered breathing and changes in quality of life with rapid maxillary expansion. Am J Orthod Dentofacial Orthop 2013*a*;144(6):860-871.

Katyal V, Pamula Y, Martin AJ, Daynes CN, Kennedy JD, Sampson WJ. Craniofacial and upper airway morphology in pediatric sleep-disordered breathing: Systematic review and meta-analysis. Am J Orthod Dentofacial Orthop 2013*b*;143(1):20-30.e3.

Katz ES, D'Ambrosio CM. Pathophysiology of pediatric obstructive sleep apnea. Proc Am Thorac Soc 2008;5(2):253-262.

Kheirandish-Gozal L. What is "abnormal" in pediatric sleep? Respir Care 2010;55(10):1366-1374.

Kheirandish-Gozal L, Dayyat EA, Eid NS, Morton RL, Gozal D. Obstructive sleep apnea in poorly controlled asthmatic children: Effect of adenotonsillectomy. Pediatr Pulmonol 2011;46(9):913-918.

Kim JH, Guilleminault C. The nasomaxillary complex, the mandible, and sleep-disordered breathing. Sleep Breath 2011;15(2):185-193.

Korayem MM, Witmans M, MacLean J, Heo G, El-Hakim H, Flores-Mir C, Major PW. Craniofacial morphology in pediatric patients with persistent obstructive sleep apnea with or without positive airway pressure therapy: A cross-sectional cephalometric comparison with controls. Am J Orthod Dentofacial Orthop 2013;144(1):78-85.

Kubba H, Bingham BJ. Endoscopy in the assessment of children with nasal obstruction. J Laryngol Otol 2001;115(5):380-384.

Lam MH, Zhang J, Li AM, Wing YK. A community study of sleep bruxism in Hong Kong children: Associations with comorbid sleep disorders and neurobehavioral consequences. Sleep Med 2011;12(7):641-645.

Lamm C, Mandeli J, Kattan M. Evaluation of home audiotapes as an abbreviated test for obstructive sleep apnea syndrome (OSAS) in children. Pediatr Pulmonol 1999;27(4):267-272.

Linder-Aronson S, Henrikson CO. Radiocephalometric analysis of anteroposterior nasopharyngeal dimensions in 6- to 12-year-old mouth breathers compared with nose breathers. ORL J Otorhinolaryngol Relat Spec 1973;35(1):19-29.

Lumeng JC, Chervin RD. Epidemiology of pediatric obstructive sleep apnea. Proc Am Thorac Soc 2008;5(2):242-252.

Major MP, Flores-Mir C, Major PW. Assessment of lateral cephalometric diagnosis of adenoid hypertrophy and posterior upper airway obstruction: A systematic review. Am J Orthod Dentofacial Orthop 2006; 130(6):700-708.

Major MP, Saltaji H, El-Hakim H, Witmans M, Major P, Flores-Mir C. The accuracy of diagnostic tests for adenoid hypertrophy: A systematic review. J Am Dent Assoc 2014*a*;145(3):247-254.

Major MP, Witmans M, El-Hakim H, Major PW, Flores-Mir C. Agreement between cone-beam computed tomography and nasoendoscopy evaluations of adenoid hypertrophy. Am J Orthod Dentofacial Orthop 2014*b*;146(4):451-459.

Marcus CL, Brooks LJ, Draper KA, Gozal D, Halbower AC, Jones J, Schechter MS, Ward SD, Sheldon SH, Shiffman RN, Lehmann C, Spruyt K; American Academy of Pediatrics. Diagnosis and management of childhood obstructive sleep apnea syndrome. Pediatrics 2012;130(3):e714-e755.

Marcus CL, Moore RH, Rosen CL, Giordani B, Garetz SL, Taylor HG, Mitchell RB, Amin R, Katz ES, Arens R, Paruthi S, Muzumdar H, Gozal D, Thomas NH, Ware J, Beebe D, Snyder K, Elden L, Sprecher RC, Willging

P, Jones D, Bent JP, Hoban T, Chervin RD, Ellenberg SS, Redline S; Childhood Adenotonsillectomy Trial (CHAT). A randomized trial of adenotonsillectomy for childhood sleep apnea. N Eng J Med 2013; 368(25):2366-2376.

Marklund M, Verbraecken J, Randerath W. Non-CPAP therapies in obstructive sleep apnoea: Mandibular advancement device therapy. Eur Respir J 2012;39(5):1241-1247.

Miles PG, Vig PS, Weyant RJ, Forrest TD, Rockette HE Jr. Craniofacial structure and obstructive sleep apnea syndrome: A qualitative analysis and meta-analysis of the literature. Am J Orthod Dentofacial Orthop 1996; 109(2):163-172.

Mitchell RB, Pereira KD, Friedman NR. Sleep-disordered breathing in children: Survey of current practice. Laryngoscope 2006;116(6):956-958.

Nazarali N, Altalibi M, Nazarali S, Major MP, Flores-Mir C, Major PW. Mandibular advancement appliances for the treatment of paediatric obstructive sleep apnea: A systematic review. Eur J Orthod 2015; 37(6):618-626.

Ng DK, Kwok KL, Poon G, Chau KW. Habitual snoring and sleep bruxism in a paediatric outpatient population in Hong Kong. Singapore Med J 2002;43(11):554-556.

Oliveti JF, Kercsmar CM, Redline S. Pre- and perinatal risk factors for asthma in inner city African-American children. Am J Epidemiol 1996; 143(6):570-577.

Osorio F, Perilla M, Doyle DJ, Palomo JM. Cone beam computed tomography: An innovative tool for airway assessment. Anesth Analg 2008; 106(6):1803-1807.

Ozbek MM, Memikoglu TU, Gögen H, Lowe AA, Baspinar E. Oropharyngeal airway dimensions and functional-orthopedic treatment in skeletal Class II cases. Angle Orthod 1998;68(4):327-336.

Pirelli P, Marullo M, Casagrande M, Tornaghi M. Espansione rapida del mascellare: Effetti sulla funzionalità respiratoria ed uditiva. Mond Ortodont 1995;20:129-135. [In Italian.]

Pirelli P, Saponara M, Guilleminault C. Rapid maxillary expansion in children with obstructive sleep apnea syndrome. Sleep 2004;27(4):761-766.

Pliska B, Lowe AA, Almeida FR. The orthodontist and the obstructive sleep apnea patient. Int J Orthod Milwaukee 2012;23(3):19-22.

Principato JJ. Upper airway obstruction and craniofacial morphology. Otolaryngol Head Neck Surg 1991;104(6):881-890.

Proffit WR. Concepts of growth and development. In Proffit WR, Fields HW Jr, Sarver DM, eds. *Contemporary Orthodontics*. 5th ed. St. Louis: Elsevier Mosby 2013:26-31.

Redline S, Tishler PV, Schluchter M, Aylor J, Clark K, Graham G. Risk factors for sleep-disordered breathing in children: Associations with obesity, race, and respiratory problems. Am J Respir Crit Care Med 1999; 159(5 Pt 1):1527-1532.

Robertson CJ. Obstructive sleep apnoea: Part I. Diagnosis, aetiology, and current treatment. N Z Dent J 1996;92(4):110-113.

Rosen CL, Larkin EK, Kirchner HL, Emancipator JL, Bivins SF, Surovec SA, Martin RJ, Redline S. Prevalence and risk factors for sleep-disordered breathing in 8- to 11-year-old children: Association with race and prematurity. J Pediatr 2003;142(4):383-389.

Ross K. Sleep-disordered breathing and childhood asthma: Clinical implications. Curr Opin Pulm Med 2013;19(1):79-83.

Ross KR, Storfer-Isser A, Hart MA, Kibler AM, Rueschman M, Rosen CL, Kercsmar CM, Redline S. Sleep-disordered breathing is associated with asthma severity in children. J Pediatr 2012;160(5):736-742.

Sivan Y, Kornecki A, Schonfeld T. Screening obstructive sleep apnoea syndrome by home videotape recording in children. Eur Resp J 1996; 9(10):2127-2131.

Sjöholm TT, Lowe AA, Miyamoto K, Fleetham JA, Ryan CF. Sleep bruxism in patients with sleep-disordered breathing. Arch Oral Biol 2000; 45(10):889-896.

Spruyt K, Gozal D. Screening of pediatric sleep-disordered breathing: A proposed unbiased discriminative set of questions using clinical severity scales. Chest 2012;142(6):1508-1515.

Stradling JR, Thomas G, Warley AR, Williams P, Freeland A. Effect of adenotonsillectomy on nocturnal hypoxaemia, sleep disturbance, and symptoms in snoring children. Lancet 1990;335(8684):249-253.

Suto Y, Matsuda E, Inoue Y. MRI of the pharynx in young patients with sleep disordered breathing. Br J Radiol 1996;69(827):1000-1004.

Tauman R, Gulliver TE, Krishna J, Montgomery-Downs HE, O'Brien LM, Ivanenko A, Gozal D. Persistence of obstructive sleep apnea syndrome in children after adenotonsillectomy. J Pediatr 2006;149(6):803-808.

Valery PC, Masters IB, Chang AB. Snoring and its association with asthma in Indigenous children living in the Torres Strait and Northern Peninsula Area. J Paediatr Child Health 2004;40(8):461-465.

Villa MP, Malagola C, Pagani J, Montesano M, Rizzoli A, Guilleminault C, Ronchetti R. Rapid maxillary expansion in children with obstructive sleep apnea syndrome: 12-month follow-up. Sleep Med 2007;8(2): 128-134.

Villa MP, Miano S, Rizzoli A. Mandibular advancement devices are an alternative and valid treatment for pediatric obstructive sleep apnea syndrome. Sleep Breath 2012;16(4):971-976.

Wang D, Clement P, Kaufman L, Derde MP. Fiberoptic examination of the nasal cavity and nasopharynx in children. Int J Pediatr Otorhinolaryngol 1992;24(1):35-44.

Weider DJ, Hauri PJ. Nocturnal enuresis in children with upper airway obstruction. Int J Pediatr Otorhinolaryngol 1985;9(2):173-182.

Witmans M, Young R. Update on pediatric sleep-disordered breathing. Pediatr Clin North Am 2011;58(3):571-589.

Wong TK. The search on an ideal disease marker for childhood obstructive sleep apnea syndrome. Sleep 2011;34(2):133-134.

Wood RE. Evaluation of the upper airway in children. Curr Opin Pediatr 2008;20(3):266-271.

Woodside DG, Linder-Aronson S, Lundström A, McWilliam J. Mandibular and maxillary growth after changed mode of breathing. Am J Orthod Dentofacial Orthop 1991;100(1):1-18.

Wysocki J, Krasny M, Skarzyński PH. Patency of nasopharynx and a cephalometric image in the children with orthodontic problems. Int J Pediatr Otorhinolaryngol 2009;73(12):1803-1809.

Yilmaz I, Caylakli F, Yilmazer C, Sener M, Ozluoglu LN. Correlation of diagnostic systems with adenoidal tissue volume: A blind prospective study. Int J Pediatr Otorhinolaryngol 2008;72(8):1235-1240.

Ysunza A, Pamplona MC, Ortega JM, Prado H. Video fluoroscopy for evaluating adenoid hypertrophy in children. Int J Pediatr Otorhinolaryngol 2008;72(8):1159-1165.

Zettergren-Wijk L, Forsberg CM, Linder-Aronson S. Changes in dentofacial morphology after adeno-/tonsillectomy in young children with obstructive sleep apnoea: A 5-year follow-up study. Eur J Orthod 2006; 28(4):319-326.

OBSTRUCTIVE SLEEP APNEA IN CHILDREN: MORE THAN JUST LARGE TONSILS

Timothy F. Hoban

ABSTRACT

Modern scientific descriptions of obstructive sleep apnea (OSA) date from the 1960s, when the syndrome first was reported in the adult population. It subsequently was discovered that the condition also is relatively common in children, but that the pathophysiology of airway obstruction, clinical consequences and optimal treatments differ considerably for children compared to adults. This chapter reviews current understanding of OSA in children, including historical aspects, clinical symptoms, comorbidities, diagnostic testing and discussion of first-line medical and surgical treatments.

KEY WORDS: obstructive sleep apnea, adenotonsillectomy, CPAP, polysomnography, pediatric

---

INTRODUCTION AND HISTORICAL ASPECTS

> *The establishment of mouth-breathing is the symptom which first attracts the attention. It is not so noticeable by day, although the child may present the vacant expression characteristic of this condition.*
>
> *At night the child's sleep is greatly disturbed; the respirations are loud and snorting, and there are sometimes prolonged pauses, followed by deep, noisy inspirations.* -Osler, 1892

> *The fact, however, that children, the victims of nasal and pharyngeal obstructions, often suffer from headaches, especially when engaged in study, and frequently evince marked inability to fix their attention on their lessons or work for any length of time, has in recent years led many to suspect that these symptoms [are] in part a reflection of some evident hampering of the cerebral functions.*
>
> -Hill, 1889

Although clinically astute descriptions of childhood obstructive sleep apnea (OSA) and its association with impaired learning and attention date from the late 18th century, the first report correlating these clinical features with polysomnographic (PSG) findings dates only from 1976, a decade after OSA first was reported in the adult population (Guilleminault et al., 1976).

Research during the subsequent 40 years has suggested that the pathophysiology of OSA differs considerably for children compared to adults. Children more commonly experience partial airway obstruction associated with arousals from sleep compared to the episodic complete airway obstruction associated with cyclical desaturations of blood oxygen (SpO2) that typically are seen in adults (Ward and Marcus, 1996). In addition, children with airway obstruction during sleep often experience clinically significant symptoms even when the frequency of respiratory disturbances is relatively low and possibly even normal by adult standards (Rosen et al., 1992).

Recognition of the fundamental differences between adult and pediatric OSA led to the establishment of pediatric-specific diagnostic criteria, summarized in Table 1 (Sateia, 2014). Diagnosis of OSA in a child requires verification of referable symptoms either during sleep (e.g., snoring) or wakefulness (e.g., learning problems) in conjunction with PSG evidence of airway obstruction (e.g., at least one obstructive respiratory disturbance per hour of sleep).

CLINICAL FEATURES OF CHILDHOOD OSA

Children with OSA usually exhibit some degree of noisy respiration during sleep, although this usually is subtler compared to the loud snoring often manifested by adults with the condition. Mouth breathing, restlessness and diaphoresis during sleep represent other common nocturnal symptoms in affected children (Brouillette and Hanson, 1984). Because children with OSA more commonly experience partial rather than complete airway obstruction during sleep, witnessed respiratory pauses seldom are observed and reported by parents.

Daytime symptoms for children with OSA are highly variable. Some children exhibit sore throat, dry mouth, grogginess or headache upon morning waking. Daytime somnolence usually is more subtle for children with OSA compared to adults and sometimes is evident only

Table 1. Diagnostic criteria for pediatric obstructive sleep apnea (OSA). Criteria A and B both must be satisfied. Adapted from The International Classification of Sleep Disorders, 3rd ed. (Sateia, 2014).

A. Presence of one or more of the following symptoms:
   1. Snoring
   2. Obstructed, labored or paradoxical respiration during sleep
   3. Hyperactivity, disturbed behavior, learning problems or sleepiness
B. PSG demonstrates one or both of the following findings:
   1. One or more obstructive apneas, hypopneas or mixed apneas/hour of sleep (excludes central apneas); or
   2. Presence of obstructive hypoventilation ($\geq$ 25% of total sleep time spent with PCO2 > 50 mm Hg) associated with one or more of the following findings:
      a. Snoring;
      b. Flattening/flow restriction of the inspiratory nasal pressure waveform; or
      c. Paradoxical thoraco-abdominal effort

intermittently during sedentary activities (e.g., riding in an automobile). Inattention, hyperactivity, impaired learning and disturbed behavior represent the most common waking symptoms of OSA during childhood, but these can be variable in severity and easily misattributed to other potential causes (e.g., attention-deficit hyperactivity disorder [ADHD]; O'Brien et al., 2003, 2004).

The physical examination often is normal for children with OSA, but predisposing anatomic features may be apparent in some children. The presence or absence of tonsillar hypertrophy can be verified easily on physical examination, where tonsillar size is graded on a scale of 1+ (tonsils hidden by tonsillar pillars) to 4+ (tonsils extending to the midline). Although the adenoids cannot be visualized directly without endoscopic evaluation, some children with adenoidal hypertrophy exhibit "adenoid facies," characterized by a drooping, elongated facial appearance with mouth breathing while awake.

Adenotonsillar hypertrophy is neither necessary nor sufficient, however, for the diagnosis of childhood OSA. Other anatomic and functional factors also can cause airway obstruction during sleep and adenotonsillar enlargement in isolation does not universally cause clinically significant airway obstruction. Craniofacial features associated with increased risk for childhood OSA include narrow or high-arched palate, mandibular or maxillary hypoplasia, low-lying soft palate and macroglossia.

Obesity also represents a common and increasingly prevalent physical finding among children with OSA. Among a group of 22 obese adolescents without specific sleep complaints, ten (45%) were found to have OSA or obstructive hypoventilation/hypercapnia (Marcus and Curtis, 1996).

The prevalence of OSA in the general pediatric population is estimated to be about 2%. Although OSA may occur at any point during childhood, it is most common between two and eight years of age, when adenotonsillar size often is maximal relative to airway size (Marcus, 2001).

A variety of medical conditions are thought to be associated with increased risk for OSA during childhood (Table 2). It is estimated that more than 33% of children with Trisomy 21 (Down syndrome) may have OSA, potentially related to the macroglossia, hypotonia and obesity that typically are associated with the condition (de Miguel-Diez et al., 2003; Ng et al., 2006). Craniofacial disorders also are associated with increased risk for childhood OSA, including conditions associated with mandibular or maxillary hypoplasia (e.g., Pierre Robin sequence), skeletal dysplasias associated with substantial craniofacial involvement (e.g., achondroplasia; Klippel-Feil syndrome, characterized by congenital fusion of any two of the seven cervical vertebrae) and disorders associated with cleft palate (Cielo and Marcus, 2015). Airway obstruction during sleep potentially can worsen following surgical repair of cleft palate (Smith et al., 2013).

TREATMENT OF CHILDHOOD OSA

Treatment of OSA in children differs considerably compared to adults, for whom continuous positive airway pressure (CPAP) and oral appliances represent the most frequent forms of therapy.

Adenotonsillectomy represents the most commonly administered treatment for affected children, although the procedure does not "cure" OSA in all cases (Hoban, 2005). One large case series reported that only 25% of children with OSA achieved complete normalization of respiratory parameters on PSG following adenotonsillectomy (Tauman et al., 2006). A larger controlled multi-center trial where 397 children with OSA were randomized either to adenotonsillectomy or watchful waiting reported

Table 2. Disorders associated with increased risk for childhood OSA.

Disorders associated with increased risk for childhood OSA
Disorders associated with micrognathia
Pierre Robin syndrome
Treacher-Collins syndrome
Russel-Silver syndrome
Disorders associated with maxillary hypoplasia
Crouzon syndrome
Apert syndrome
Pfeiffer syndrome
Disorders frequently associated with cleft palate
Velocardiofacial syndrome
Pierre Robin syndrome
Apert syndrome
Other medical and genetic disorders associated with risk for childhood OSA
Obesity
Down syndrome
Prader Willi syndrome
Skeletal dysplasias (e.g., achondroplasia, Klippel-Feil syndrome)

more frequent normalization of PSG findings in children receiving surgical treatment (79%) compared to children assigned to watchful waiting (46%; Marcus et al., 2013). Factors associated with increased risk for persistent OSA following adenotonsillectomy include high preoperative severity of OSA, obesity and atypical upper airway anatomy (e.g., low-lying soft palate, narrow or high-arched hard palate, or maxillary/mandibular hypoplasia).

Effective treatment of childhood OSA *via* adenotonsillectomy often is associated with tangible improvements in attention, learning, behavior, or quality of life measures in addition to alleviation of obstructive symptoms during sleep (Tran et al., 2005; Chervin et al., 2006). Several reports suggest that excessive healthcare costs and utilization among children with OSA may improve following adenotonsillectomy treatment (Reuveni et al., 2002; Tarasiuk et al., 2004).

The potential benefits of adenotonsillectomy for treatment of childhood OSA must be balanced against the risks and side effects of the procedure. Post-operative side effects typically are mild and self-limited in healthy, older children. Those who are young, have severe OSA or have associated medical or craniofacial disorders are at increased risk for more frequent and severe complications, which may include bleeding, infection or respiratory compromise (McColley et al., 1992; Schechter et al., 2002). Children in these high-risk groups often are hospitalized for close post-operative monitoring rather than receiving outpatient surgery.

CPAP represents the most common non-surgical treatment for children with OSA. It is used most often for children who are not appropriate candidates for adenotonsillectomy and for whom OSA persists despite adenotonsillectomy. It is considered a first-line treatment for children of all ages, although it has not been approved formally by the United States Food and Drug Administration for use in children weighing less than 30 kg. A variety of CPAP masks are available in sizes suitable for children, including nasal, oral-nasal and full-face masks.

Side effects of CPAP treatment for children generally are mild and self-limited. Mild irritation of skin under the CPAP mask is not uncommon, but typically responds to cleaning or replacement of the mask cushion, use of mask liners, loosening of mask fit or transition to an alternate mask interface. Nasal congestion or dryness often respond to use of a CPAP humidifier (Marcus et al., 2006). Serious side effects are uncommon, but skin breakdown may occur when irritation from a CPAP mask is not treated adequately. Acquired mid-face hypoplasia secondary to pressure effects from the CPAP mask has been reported as a rare complication of long-term treatment (Li et al., 2000).

As is the case for adults, achieving long-term adherence to CPAP treatment of OSA sometimes is challenging for children (Marcus et al., 2006). Long-term adherence is achieved most consistently in children who understand the need for CPAP, recognize the benefits of therapy and receive appropriate support and assistance from parents to maintain regular use. Adherence is most challenging for children with developmental disabilities or behavioral disorders, or when the child and his/her parents do not understand or accept the importance of maintaining consistent use. The likelihood of successful long-term adherence to CPAP treatment can be improved through use of age-appropriate desensitization and behavioral reinforcement techniques (Rains, 1995; Massa et al., 2002).

Successful adherence to CPAP treatment of childhood OSA is associated with positive neurobehavioral outcomes, including improvements in daytime attention, behavior and sleepiness (Marcus et al., 2006, 2012).

In cases where neither adenotonsillectomy nor CPAP represent feasible treatment options, alternative modes of therapy can be considered. *Supplemental oxygen* by nasal cannula is considered a partially effective treatment which often alleviates the severity of oxygen desaturation with less impact upon the frequency or severity of apneic events (Marcus et al., 1995). Sleep-related hypoventilation/hypercapnia represents an occasional complication of this treatment, so it is recommended that oxygen therapy for childhood OSA be titrated carefully during PSG with CO2 monitoring.

Several studies have reported that *nasal steroids* may alleviate upper airway obstruction in some children with OSA (Alexopoulos et al., 2004; Mansfield et al., 2004). A double-blind, randomized, crossover trial of intra-nasal budesonide *versus* placebo in children with mild OSA reported normalization of PSG measures for 54% of subjects following six weeks of active treatment (Kheirandish-Gozal and Gozal, 2008). It remains uncertain, however, to what extent nasal steroids are effective and safe for longer-term treatment of childhood OSA.

Alternative surgical treatments for OSA are approached cautiously in children due to the paucity of pediatric-specific data regarding safety and efficacy. *Uvulopalatopharyngoplasty* (UPPP) is performed for some children with low-lying soft palates or redundant pharyngeal soft tissue—sometimes concurrently with adenotonsillectomy—with some reports of clinical or PSG improvements in sleep (Kerschner et al., 2002). For children with significant micrognathia, *distraction osteogenesis* or other *mandibular advancement procedures* sometimes are performed (Rachmiel et al., 2012). When significant maxillary hypoplasia is present, *Le Fort advancement procedures* sometimes alleviate upper airway obstruction during sleep while correcting cosmetic deformity (Nout et al., 2012).

Treatment of childhood OSA using orthodontic procedures (e.g. *rapid maxillary expansion*; Table 3) or *mandibular advancing oral appliances* are discussed in a separate chapter within this monograph (Conley, 2018).

CONCLUSIONS

The pathophysiology, clinical manifestations and treatment of OSA in children are distinctly different compared to the adult population.

Table 3. First-line and alternative treatments for childhood OSA.

Treatments
First-line treatments
Adenotonsillectomy
CPAP
Alternative medical treatments
Oxygen by nasal cannula
Nasal steroids
Alternative surgical treatments
Uvulopalatopharyngoplasty (UPPP)
Maxillary or mandibular advancements procedures
Alternative orthodontic/dental treatments
Maxillary expansion
Mandibular advancing oral appliances

The fact that adenotonsillectomy does not cure the condition in all children underscores the importance of identifying and treating other potential causes of upper airway obstruction, in addition to adenotonsillar hypertrophy.

Although adenotonsillectomy and CPAP represent widely available and reasonably effective treatments for OSA in children, these therapies are not feasible or always fully effective, particularly in children with associated medical, craniofacial or neurodevelopmental disorders. Further work remains necessary to identify a broader range of effective medical, surgical, orthodontic and functional dental therapies for use in children with OSA.

REFERENCES

Alexopoulos EI, Kaditis AG, Kalampouka E, Kostadima E, Angelopoulos NV, Mikraki V, Skenteris N, Gourgoulianis K. Nasal corticosteroids for children with snoring. Pediatr Pulmonol 2004;38(2):161-167.

Brouillette R, Hanson D, David R, Klemka L, Szatkowski A, Fernbach S, Hunt C. A diagnostic approach to suspected obstructive sleep apnea in children. J Pediatr 1984;105(1):10-14.

Chervin RD, Ruzicka DL, Giordani BJ, Weatherly RA, Dillon JE, Hodges EK, Marcus CL, Guire KE. Sleep-disordered breathing, behavior, and cognition in children before and after adenotonsillectomy. Pediatrics 2006;117(4):e769-e778.

Cielo CM, Marcus CL. Obstructive sleep apnoea in children with craniofacial syndromes. Paediatr Respir Rev 2015;16(3):189-196.

Conley RS. Managing obstructive sleep apnea with oral appliances. In: McNamara JA Jr, Shelgikar AV, eds. *Sleep Apnea: What Every Clinician (and Patient) Should Know.* Craniofacial Growth Series, Center for Human Growth and Development, The University of Michigan, Ann Arbor, MI 2018;54:131-153.

de Miguel-Diez J, Villa-Asensi JR, Alvarez-Sala JL. Prevalence of sleep-disordered breathing in children with Down syndrome: Polygraphic findings in 108 children. Sleep 2003;26(8):1006-1009.

Guilleminault C, Eldridge FL, Simmons FB, Dement WC. Sleep apnea in eight children. Pediatrics 1976;58(1):23-30.

Hill W. On some causes of backwardness and stupidity in children: And the relief of these symptoms in some instances by nasopharyngeal scarification. Br Med J 1889;2(1500):711-712.

Hoban TF. Obstructive sleep apnea in children. Curr Treat Options Neurol 2005;7(5):353-361.

Kerschner JE, Lynch JB, Kleiner H, Flanary VA, Rice TB. Uvulopalatopharyngoplasty with tonsillectomy and adenoidectomy as a treatment for obstructive sleep apnea in neurologically impaired children. Int J Pediatr Otorhinolaryngol 2002;62(3):229-235.

Kheirandish-Gozal L, Gozal D. Intranasal budesonide treatment for children with mild obstructive sleep apnea syndrome. Pediatrics 2008;122(1):e149-e155.

Li KK, Riley RW, Guilleminault C. An unreported risk in the use of home nasal continuous positive airway pressure and home nasal ventilation in children: Mid-face hypoplasia. Chest 2000;117(3):916-918.

Mansfield LE, Diaz G, Posey CR, Flores-Neder J. Sleep disordered breathing and daytime quality of life in children with allergic rhinitis during treatment with intranasal budesonide. Ann Allergy Asthma Immunol 2004;92(2):240-244.

Marcus CL. Sleep-disordered breathing in children. Am J Respir Crit Care Med 2001;164(1):16-30.

Marcus CL, Carroll JL, Bamford O, Pyzik P, Loughlin GM. Supplemental oxygen during sleep in children with sleep-disordered breathing. Am J Respir Crit Care Med 1995;152(4 Pt 1):1297-1301.

Marcus CL, Curtis S, Koerner CB, Joffe A, Serwint JR, Loughlin GM. Evaluation of pulmonary function and polysomnography in obese children and adolescents. Pediatr Pulmonol 1996;21(3):176-183.

Marcus CL, Moore RH, Rosen CL, Giordani B, Garetz SL, Taylor HG, Mitchell RB, Amin R, Katz ES, Arens R, Paruthi S, Muzumdar H, Gozal D, Thomas NH, Ware J, Beebe D, Snyder K, Elden L, Sprecher RC, Willging P, Jones D, Bent JP, Hoban T, Chervin RD, Ellenberg SS, Redline S; Childhood Adenotonsillectomy Trial (CHAT). A randomized trial of adenotonsillectomy for childhood sleep apnea. N Engl J Med 2013;368(25):2366-2376.

Marcus CL, Radcliffe J, Konstantinopoulou S, Beck SE, Cornaglia MA, Traylor J, DiFeo N, Karamessinis LR, Gallagher PR, Meltzer LJ. Effects of positive airway pressure therapy on neurobehavioral outcomes in children with obstructive sleep apnea. Am J Respir Crit Care Med 2012;185(9):998-1003.

Marcus CL, Rosen G, Ward SL, Halbower AC, Sterni L, Lutz J, Stading PJ, Bolduc D, Gordon N. Adherence to and effectiveness of positive airway pressure therapy in children with obstructive sleep apnea. Pediatrics 2006;117(3):e442-e451.

Massa F, Gonsalez S, Laverty A, Wallis C, Lane R. The use of nasal continuous positive airway pressure to treat obstructive sleep apnoea. Arch Dis Child 2002;87(5):438-443.

McColley SA, April MM, Carroll JL, Naclerio RM, Loughlin GM. Respiratory compromise after adenotonsillectomy in children with obstructive sleep apnea. Arch Otolaryngol Head Neck Surg 1992;118(9):940-943.

Ng DK, Hui HN, Chan CH, Kowk KL, Chow PY, Cheung JM, Leung SY. Obstructive sleep apnoea in children with Down syndrome. Singapore Med J 2006;47(9):774-779.

Nout E, Bannink N, Koudstaal MJ, Veenland JF, Joosten KF, Poublon RM, van der Wal KG, Mathijssen IM, Wolvius EB. Upper airway changes in syndromic craniosynostosis patients following midface or monobloc advancement: Correlation between volume changes and respiratory outcome. J Craniomaxillofac Surg 2012;40(3):209-214.

O'Brien LM, Holbrook CR, Mervis CB, Klaus CJ, Bruner JL, Raffield TJ, Rutherford J, Mehl RC, Wang M, Tuell A, Hume BC, Gozal D. Sleep and neurobehavioral characteristics of 5- to 7-year-old children with parentally reported symptoms of attention-deficit/hyperactivity disorder. Pediatrics 2003;111(3):554-563.

O'Brien LM, Mervis CB, Holbrook CR, Bruner JL, Bmith NH, McNally N, McClimment MC, Gozal D. Neurobehavioral correlates of sleep-disordered breathing in children. J Sleep Res 2004;13(2):165-172.

Osler W. Chroninc tonsillitis. *The Principles and Practice of Medicine.* New York; Appleton 1892:335-339.

Rachmiel A, Srouji S, Emodi O, Aizenbud D. Distraction osteogenesis for tracheostomy dependent children with severe micrognathia. J Craniofac Surg 2012;23(2):459-463.

Rains JC. Treatment of obstructive sleep apnea in pediatric patients: Behavioral intervention for compliance with nasal continuous positive airway pressure. Clin Pediatr 1995;34(10):535-541.

Reuveni H, Simon T, Tal A, Elhayany A, Tarasiuk A. Health care services utilization in children with obstructive sleep apnea syndrome. Pediatrics 2002;110(1 Pt 1):68-72.

Rosen CL, D'Andrea L, Haddad GG. Adult criteria for obstructive sleep apnea do not identify children with serious obstruction. Am Rev Respir Dis 1992;146(5 Pt 1):1231-1234.

Sateia MJ. International classification of sleep disorders. Third edition: Highlights and modifications. Chest 2014;146(5):1387-1394.

Schechter MS; Section on Pediatric Pulmonology, Subcommittee on Obstructive Sleep Apnea Syndrome. Technical report: Diagnosis and management of childhood obstructive sleep apnea syndrome. Pediatrics 2002;109(4):e69.

Smith D, Abdullah SE, Moores A, Wynne DM. Post-operative respiratory distress following primary cleft palate repair. J Laryngol Otol 2013; 127(1):65-66.

Tarasiuk A, Simon T, Tal A, Reuveni H. Adenotonsillectomy in children with obstructive sleep apnea syndrome reduces health care utilization. Pediatrics 2004;113(2):351-356.

Tauman R, Gulliver TE, Krishna J, Montgomery-Downs HE, O'Brien LM, Ivanenko A, Gozal D. Persistence of obstructive sleep apnea syndrome in children after adenotonsillectomy. J Pediatr 2006;149(6):803-808.

Tran KD, Nguyen CD, Weedon J, Goldstein NA. Child behavior and quality of life in pediatric obstructive sleep apnea. Arch Otolaryngol Head Neck Surg 2005;131(1):52-57.

Ward SL, Marcus CL. Obstructive sleep apnea in infants and young children. J Clin Neurophysiol 1996;13(3):198-207.

PEDIATRIC OBSTRUCTIVE SLEEP APNEA: THE ORTHODONTIC PERSPECTIVE

Benjamin Pliska

ABSTRACT

Obstructive sleep apnea (OSA) is a relatively common condition in children that causes significant negative health effects during a critical period of physical and neurocognitive development. Nasal obstruction and the corresponding sleep-disordered breathing traditionally has thought to be a significant cause of altered dentofacial growth and development; however, evidence of a strong association for the majority of patients does not hold up under contemporary scrutiny. The orthodontic correction of underlying skeletal discrepancies—specifically maxillary constriction and retrognathia—show promise as alternative treatments for OSA in select patients. Future research will delineate in which children these forms of orthodontic procedures will be most effective, serving as a first-line or adjunctive therapy for obstructive sleep disorders. At present, management of a multi-factorial disease (e.g., OSA) will be managed best by a multi-disciplinary approach including the specialty of orthodontics.

KEY WORDS: growth, development, maxillary expansion, obstructive sleep apnea

---

INTRODUCTION

Obstructive sleep apnea (OSA) is a sleep disorder characterized by recurrent, partial or complete episodes of upper airway obstruction, commonly associated with intermittent hypoxemia and sleep fragmentation (American Thoracic Society, 1996). The potential consequences of OSA are serious and can impact quality of life, neurocognitive development and cardio-respiratory health dramatically in children. OSA-associated symptoms often reported by parents and patients include daytime sleepiness, morning headaches, loud and/or labored breathing during sleep, snoring, altered mood, impaired learning and hyperactivity (Marcus et al., 2012).

The best data available from international studies report a prevalence rate of 1.2 to 5.7% in children (O'Brien et al., 2003; Bixler et al., 2009; Li et al., 2010), meaning that most orthodontists will encounter new patients with OSA several times a year; thus, the need for improved awareness of the problem within the profession is obvious.

The deficient function arising from the partial or complete collapse of the upper airway during sleep is a dynamic and complex process. OSA is a multi-factorial disease with interrelated pathophysiological factors that contribute either to a decreased caliber of the airway or increased upper airway collapsibility, or both. As such, OSA in children can result from a combination of factors including inflammation and altered neuromuscular tone of the upper airway, craniofacial disproportion and enlarged tonsillar tissues (Sinha and Guilleminault, 2010).

Hypertrophic tonsils and adenoids are the most common and perhaps well-recognized anatomic risk factors of OSA in children. Due to early growth that outpaces that of the surrounding hard and soft tissues, the adenoids and tonsils are found to be largest relative to the surrounding anatomy between the ages of four and six years of age (Handelman and Osborne, 1976; Linder-Aronson and Leighton, 1983). This early lymphatic growth results in a marked reduction in volume of the nasopharynx and corresponds to the same age range at which OSA most frequently is seen in children. Later in early adolescence, the volume of the upper airway increases due to the concurrent increase in vertical skeletal growth and regression of the lymphoid tissue (Abramson et al., 2009).

As childhood obesity rates continue to inch upward in developed nations, obesity is becoming a significant contributor to the pathophysiology of the disease. The associated fatty infiltration of the soft tissues of the upper airway leads to a narrowing of the airway, while subcutaneous fat deposits under the chin and anterior neck increase pharyngeal collapsibility (Dayyat et al., 2007). Similar to adenoid and tonsil hypertrophy, numerous studies have identified obesity as an independent risk factor for both snoring and OSA (Marcus et al., 2012), while others have proposed obesity-related OSA as a unique phenotype of the disease in children (Dayyat et al., 2007).

It also is clear that any significant aberration in overall craniofacial anatomy or neuromuscular tone will lead to a higher prevalence of OSA, as observed in patients with syndromes associated with maxillary hypoplasia (e.g., Apert's, Crouzon, trisomy 21 and cleft palate; Muntz et al., 2008; Alsaadi et al., 2013; Hill et al., 2016). Similarly, Treacher Collins syndrome or Pierre-Robin sequence patients—both of which are characterized by mandibular hypoplasia—also will have greater rates of OSA than healthy non-syndromic populations will (Anderson et al., 2011). In addition to the challenges involved with the orthodontic management of these patients, treating clinicians should carry a high level of suspicion and a low threshold for referral to assess and manage probable sleep-related respiratory disturbances.

CRANIOFACIAL MORPHOLOGY AS A RISK FACTOR FOR OSA

The morphology or position of craniofacial bones can lead to a narrower airway, either indirectly through the modification of the normal insertion points of the muscles of the upper airway and the consequent impairment of their function, or directly by defining a narrower restrictive skeletal outer limit to the upper airway (Cappabianca et al., 2012). However, the relationship between craniofacial morphology and abnormal mode of respiration long has been a point of controversy in orthodontics (Vig, 1998); unlike in the cleft and syndromic populations, it is unclear to what extent craniofacial morphology contributes to OSA in otherwise healthy patients.

An equally challenging question: is OSA itself a significant or common cause of malocclusion or rather an abnormal dentofacial development in orthodontic patients? Classically in orthodontics, there has thought to have been a linear cause-and-effect relationship between impaired nasal respiration and altered facial morphology. This relationship is due to the postural changes that are observed with nasal obstruction, where the head tilts and extends and the mandible and tongue lower to facilitate oral breathing.

As the tongue no longer is situated in the palatal vault, an imbalance in the resting forces on the arches is created, with a greater amount of palatally-directed force being present, which leads to constriction of the maxilla and dentition. As the mandible also assumes a lowered position, the posterior teeth overerupt into the increased interocclusal space. The resulting characteristics together have been termed "adenoid facies"; this term is used commonly to described malocclusions thought to have an airway related etiology (Fig. 1).

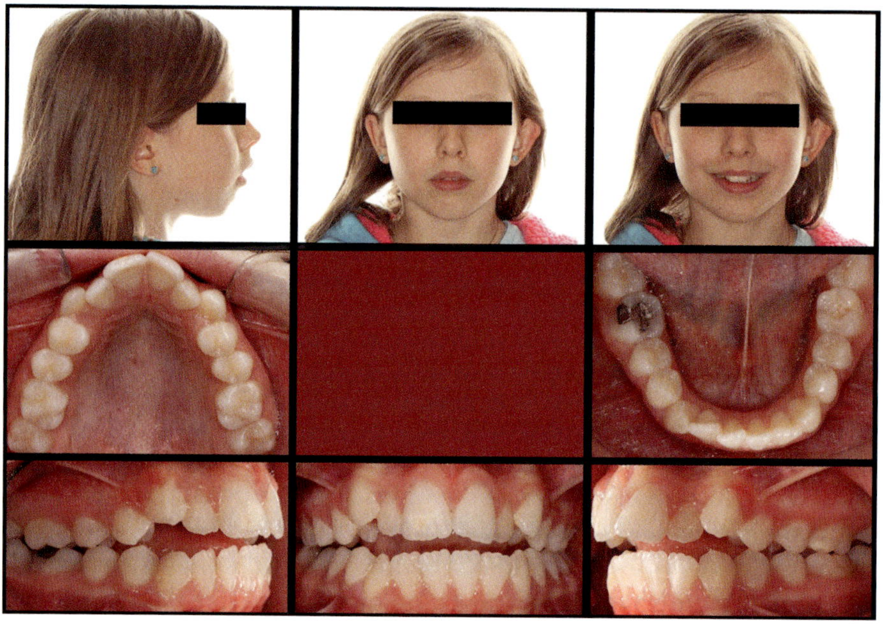

Figure 1. An 11-year-old patient presented with the characteristics classically associated with impaired nasal respiration and related sleep-disordered breathing. Significant findings include an increased lower face height and convex profile, anterior open bite, Class II malocclusion and transverse constriction of the maxillary arch. Note this patient and family reported no OSA-associated symptoms and had a negative history of sleep-disordered breathing or treatment.

Much of our early ideas of malocclusion as it relates to airway issues stems from the work of Linder-Aronson and Woodside, to be discussed below. Their cephalometric observations of various cohorts of Swedish nasally-obstructed children who had undergone adenoid and/or tonsillectomy procedures led to the theory that many malocclusions previously thought to be predisposed genetically were rather of neuromuscular origins.

The first key tenet of their work was that children with chronic nasal obstruction—often due to hypertrophic adenoids—may present with altered facial morphology. This observation was detailed first in Linder-Aronson's landmark paper (1970), a study that compared the average cephalometric tracings of a group of 81 children with nasal obstruction to that of a group of age-matched controls. The obstructed group, all of whom were scheduled to undergo adenoidectomy, presented with significantly longer face heights, as well as steeper mandibular plane angles and greater amounts of retrognathia.

The growth of these children subsequently was followed over a five-year period and again compared to the growth changes of matched controls. Following adenoidectomy and a change in breathing pattern, some children saw a change to a more horizontal direction or "normalization" of mandibular growth over the follow-up period (Linder-Aronson et al., 1986). This finding led to the second key tenet of their work, which was that significant changes in dental and skeletal patterns of malocclusion could be achieved when a change from mouth to nasal breathing could be made in conjunction with orthodontic therapy. At the time, this was welcome information as a strategy for treating the challenging malocclusions that are associated with excessive vertical growth.

Perhaps most confounding of their findings is that not all children responded in the same manner—either with maladaptive postural changes of the tongue and mandible when there initially is nasal obstruction or sleep-disordered breathing, or with an improvement in the mode of breathing or a positive change in direction of growth following treatment. This inconsistency is exemplified best by the Linder-Aronson and Woodside statement (2000): "The neuromuscular suspension of the mandible is a highly sensitive mechanism that responds with altered mandibular posture *in some cases* of chronic nasopharyngeal obstruction" (emphasis added).

Unfortunately, this fact has been overlooked largely when these studies subsequently are referred to, leading many orthodontists to believe that disordered breathing as a cause of malocclusion is the rule rather than the (perhaps rare) exception. If, in fact, it is the individual's neuromuscular response either to the initial impaired breathing during sleep or mode of breathing following treatment that drives the dentofacial morphology, diagnostic factors (e.g., nasopharyngeal obstruction or OSA) may be relatively unimportant in this population. The problem remains, however, that we have no clear way of easily identifying the individual patients in whom this adverse neuromuscular adaptation has occurred.

An association of altered morphology with OSA in children often has been reported in the literature and most commonly includes the dentofacial characteristics related to a long, narrow face. Studies of sleep disordered-breathing children including intra-oral evaluations in their assessments have described a constricted maxilla with a high palate, posterior dental crossbite and increased overjet, all of which have a higher prevalence in children with sleep-breathing abnormalities (Zucconi et al., 1999; Marino et al., 2009; Pirilä-Parkkinen et al., 2009).

Cephalometric investigations report that increased OSA severity has been found to correlate significantly with both an increased mandibular plane angle and an inferior hyoid bone position, as well as decreased posterior airway space in children (Finkelstein et al., 2000; Ozdemir et al., 2004; Ping-Ying et al., 2012). The influence of the lower jaw appears to be complex, as OSA children have been shown to have a retrusive mandibular position (Deng and Gao, 2012), but not a smaller mandible in terms of absolute size compared to normal controls (Schiffman et al., 2004).

The cephalometric characteristics of OSA children have been summarized in two separate systematic reviews, with OSA diagnosis made either from the gold standard of overnight polysomnography (PSG; Flores-Mir et al., 2013), or a combination of PSG and less reliable parental questionnaires (Katyal et al., 2013). The meta-analysis of the included studies confirmed that compared to controls, OSA children had statistically significant increases in measures of facial convexity and vertical growth. The differences found were not large, however, with the increase in ANB angle reported at 1.4° and increase in the angle of the mandibular plane of 4.2°, amounts that rarely would be considered significant clinically. These results led Katyal and colleagues (2013) to state that evidence for a direct causal relationship between craniofacial structure and pediatric sleep-disordered breathing is unsupported by this meta-analysis.

Much of the existing literature on the clinical dentofacial characteristics of children with OSA largely has been limited by studies that predominantly are retrospective, involve small sample sizes and infrequently use overnight PSG for OSA diagnosis. Better data are becoming available, however, as craniofacial morphology increasingly is being recognized as a confounding factor in the etiology and management of OSA, leading to improved collaboration between medical and dental specialists.

One such example is the study by Smith and associates (2016) that involved taking maxillary arch impressions of a cohort of 42 young children with PSG-diagnosed OSA during their adenotonsillectomy procedure. The hypothesis that was tested was whether these children would have a reduction in the transverse dimension of the maxilla due to their disordered breathing.

Maxillary dental casts of this treatment group were compared to casts of a group of 19 non-snoring children recruited from the otolaryngology clinic of the same hospital who served as controls. The authors found statistically significant differences between groups in interdental distances between the first and second deciduous molars, as well as the first permanent molars, but not in palatal height. However, the differences in intermolar width averaged roughly 1 mm, smaller than the standard deviations reported and unlikely to be of much clinical significance.

Perhaps the most expansive clinical investigation into the relationship of dentofacial morphology in children with OSA is that of the Pediatric Dental Sleep Apnea Network, which consisted of a team of researchers from three Canadian universities (http://pdsa.ca). The aim of the trial was to determine the prevalence of malocclusion in children with sleep-disordered breathing. The protocol involved performing a clinical orthodontic examination of children scheduled to undergo overnight PSG at the respective hospital-based sleep clinic. This collaborative investigation then would allow for a prospective and direct comparison of clinical dentofacial characteristics with an objective measurement and diagnosis of OSA in the recruited patients, overcoming many of the limitations of previous investigations relating facial and dental morphology to sleep-disordered breathing.

As of 2016, a total of 282 patients with an average age of nine years and 64% of whom were males have been recruited into the trial. The preliminary data of these patients are presented in Table 1, which has the patients separated into two groups based on their Apnea-Hypopnea Index (AHI). The major finding of this trial so far has been that children diagnosed with OSA—as indicated by having an AHI greater than 2/hour—do not present with higher rates of malocclusion or facial characteristics associated with abnormal respiration (e.g., adenoid facies), nor does malocclusion increase with increasing severity of OSA (unpublished data).

Table 1. Prevalence of specific dentofacial characteristics in a sample of children undergoing PSG. No statistically significant difference in prevalence rates was found between groups.

	Non-OSA AHI < 2 (n = 141)	**OSA AHI ≥ 2 (n = 141)**
Retrognathia	34.8%	36.9%
Posterior crossbite	14.9%	17.7%
Narrow palate	23.4%	19.3%
Anterior open bite	7.7%	6.7%
Overjet > 5 mm	21.4%	18.0%

To be clear, these data do not indicate that sleep-disordered breathing cannot be an etiologic factor in malocclusion, but rather that this cause-and-effect relationship likely occurs in a minority of patients if it occurs at all. Perhaps more significant is that recent orthodontic data align with the medical literature, which supports the complex pathophysiology of sleep-disordered breathing.

The problem is not solely an anatomical one; tonsil size has been shown to be a poor predictor of OSA severity (Nolan and Brietzke, 2011); similarly, a thorough clinical history of symptoms, combined with an examination of clinical signs, does not discriminate different levels of disease severity (Mitchell et al., 2015). Additionally, many children have significant tonsil and adenoid hypertrophy, yet few develop breathing problems and some patients may suffer from OSA in the absence of any hypertrophic tissues at all (Arens and Marcus, 2004).

When this same concept is applied to dentofacial morphology, it should be no surprise that children often may present with different malocclusions and facial forms and yet have similar levels of disease severity (Fig. 2). For many children, it is not tissue hypertrophy or altered anatomy, but rather subtle abnormalities in upper airway neuromuscular function that are the underlying primary cause of OSA (Marcus et al., 1994).

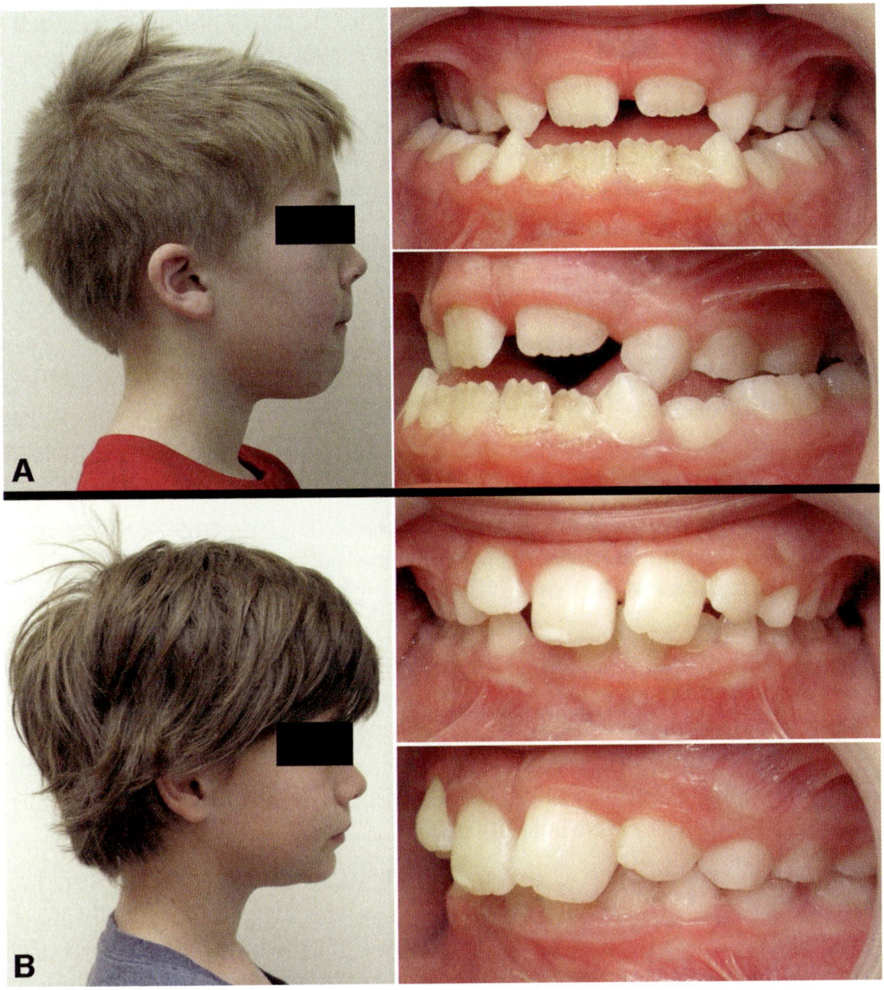

Figure 2. Two eight-year-old boys presented with markedly different dentofacial characteristics and yet both had the same moderate level of OSA diagnosed following PSG. *A*: Note the straight profile, increased lower face height and significant maxillary hypoplasia as indicated by the bilateral posterior crossbite and negative overjet. *B*: A mildly convex profile with normal transverse coordination of the dental arches. Moderate anterior crowding is present with normal anteroposterior occlusal relationship in the mixed dentition.

The challenge remains that we are not able yet to determine in any particular individual what is the contribution of neuromuscular deficiency compared to anatomical limitations to the etiology of the sleep-disordered breathing.

It also may be that a far too simplistic approach has been used to assess the relationship between dentofacial morphology and sleep-disordered breathing. The low prevalence of adenoid facies-type characteristics in patients with diagnosed OSA would indicate that it likely is not a direct cause-and-effect relationship of mouth breathing leading to a postural change of the mandible and an associated malocclusion, as first described by Linder-Aronson (1970).

Fortunately, we as a profession are starting to look at alternative explanations to link sleep-disordered breathing with dentofacial growth and development. One such area of emerging research is examining an epigenetic mechanism of the relationship. Investigators evaluating genotyping differences in mandibular growth have found links in various polymorphisms in the growth hormone receptor gene and mandibular morphology (Hartsfield et al., 2013).

This relationship becomes relevant to OSA as the release of growth hormone and its mediators can be interrupted with fragmented sleep, while mandibular ramus height has been reported to be deficient with growth hormone deficiency and also to rebound significantly, along with growth of the masseter muscle if growth hormone is administered (Peltomäki, 2007). The epigenetic theory is that genetic differences in how a patient's mandibular growth responds to alteration in growth hormone levels are responsible for the variations observed in craniofacial form. This theory seems to be a plausible explanation of why there is such a wide individual response in craniofacial changes in the presence of OSA and an important area of future investigation.

To summarize the relationship between OSA and dentofacial morphology in children, it appears that there is a lack of strong evidence supporting abnormal respiration or OSA as a frequent or common cause of malocclusion, also likely including those that present with the features often associated with adenoid facies (Fig. 3). Conversely, the extent to which craniofacial form contributes or exacerbates OSA symptoms in otherwise healthy children is not understood completely and should be an area of continued research.

TREATMENT OPTIONS FOR OSA IN CHILDREN

The primarily recognized etiologic factor for OSA in children is hypertrophic tonsils and adenoids. Hence, the most common treatment for

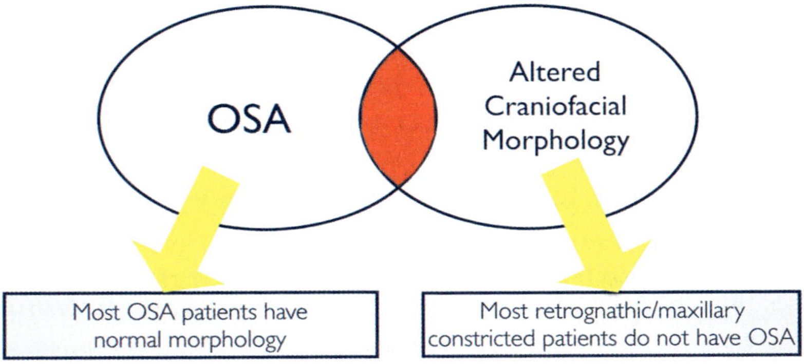

Figure 3. Relationship between OSA and altered craniofacial morphology in children.

this problem remains adenotonsillectomy, the surgical removal of enlarged tonsils and adenoids. A recent review of the national Canadian hospital discharge database revealed that 70.5% of OSA patients between ages 1 to 9 and 49.8% of patients between ages 10 to 19 received a therapeutic intervention that involved the removal of adenotonsillar tissue (Spurr et al., 2011).

This surgery, however, frequently is associated with post-operative throat pain and pain on swallowing. The pain can be severe enough for the patient to restrict intake of liquids, resulting in dehydration and potential hospitalization, particularly in very young children. Other post-operative complications include post-operative hemorrhage, sometimes requiring further surgery, and respiratory complications. Residual OSA following adenotonsillectomy is not uncommon, occurring in an estimated 13% (Ye et al., 2010) to 29% (Mitchell, 2007) of children, with rates varying depending on the specific patient population or patient characteristics. Children who are obese, of African-American ethnicity, or who have more severe OSA are less likely to have their AHI normalize following surgery (Marcus et al., 2013).

If adenotonsillectomy is not indicated or if residual symptoms remain following surgery, then patients may be prescribed positive airway pressure (PAP) therapy. While typically completely effective at resolving obstructive respiratory episodes when worn, PAP therapy often suffers from poor patient adherence. As such, this form of treatment often is limited to medically compromised or syndromic children.

The use of intranasal steroid sprays (Zhang et al., 2008) or oral anti-inflammatories (e.g., leukotriene-receptor antagonists; Goldbart et al., 2005) more recently have been promoted as alternative and less invasive forms of treatment in select patients. While medical therapy is promising, the effects are modest and perhaps best reserved for patients with an unclear history of symptoms or for cases of mild sleep-disordered breathing severity.

The current paradigm describing the pathophysiology of the disease is that children with OSA have subtle abnormalities in upper airway neuromuscular function or structure (Marcus et al., 1994). When these variations in structure or function are present, tonsil and adenoid hypertrophy may precipitate OSA symptoms, but the hypertrophy may not be the underlying primary cause of the disease for many children. In a similar fashion, it may be hypothesized that altered craniofacial morphology may increase the likelihood of developing OSA in susceptible individuals who have a limited adaptive capacity due to an underlying abnormality in their neuromuscular tone during sleep.

The multi-factorial nature of OSA suggests that there will not be a universal treatment solution for all children. Currently, the specific treatment employed depends on the specific risk factor most prominent in a particular patient with sleep-disordered breathing; in most children, primary factors remain hypertrophic tonsils and adenoids. While adenotonsillectomy generally is an effective treatment, the underlying abnormalities remain after surgery for patients with craniofacial disproportions; thus, incomplete resolution following surgery should be expected.

In cases where abnormal craniofacial morphology is diagnosed in a child with sleep-disordered breathing, it would be reasonable to assume that this abnormal morphology may be a contributing factor in the disease. Orthodontic therapy for OSA could be considered in these patients. To date, treatment of two specific anatomic deficiencies—maxillary transverse constriction and mandibular retrognathia—have been studied as forms of orthodontic intervention in the pediatric OSA population.

MAXILLARY EXPANSION TREATMENT OF PEDIATRIC OSA

Maxillary expansion is a non-invasive and common form of orthodontic treatment used to correct transverse discrepancies in the upper jaw. This treatment typically involves the bonding of an expansion device to the upper posterior teeth, which then is activated at specific intervals to apply a force across the midpalatal suture. Over a period of two to four weeks and 7 to 14 mm of expansion, the resulting distraction osteogenesis produces a larger maxilla with increased arch perimeter and intermolar width.

Other indirect effects potentially beneficial for respiration can include a decrease in nasal resistance (De Felippe et al., 2009) and increase in volume of the nasopharynx and nasal cavity (Smith et al., 2012). The technique is limited in magnitude by the lower jaw, as the upper and lower dental arches typically must be coordinated in terms of width following expansion. The technique also relies on a patent midpalatal suture; therefore, a patient's age of 13 to 14 years typically is the upper limit for this non-surgical expansion.

Pirelli and colleagues (2004) were the first to report the use of maxillary expansion for the treatment of pediatric OSA. In their observational study, which lacked a control group, 31 non-obese children with an average age of 8.7 years and no adenoid or tonsillar hypertrophy were treated with maxillary expansion. Following four months of treatment, the patients' average pre-treatment AHI of 12.2/hour was reduced to < 1/hour and the patients were considered free of OSA.

Similarly, Villa and associates (2011) found promising results in a small cohort of eight children who were followed over a three-year period. The children initially were aged 6.6 years and had some form of tonsillar or adenoid hypertrophy at the start of the study. All children were treated with maxillary expansion and their initial AHI was reduced from 6.3/hour to 2.4/hour. This reduction in AHI was found to remain stable at the three-year follow up. Though the AHI was not corrected fully to ideal levels (< 1/hour), the investigators reported that all clinical symptoms had resolved by the end of the treatment period.

From these initial studies on maxillary expansion, which interestingly all originated from Italian research groups, several others since have been published and most recently have been summarized in a scoping systematic review (Camacho et al., 2017). From 17 different publications, maxillary expansion treatment of a total of 314 OSA patients averaging 7.6 years in age were aggregated. On average for these children, the AHI was reduced from 8.9 ± 7.0 to 2.7 ± 3.3/hour, which corresponded to a mean reduction in AHI of 70% (Camacho et al., 2017). Clearly, these are impressive numbers, especially as most of the included children also had hypertrophic tonsils and adenoids, making many of them candidates for adenotonsillectomy surgery.

Consideration is warranted, however, when reviewing the effects of maxillary expansion for OSA. The authors of this recent review acknowledged from further analysis that there was a high risk of publication bias in the existing literature, indicating that instances of negative results were less likely to be reported. Indeed, under closer inspection, within the existing literature there are several instances of children either with limited improvement in OSA symptoms following maxillary expansion treatment (Marino et al., 2012), or having a deterioration and worsening of their disease (Villa et al., 2013).

Furthermore, to date, all published studies examining maxillary expansion either have been case reports or series lacking a control group for comparison, meaning it is unknown how many of the included patients would have seen a spontaneous resolution of their symptoms in the absence of any therapy. This comment is not an insignificant point, as it has been observed in a recent large scale randomized control trial of adenotonsillectomy treatment in children that 46% of the control group saw a full regression of their OSA over a five-month period (Marcus et al., 2013).

As stated previously, the increase in the transverse dimension of the maxilla during orthodontic expansion is limited by the width of the mandibular dentition, as patients should finish treatment with a functional occlusion. Would the results be improved if a greater amount of expansion was attainable? This was the question investigated in Quo and associates' publication (2017) in which researchers reported on a retrospective case series of children who underwent bimaxillary expansion. The expansion protocol employed for these patients included the typical banded maxillary expansion appliance, but with the addition of a banded expander in the mandibular arch to provide buccal tipping of the posterior dentition. In this way, the authors could achieve greater than 6 mm of maxillary expansion and maintain a functional occlusion at the end of treatment.

All patients had been diagnosed with OSA from overnight PSG, which then was repeated following orthodontic expansion. The results for 45 patients, averaging 7.6 ± 2.8 years of age, were reported (abbreviated and shown in Table 2) and serve to highlight the unpredictable nature of maxillary expansion therapy on sleep breathing parameters. While patients with severe OSA saw a dramatic reduction in AHI, which for the group averaged 50%, the mild OSA group had their AHI increase more than double on average. In fact, of the 45 study patients, 15 were observed to have their AHI increase following the expansion treatment. Unfortunately, due to the small group sample sizes, the authors were unable to determine characteristics that may have predicted an improvement or deterioration of sleep-breathing parameters.

To summarize the literature on maxillary expansion as a treatment option for pediatric patients with OSA, we can see examples of expansion dramatically improving symptoms and objective measures of the disease despite the presence of hypertrophic tonsils and adenoids. The treatment results persist over the long term (greater than three years), effectively eliminating residual disease when traditional adenotonsillectomy has proven unsuccessful.

Balancing this great potential for the select patients who present with maxillary constriction, however, is the reality that overall the quality of evidence is not high and further research in the form of controlled randomized clinical trials is needed to validate this form of therapy further. Until that time, caution is warranted as consistent improvement in children and the tenet of "doing no harm" is not a certainty due to complex pathophysiology of the disease. If maxillary expansion is performed with the main intent of improving symptoms of OSA, then it would be prudent to carry out treatment in consultation with the physician managing the child's care.

Table 2. Treatment effects on the AHI of 45 OSA children treated with a bimaxillary expansion protocol, divided into groups based on initial disease severity. Note the significant increase in AHI, indicating a worsening of OSA in the mild group. Adapted from Quo et al., 2017.

AHI	Pre-expansion	Post-expansion	P
Mild (n = 12)	2.9 (0.1-4.0)	6.1 (0.0-30.8)	0.03
Moderate (n = 17)	7.1 (5.2-10)	6.1 (0.0-20.2)	0.25
Severe (n = 16)	22.0 (10.4-32.5)	10.3 (3.7-25.3)	0.001

MANDIBULAR ADVANCEMENT TREATMENT OF PEDIATRIC OSA

Mandibular retrognathia is the second anatomical deficiency that has been treated orthodontically for the purposes of improving OSA symptoms. Unlike the extensive literature of the effectiveness of treating adults with mandibular advancement devices, however, there is much less data available for pediatric patients. Moreover, the mechanism by which treatment occurs also varies between children and adults. In adult OSA patients, the main treatment effect of wearing an oral appliance to advance the mandible is to increase the caliber of the upper airway, particularly the oropharynx (Chan et al., 2010).

It is interesting to note that with anterior protrusion of the mandible, the greatest amount of volume increase occurs in the transverse dimension of the upper airway. This treatment effect occurs only as long as the appliance is worn by the patient as they sleep, however. In contrast, the objective in growing OSA patients with retrognathia is to improve the position of the mandible and associated soft tissues of the upper airway permanently through growth modification. This growth modification, of course, is used commonly in orthodontics for the treatment of Class II malocclusions, though the goal here is to improve the caliber and decrease the collapsibility of the upper airway.

The supporting data for the effectiveness of mandibular advancement treatment in children also have been assembled in a systematic review and meta-analysis (Huynh et al., 2016). Following a scoping review of the published literature and application of screening criteria by the authors, two studies with a combined total of 39 children averaging six years of age were assessed further for the meta-analysis. The pooled data revealed an average reduction in AHI of 5.1/hour from treatment, demonstrating the potential of this form of therapy in patients over a wide age range of 5 to 12 years and, in some cases, despite the presence of hypertrophic tonsils and adenoids. Due to the small number of patients with reported treatment data, caution is warranted when considering generalizing the results. Furthermore, unlike in adults, mandibular advancement with growth modification is limited to retrognathic patients who otherwise would require treatment of a posteriorly positioned lower jaw and dentition.

Although orthodontics as a form of therapy for OSA shows potential in select children with specific dentofacial abnormalities (Fig. 4), it

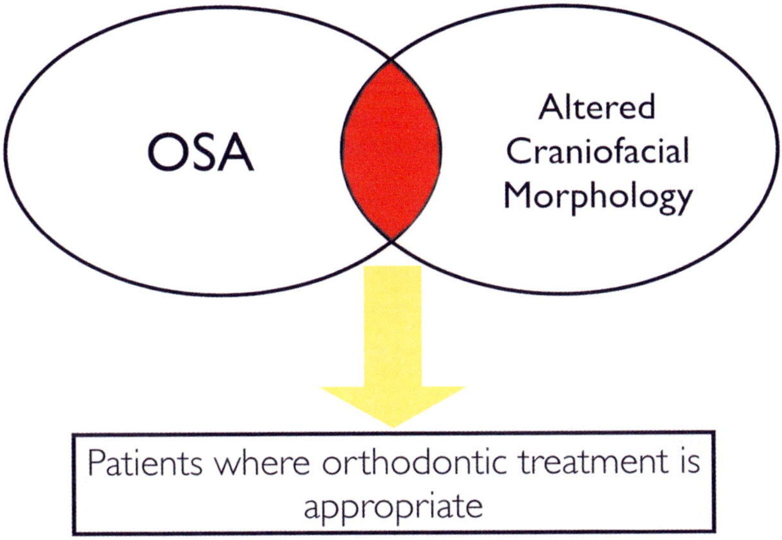

Figure 4. Schematic of the type of child who has both OSA and a deficient mandible or constricted maxilla, where orthodontic treatment with the intention of improving both problems should be considered.

has not been adapted widely; the extent to which this approach may be applicable in the everyday medical management of pediatric patients remains to be determined for many different patient populations. Indeed, most children presenting for management of suspected obstructive sleep issues may not have any obvious signs of altered craniofacial morphology, which makes collaboration among medicine and orthodontics of key importance.

With the goal of forming a better understanding of the number of patients in which specific forms of orthodontic treatment is indicated, the prevalence and severity of malocclusion in a cohort of children referred to an otolaryngology department for the management of sleep-disordered breathing recently has been reported (Pliska et al., 2017). It is this type of clinical setting in which treatment decisions for OSA management often are made, making it the ideal environment to determine the potential applicability of orthodontic intervention.

Following the review of 110 consecutively assessed children, the authors reported that maxillary expansion—as indicated by the presence of a posterior crossbite—and mandibular advancement—as indicated by an overjet greater than 6 mm—were indicated in 15.5% and 4.8% of the sample, respectively. If these prevalence rates and criteria for intervention are taken together, ultimately one in five children with OSA possibly could be managed by orthodontic treatment.

CONCLUSIONS

Clearly, there is a role for orthodontic assessment and treatment of children who suffer from OSA. Future research will determine to what extent altered dentofacial morphology plays in the etiology of the complex pathophysiology of the disease and will clarify in which patients orthodontic intervention can be applied most successfully.

REFERENCES

Abramson Z, Susarla S, Troulis M, Kaban L. Age-related changes of the upper airway assessed by 3-dimensional computed tomography. J Craniofac Surg 2009;20(Suppl 1):657-663.

Alsaadi MM, Iqbal SM, Elgamal EA, Salih MA, Gozal D. Sleep-disordered breathing in children with craniosynostosis. Sleep Breath 2013;17(1): 389-393.

American Thoracic Society. Standards and indications for cardiopulmonary sleep studies in children. Am J Respir Crit Care Med 1996;153(2): 866-878.

Anderson IC, Sedaghat AR, McGinley BM, Redett RJ, Boss EF, Ishman SL. Prevalence and severity of obstructive sleep apnea and snoring in infants with Pierre Robin sequence. Cleft Palate Craniofac J 2011;48(5): 614-618.

Arens R, Marcus CL. Pathophysiology of upper airway obstruction: A developmental perspective. Sleep 2004;27(5):997-1019.

Bixler EO, Vgontzas AN, Lin HM, Liao D, Calhoun S, Vela-Bueno A, Fedok F, Vlasic V, Graff G. Sleep disordered breathing in children in a general population sample: Prevalence and risk factors. Sleep 2009;32(6): 731-736.

Camacho M, Chang ET, Song SA, Abdullatif J, Zaghi S, Pirelli P, Certal V, Guilleminault C. Rapid maxillary expansion for pediatric obstructive sleep apnea: A systematic review and meta-analysis. Laryngoscope 2017;127(7):1712-1719.

Cappabianca S, Iaselli F, Negro A, Basile A, Reginelli A, Grassi R, Rotondo A. Magnetic resonance imaging in the evaluation of anatomical risk factors for pediatric obstructive sleep apnoea-hypopnoea: A pilot study. Int J Pediatr Otorhinolaryngol 2013;77(1):69-75.

Chan AS, Sutherland K, Schwab RJ, Zeng B, Petocz P, Lee RW, Darendeliler MA, Cistulli PA. The effect of mandibular advancement on upper airway structure in obstructive sleep apnoea. Thorax 2010;65(8):726-732.

Dayyat E, Kheirandish-Gozal L, Gozal D. Childhood obstructive sleep apnea: One or two distinct disease entities? Sleep Med Clin 2007;2(3): 433-444.

De Felippe NL, Bhushan N, Da Silveira AC, Viana G, Smith B. Long-term effects of orthodontic therapy on the maxillary dental arch and nasal cavity. Am J Orthod Dentofacial Orthop 2009;136(4):490.e1-e8.

Deng J, Gao X. A case-control study of craniofacial features of children with obstructed sleep apnea. Sleep Breath 2012;16(4):1219-1227.

Finkelstein Y, Wexler D, Berger G, Nachmany A, Shapiro-Feinberg M, Ophir D. Anatomical basis of sleep-related breathing abnormalities in children with nasal obstruction. Arch Otolaryngol Head Neck Surg 2000;126(5):593-600.

Flores-Mir C, Korayem M, Heo G, Witmans M, Major MP, Major PW. Craniofacial morphological characteristics in children with obstructive sleep apnea syndrome: A systematic review and meta-analysis. J Am Dent Assoc 2013;144(3):269-277.

Goldbart AD, Goldman JL, Veling MC, Gozal D. Leukotriene modifier therapy for mild sleep-disordered breathing in children. Am J Respir Crit Care Med 2005;172(3):364-370.

Handelman CS, Osborne G. Growth of the nasopharynx and adenoid development from one to eighteen years. Angle Orthod 1976;46(3):243-259.

Hartsfield JK, Morford LA, Otero LM, Fardo DW. Genetics and non-syndromic facial growth. J Pediatr Genet 2013;2(1):9-20.

Hill CM, Evans HJ, Elphick H, Farquhar M, Pickering RM, Kingshott R, Martin J, Reynolds J, Joyce A, Rush C, Gavlak JC, Gringas P. Prevalence and predictors of obstructive sleep apnoea in young children with Down syndrome. Sleep Med 2016;27-28:99-106.

Huynh NT, Desplats E, Almeida FR. Orthodontics treatments for managing obstructive sleep apnea syndrome in children: A systematic review and meta-analysis. Sleep Med Rev 2016;25:84-94.

Katyal V, Pamula Y, Martin AJ, Daynes CN, Kennedy JD, Sampson WJ. Craniofacial and upper airway morphology in pediatric sleep-disordered breathing: Systematic review and meta-analysis. Am J Orthod Dentofacial Orthop 2013;143(1):20-30.e3.

Li AM, So HK, Au CT, Ho C, Lau J, Ng SK, Abdullah VJ, Fok TF, Wing YK. Epidemiology of obstructive sleep apnoea syndrome in Chinese children: A two-phase community study. Thorax 2010;65(11):991-997.

Linder-Aronson S. Adenoids: Their effect on mode of breathing and nasal airflow and their relationship to characteristics of the facial skeleton and the dentition: A biometric, rhino-manometric and cephalometro-radiographic study on children with and without adenoids. Acta Otolaryngol Suppl 1970;265:1-132.

Linder-Aronson S, Leighton BC. A longitudinal study of the development of the posterior nasopharyngeal wall between 3 and 16 years of age. Eur J Orthod 1983;5(1):47-58.

Linder-Aronson S, Woodside DG. *Excess Face Height Malocclusion.* Quintessence Publishing Co. 2000.

Linder-Aronson S, Woodside DG, Lundström A. Mandibular growth direction following adenoidectomy. Am J Orthod 1986;89(4):273-284.

Marcus CL, Brooks LJ, Draper KA, Gozal D, Halbower AC, Jones J, Schechter MS, Ward SD, Sheldon SH, Shiffman RN, Lehmann C, Spruyt K; American Academy of Pediatrics. Diagnosis and management of childhood obstructive sleep apnea syndrome. Pediatrics 2012;130(3): e714-e755.

Marcus CL, McColley SA, Carroll JL, Loughlin GM, Smith PL, Schwartz AR. Upper airway collapsibility in children with obstructive sleep apnea syndrome. J Appl Physiol 1994;77(2):918-924.

Marcus CL, Moore RH, Rosen CL, Giordani B, Garetz SL, Taylor HG, Mitchell RB, Amin R, Katz ES, Arens R, Paruthi S, Muzumdar H, Gozal D, Thomas NH, Ware J, Beebe D, Snyder K, Elden L, Sprecher RC, Willging P, Jones D, Bent JP, Hoban T, Chervin RD, Ellenberg SS, Redline S; Childhood Adenotonsillectomy Trial (CHAT). A randomized trial of adenotonsillectomy for childhood sleep apnea. N Eng J Med 2013;368 (25):2366-2376.

Marino A, Malagnino I, Ranieri R, Villa MP, Malagola C. Craniofacial morphology in preschool children with obstructive sleep apnoea syndrome. Eur J Paediatr Dent 2009;10(4):181-184.

Marino A, Ranieri R, Chiarotti F, Villa MP, Malagola C. Rapid maxillary expansion in children with obstructive sleep apnoea syndrome (OSAS). Eur J Paediatr Dent 2012;13(1):57-63.

Mitchell RB. Adenotonsillectomy for obstructive sleep apnea in children: Outcome evaluated by pre- and postoperative polysomnography. Laryngoscope 2007;117(10):1844-1854.

Mitchell RB, Garetz S, Moore RH, Rosen CL, Marcus CL, Katz ES, Arens R, Chervin RD, Paruthi S, Amin R, Elden L, Ellenberg SS, Redline S. The use of clinical parameters to predict obstructive sleep apnea syndrome severity in children: The Childhood Adenotonsillectomy (CHAT) study randomized clinical trial. JAMA Otolaryngol Head Neck Surg 2015; 141(2):130-136.

Muntz H, Wilson M, Park A, Smith M, Grimmer JF. Sleep disordered breathing and obstructive sleep apnea in the cleft population. Laryngoscope 2008;118(2):348-353.

Nolan J, Brietzke SE. Systematic review of pediatric tonsil size and polysomnogram-measured obstructive sleep apnea severity. Otolaryngol Head Neck Surg 2011;144(6):844-850.

O'Brien LM, Holbrook CR, Mervis CB, Klaus CJ, Bruner JL, Raffield TJ, Rutherford J, Mehl RC, Wang M, Tuell A, Hume BC, Gozal D. Sleep and neurobehavioral characteristics of 5- to 7-year-old children with parentally reported symptoms of attention-deficit/hyperactivity disorder. Pediatrics 2003;111(3):554-563.

Ozdemir H, Altin R, Söğüt A, Cinar F, Mahmutyazicioğlu K, Kart L, Uzun L, Davşanci H, Gündoğdu S, Tomaç N. Craniofacial differences according to AHI scores of children with obstructive sleep apnoea syndrome: Cephalometric study in 39 patients. Pediatr Radiol 2004;34(5):393-399.

Peltomäki T. The effect of mode of breathing on craniofacial growth: Revisited. Eur J Orthod 2007;29(5):426-429.

Ping-Ying Chiang R, Lin CM, Powell N, Chiang YC, Tsai YJ. Systematic analysis of cephalometry in obstructive sleep apnea in Asian children. Laryngoscope 2012;122(8):1867-1872.

Pirelli P, Saponara M, Guilleminault C. Rapid maxillary expansion in children with obstructive sleep apnea syndrome. Sleep 2004;27(4):761-766.

Pirilä-Parkkinen K, Pirttiniemi P, Nieminen P, Tolonen U, Pelttari U, Löppönen H. Dental arch morphology in children with sleep-disordered breathing. Eur J Orthod 2009;31(2):160-167.

Pliska BT, Lee J, Chadha NK. Prevalence of malocclusion in children with sleep-disordered breathing. J Dent Sleep Med 2017;4(2):41-44.

Quo SD, Hyunh N, Guilleminault C. Bimaxillary expansion therapy for pediatric sleep-disordered breathing. Sleep Med 2017;30:45-51.

Schiffman PH, Rubin NK, Dominguez T, Mahboubi S, Udupa JK, O'Donnell AR, McDonough JM, Maislin G, Schwab RJ, Arens R. Mandibular dimensions in children with obstructive sleep apnea syndrome. Sleep 2004;27(5):959-965.

Sinha D, Guilleminault C. Sleep disordered breathing in children. Indian J Med Res 2010;131(2):311-320.

Smith DF, Dalesio NM, Benke JR, Petrone JA, Vigilar V, Cohen AP, Ishman SL. Anthropometric and dental measurements in children with obstructive sleep apnea. J Clin Sleep Med 2016;12(9):1279-1284.

Smith T, Ghoneima A, Stewart K, Liu S, Eckert G, Halum S, Kula K. Three-dimensional computed tomography analysis of airway volume changes after rapid maxillary expansion. Am J Orthod Dentofacial Orthop 2012;141(5):618-626.

Spurr KF, Morrison DL, Graven MA, Webber A, Gilbert RW. Characterizing obstructive sleep apnea and its management in paediatric and adolescent patients hospitalized in Canada. Canadian J Respir Ther 2011; 47(2):23-28.

Vig KW. Nasal obstruction and facial growth: The strength of evidence for clinical assumptions. Am J Orthod Dentofacial Orthop 1998;113(6): 603-611.

Villa MP, Castaldo R, Miano S, Paolino MC, Vitelli O, Tabarrini A, Mazzotta AR, Cecili M, Barreto M. Adenotonsillectomy and orthodontic therapy in pediatric obstructive sleep apnea. Sleep Breath 2014;18(3):533-539.

Villa MP, Rizzoli A, Miano S, Malagola C. Efficacy of rapid maxillary expansion in children with obstructive sleep apnea syndrome: 36 months of follow-up. Sleep Breath 2011;15(2):179-184.

Ye J, Liu H, Zhang GH, Li P, Yang QT, Liu X, Li Y. Outcome of adenotonsillectomy for obstructive sleep apnea syndrome in children. Ann Otol Rhinol Laryngol 2010;119(8):506-513.

Zhang L, Mendoza-Sassi RA, César JA, Chadha NK. Intranasal corticosteroids for nasal airway obstruction in children with moderate to severe adenoidal hypertrophy. Cochrane Database Syst Rev 2008;16(3): CD006286.

Zucconi M, Caprioglio A, Calori G, Ferini-Strambi L, Oldani A, Castronovo C, Smirne S. Craniofacial modifications in children with habitual snoring and obstructive sleep apnoea: A case-control study. Eur Resp J 1999;13(2):411-417.

OBSTRUCTIVE SLEEP APNEA AND PRIMARY SNORING IN THE PEDIATRIC PATIENT

Bhoomika Ahuja

ABSTRACT

Sleep-disordered breathing (SDB) refers to a continuum of breathing disorders ranging from primary snoring at the mild end to obstructive sleep apnea (OSA) at the severe end. The associated co-morbidities and diagnostic methods for primary snoring (PS) and OSA are discussed. For a long time, PS has been classified as a benign form of SDB and no treatment usually is recommended for these patients. With increasing evidence suggesting otherwise, we look at long-term systemic effects and disease progression for habitual snorers.

KEY WORDS: obstructive sleep apnea, snoring, intermittent hypoxia, children, sleep-disordered breathing

---

INTRODUCTION

Sleep-disordered breathing (SDB) in children refers to a spectrum of sleep-related breathing disorders. Obstructive sleep apnea (OSA) is classified to be on the severe end of this spectrum. OSA is a sleep-related breathing disorder in which the subject experiences periodic episodes of full or partial obstruction of the airway resulting in oxygen desaturation or cortical arousals during sleep. On the mild end of the SDB spectrum is primary snoring (PS), which is characterized by habitual snoring and restricted airflow without the presence of intermittent hypoxia or gas exchange abnormalities. However, patients with PS may exhibit respiratory related arousals from sleep, as seen in OSA.

INCIDENCE IN THE PEDIATRIC POPULATION

The incidence of OSA in the pediatric population has been reported to be from 1-5% (Marcus et al., 2012). The peak incidence is seen at pre-school age from 2 to 5 years (Farber, 2002).

Studies reporting on PS suggest that infants and young children have an incidence rate of snoring from 5-27%. PS also is known to affect 10-12% of adolescents (Nieminen et al., 2000; Lu et al., 2003; Shin et al., 2003). While incidence varies with age, race, gender and environmental factors, both PS and OSA are underdiagnosed in the pediatric population.

Comorbidities

OSA has several associated systemic and behavioral problems in the pediatric population. These associated problems can be categorized broadly as neurocognitive, cardiovascular and metabolic. However, these problems are not just limited to children whose SDB is at the severe end of the spectrum. Children with PS also show many of the associated co-morbidities seen in children with OSA.

Neurocognitive

Children suffering from OSA often show behavior problems, which can occur even in the absence of obesity. Other neurocognitive problems include hyperactivity, attention deficit and impaired performance at school (Rudnick and Mitchell, 2007).

A study among high school students showed that poor performers had a 35% greater odds of being habitual snorers when compared to students who showed good school performance (Shin et al., 2003). Another study also found significant correlation between habitual snoring and poor school performance in primary school children in the absence of intermittent hypoxia (Urschitz et al., 2003). When tested for correlation, intermittent hypoxia did not show independent correlation to poor school performance. Thus, it is possible that the neurocognitive impairments seen in OSA and PS patients might be related to respiratory arousals rather than to intermittent hypoxia.

Cardiovascular

Cardiovascular disorders are a common risk associated with OSA in adults. Fibrinogen, a predictor of cardiovascular disease in adults, has been found in children with sleep apnea (Kaditis et al., 2004), with the levels of fibrinogen not being correlated to the severity of the Apnea-Hypopnea Index (AHI). Children with OSA present an 11-fold increased risk of left ventricular hypertrophy. Abnormality of left ventricular architecture has been found in 15% of children with PS and 39% of children with OSA (Amin et al., 2002). Non-obese, pre-pubertal children with PS also show increased daytime and nighttime blood pressure compared to non-snorers (Li et al., 2009).

Metabolic

In recent years, metabolic disorders have been studied in OSA patients and *vice versa* to examine the possible relationship between these conditions. Not only do clinical studies show that OSA is associated with impaired glucose metabolism (Tamura et al., 2008), but there have been animal studies that show this association independent of obesity at a young age. Lean mice treated with intermittent hypoxia have shown decreased insulin sensitivity (Iiyori et al., 2007), demonstrating that chronic intermittent hypoxia might have a role in inducing insulin resistance and metabolic disorder. Studies also have shown that rats treated with intermittent hypoxia show increased levels of blood glucose and decreased levels of blood insulin (Pae et al., 2013). The pathophysiological pathway of association between SDB and metabolic disturbances is yet to be understood.

DIAGNOSIS

OSA and PS cannot be differentiated solely based on clinical examination, radiographic study and patient questionnaire. Diagnosis typically is made with an overnight sleep study using a polysomnograph (PSG) in a sleep lab. The AHI represents the frequency of apneas and hypopneas per hour of sleep during the PSG. In children, an AHI of ≥ 1/hour is considered to be diagnostic for sleep apnea.

For both adults and children, OSA is an underdiagnosed and largely unrecognized disorder. In the adult population, the common phenotype associated with OSA is a male who is overweight with a sedentary lifestyle (Biggs, 2014). However, in children who do not present with obesity, OSA is characterized mostly by adenotonsillar hypertrophy and small pharyngeal space (Nixon and Brouillette, 2005).

The PSG is the gold standard for diagnosis of SDB. However, the complexity of the overnight procedure, high cost and limited availability of accredited sleep laboratories make it a challenge to diagnose SDB in children adequately. Hence, OSA and PS frequently go undiagnosed and untreated.

While adenotonsillectomy (AT) is the preferred treatment and significantly improves OSA, residual post-operative OSA may go undetected due to post-surgical sleep studies rarely being conducted in clinical practice. Other markers of residual OSA have been studied, though none of these are incorporated into routine clinical practice at present. Bhattacharjee and colleagues (2016) showed that high-sensitivity C-reactive protein serum levels might be useful as a biomarker in predicting residual OSA after AT in children.

A study aimed at finding new modalities for diagnosis in children with OSA and PS used surface EMG to study activity of the genioglossus (GG) during various tasks (e.g., deep breathing, quiet breathing and maximum protrusion; Ahuja, 2017). The tasks were performed during wakefulness. The aim of this study was to differentiate between muscle activity during wakefulness to find a method that makes a reliable preliminary diagnosis without having the pediatric patients go through an overnight sleep study. OSA patients showed higher GG activity during quiet breathing (Fig. 1) and deep breathing (Fig. 2). GG activity for subjects with OSA and PS fatigued faster than healthy controls (Fig. 3). While these studies bring new possible modalities for diagnosis to light, further studies are needed to determine the accuracy, reliability and feasibility of these proposed diagnostic methods.

PEDIATRIC SNORERS: AT-RISK GROUP

PS may be the beginning of the severity spectrum of SDB, but it has associated comorbidities that are of alarming concern. A long-term follow-up of children with PS showed that 37% of habitual snorers progressed to having OSA; about 7% of these habitual snorers developed moderate-to-severe OSA (Li et al., 2013). Children with PS show cardiovascular abnormalities including increased systolic and diastolic blood pressure. Children who show higher systolic blood pressure are at an increased risk of hypertension and metabolic syndrome in adult life (Sun et al., 2007).

Additionally, PS affects the growth and neurocognitive development in childhood, causing developmental delays and behavior problems, as well as deficiencies in school performance in primary and high school children.

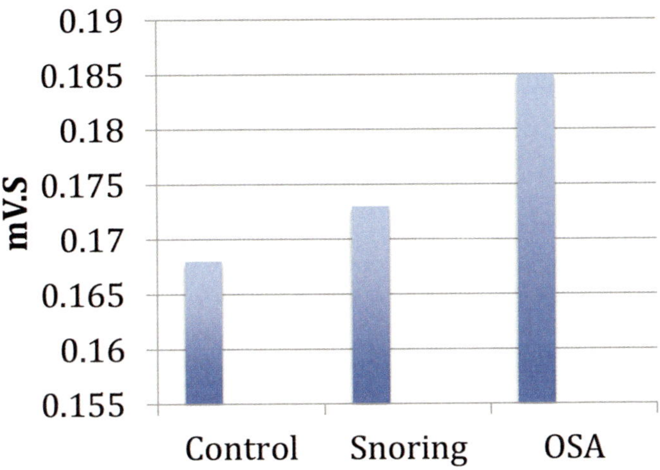

Figure 1. Genioglossus muscle activity measured by electromyography (EMG) during quiet breathing in wakefulness in OSA patients, snoring and control groups. OSA and snoring subjects show higher EMG activity of the genioglossus muscle as compared to controls with no sleep-disordered breathing. mV.S. = millivolts per second.

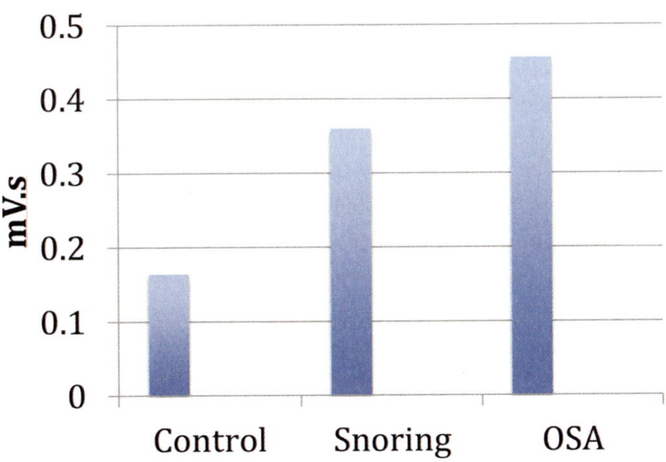

Figure 2. Genioglossus muscle activity measured by EMG during deep breathing in wakefulness in OSA patients, snoring and control groups. OSA and snoring subjects show higher EMG activity of the genioglossus muscle as compared to controls with no sleep-disordered breathing.

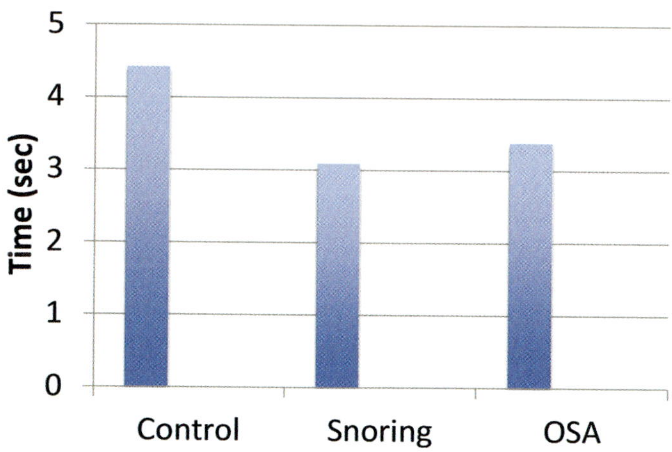

Figure 3. EMG showing fatigability of the genioglossus muscle in OSA, snoring and control groups. Children with OSA and snoring show increased fatigability of the GG than the control subjects. The exact pathophysiology of how PS leads to its associated neurocognitive and cardiovascular comorbidities is not understood fully. It is possible that the current diagnostic modalities are not sensitive enough to detect sub-threshold oxygen desaturation. PS patients also might experience frequent micro-arousals (< 3 seconds) not currently scored on PSG and more sleep fragmentation than normal healthy controls. These alterations might play a role in the manifestation of associated comorbidities. More importantly, the increased risk of progression to OSA in primary snorers makes it important that this condition no longer should be considered entirely benign.

REFERENCES

Ahuja B, Park J, Chae T, Albinali R, Patel D, Pae EK. Genioglossus fatigue in children with obstructive sleep apnea and primary snoring. Unpublished data, 2017.

Amin RS, Kimball TR, Bean JA, Jeffries JL, Willging JP, Cotton RT, Witt SA, Glascock BJ, Daniels SR. Left ventricular hypertrophy and abnormal ventricular geometry in children and adolescents with obstructive sleep apnea. Am J Respir Crit Care Med 2002;165(10):1395-1399.

Bhattacharjee R, Kheirandish-Gozal L, Kaditis AG, Verhulst SL, Gozal D. C-reactive protein as a potential biomarker of residual obstructive sleep apnea following adenotonsillectomy in children. Sleep 2016;39(2): 283-291.

Biggs SN, Nixon GM, Home RS. The conundrum of primary snoring in children: What are we missing in regards to cognitive and behavioural morbidity? Sleep Med Rev 2014;18(6):463-475.

Farber JM. Clinical practice guideline: Diagnosis and management of childhood obstructive sleep apnea syndrome. Pediatrics 2002;110(6): 1255-1257.

Iiyori N, Alonso LC, Li J, Sanders MH, Garcia-Ocana A, O'Doherty RM, Polotsky VY, O'Donnell CP. Intermittent hypoxia causes insulin resistance in lean mice independent of autonomic activity. Am J Respir Crit Care Med 2007;175(8):851-857.

Kaditis AG, Alexopoulos EI, Kalampouka E, Kostadima E, Angelopoulos N, Germenis A, Zintzaras E, Gourgoulianis K. Morning levels of fibrinogen in children with sleep-disordered breathing. Eur Respir J 2004;24(5): 790-797.

Li AM, Au CT, Ho C, Fok TF, Wing YK. Blood pressure is elevated in children with primary snoring. J Pediatr 2009;155(3):362-368.

Li AM, Zhu Y, Au CT, Lee DL, Ho C, Wing YK. Natural history of primary snoring in school-aged children: A 4-year follow-up study. Chest 2013;143(3):729-735.

Lu LR, Peat JK, Sullivan CE. Snoring in preschool children: Prevalence and association with nocturnal cough and asthma. Chest 2003;124(2):587-593.

Marcus CL, Brooks LJ, Draper KA, Gozal D, Halbower AC, Jones J, Schechter MS, Sheldon SH, Spruyt K, Ward SD, Lehmann C, Shiffman RN; American Academy of Pediatrics. Diagnosis and management of childhood obstructive sleep apnea syndrome. Pediatrics 2012;130(3):576-584.

McSharry D, O'Connor C, McNicholas T, Langran S, O'Sullivan M, Lowery M, McNicholas WT. Genioglossus fatigue in obstructive sleep apnea. Respir Physiol Neurobiol 2012;183(2):59-66.

Nieminen P, Tolonen U, Löppönen H. Snoring and obstructive sleep apnea in children: A 6-month follow-up study. Arch Otolaryngol Head Neck Surg 2000;126(4):481-486.

Nixon GM, Brouillette RT. Sleep: 8. Pediatric obstructive sleep apnoea. Thorax 2005;60(6):511-516.

Pae EK, Ahuja B, Kim M, Kim G. Impaired glucose homeostasis after a transient intermittent hypoxic exposure in neonatal rats. Biochem Biophys Res Commun 2013;441(3):637-642.

Rudnick EF, Mitchell RB. Behavior and obstructive sleep apnea in children: Is obesity a factor? Laryngoscope 2007;117(8):1463-1466.

Shin C, Joo S, Kim J, Kim T. Prevalence and correlates of habitual snoring in high school students. Chest 2003;124(5):1709-1715.

Sun SS, Grave GD, Siervogel RM, Pickoff AA, Arslanian SS, Daniels SR. Systolic blood pressure in childhood predicts hypertension and metabolic syndrome later in life. Pediatrics 2007;119(2):237-246.

Tamura A, Kawano Y, Watanabe T, Kadota J. Relationship between the severity of obstructive sleep apnea and impaired glucose metabolism in patients with obstructive sleep apnea. Respir Med 2008;102(10): 1412-1416.

Urschitz MS, Guenther A, Eggebrecht E, Wolff J, Urschitz-Duprat PM, Schlaud M, Poets CF. Snoring, intermittent hypoxia and academic performance in primary school children. Am J Respir Crit Care Med 2003; 168(4):464-468.

MANAGING OBSTRUCTIVE SLEEP APNEA WITH ORAL APPLIANCES

R. Scott Conley

ABSTRACT

INTRODUCTION: Obstructive sleep apnea (OSA) is a multi-factorial disease that can increase risk for numerous comorbidities if left untreated. Dentists and dental specialists can provide lifesaving screening assessments and refer patients to sleep medicine specialists for definitive diagnosis and team-based treatment planning. The gold standard treatment for OSA remains positive air pressure (PAP), but following referral from a sleep physician, the dental community can provide successful oral appliance (OA) therapy. To do so, dentists must be knowledgeable of and adhere to recent American Dental Association (ADA) guidelines and recommendations so that they continue to operate within their legally defined scope of care. **MATERIALS AND METHODS**: A critical review of the dental and medical literature regarding oral appliance therapy for OSA was performed. While systematic review is preferred, due to short study durations, limited numbers of patients and the lack of well-controlled prospective randomized trials, an appropriate systematic review could not be performed. **CONCLUSIONS**: While no single design of mandibular advancement appliance (MAD) is successful in every patient, the preponderance of the evidence suggests that MADs can be a highly effective treatment for selected OSA patients. To date, attempts to predict patients accurately who will respond favorably from those who will not have been unsuccessful. Therefore, dentists who elect to provide OA therapy must follow up on their own patients to manage the dental side effects that may occur, as well as refer the patient back to the sleep medicine team for definitive and objective treatment response assessment.

KEY WORDS: obstructive sleep apnea, oral appliance therapy, efficacy, evidence-based dentistry, inter-professional care

---

INTRODUCTION

Obstructive sleep apnea (OSA) is part of a spectrum of sleep-disordered breathing (SDB) conditions (Young et al., 1993). Over time, OSA has transitioned from being an obscure, little-known medical condition to one that is known widely both by the medical community and the lay public (Strollo and Rogers, 1996; Woods et al., 2015). While OSA often is thought only to affect overweight males in the mid-to-late age range; greater understanding of the disease has established that OSA affects a broader segment of the public including women and children (Bixler et al., 2001).

Symptoms of OSA include excessive daytime sleepiness, falling asleep unintentionally, difficulty staying focused, depression, headaches and nausea. However, other sleep disorders—including central sleep apnea (CSA), mixed sleep apnea, restless leg syndrome, narcolepsy, insomnia and many others—exhibit similar clinical presentations (Ohayon et al., 2002; Koo, 2015; Donovan and Kapur, 2016). Because treatment protocols are different for each sleep disorder, patients must undergo a proper diagnostic protocol so that disease-specific treatment can be implemented. While well intentioned, the dentist or physician who provides treatment without a definitive diagnosis may harm rather than help the patient.

DIAGNOSTIC PROTOCOLS

When patients present to the dental or dental specialist's office, the first level of information collected must be the health history. Unfortunately, large numbers of OSA patients are not aware that they are affected. If the patient's sleep partner comes to the office, s/he may report frequent loud snorts or gasps during sleep, but this information often is not available.

To screen for excessive daytime sleepiness, dentists and dental specialists can implement the Epworth Sleepiness Scale into their routine adult health history (Rosenthal and Dolan, 2008). The survey requires patients to self-report answers to eight short questions regarding how likely they are to fall asleep in a given situation. The survey uses a 0 to 3 scale with "0" denoting no chance of falling asleep and "3" denoting they are highly likely to fall asleep during that particular activity; the answers yield a total score ranging from 0 to 24. Items surveyed include common every-day activities (e.g., sitting and reading, watching television and being a passenger in a car). Elevated scores ($\geq$ 10) indicate excessive daytime sleepiness, but this is not unique to a specific etiology. While useful as a screening tool, the test is unable to distinguish between various sleep disorders and, therefore, patients with elevated scores must be referred to a sleep specialist to obtain a definitive diagnosis of OSA.

The gold standard for diagnosis of OSA is an attended overnight polysomnogram (PSG), which is read and scored by a physician (Kapur et al., 2017). It is recommended that the sleep study interpretation be performed by a board-certified sleep medicine specialist. An attended PSG provides the most accurate diagnosis because the patient is observed directly during his/her sleep, with all required parameters being monitored continuously. Should the patient dislodge one or more of the sensors, the sleep technologist can enter the room and re-attach it, which helps maintain the integrity of the recorded data.

In some instances, home sleep apnea tests are used to evaluate for OSA. Because patients must place their own sensors properly, at-home testing is less reliable, less accurate and less complete (Dawson et al., 2015). Home sleep apnea testing is indicated for patients with a high clinical suspicion of OSA and lack of medical comorbidities (e.g., severe heart failure, significant pulmonary disease, neuromuscular disease, comorbid sleep disorder). Since home sleep apnea tests record fewer channels and are less sensitive, a negative home sleep apnea test always should be followed by an attended PSG.

OSA is distinguished from CSA by the presence of respiratory effort during apneas and hypopneas. With OSA, the patient attempts to breathe, as demonstrated by contraction of the muscles of the chest wall, but the airway is obstructed and prevents the air from reaching the lungs (Strollo and Rogers, 1996). CSA patients fail to receive the necessary input from the brainstem to breathe and, as a result, they do not demonstrate chest wall contraction (Donovan and Kapur, 2016). Because home sleep apnea tests do not have brainwave monitoring, it can be difficult to characterize central apneas and hypopneas fully (Kapur et al., 2017). At-home testing models often provide insufficient data to discern the true cause of the respiratory disturbance (Dawson et al., 2015).

The severity of OSA is measured by the apnea-hypopnea index (AHI), which sums the number of breathing cessations (apnea) and breathing reductions (hypopnea) that occur per hour of sleep. Adult patients are classified as shown in Table 1 (Escourrou et al., 2015):

Table 1. The official American Association of Sleep Medicine (AASM) severity scale for adult patients with sleep apnea. Patients can range from normal to severe based on the number of breathing reductions or cessations per hour of sleep.

CLASSIFICATION	AHI
Normal	≤ 5
Mild	5-15
Moderate	15-29
Severe	> 30

Other common items measured during a PSG include the time taken to fall asleep, average oxygen saturation and minimum oxygen saturation (Kapur et al., 2017). Patients with mild AHI but severely reduced oxygen saturation may be classified by the more severe of the two criteria. Dentists who suspect OSA perform the Epworth scale in their office and refer their patients for a PSG provide an invaluable service. Their patients not only receive a definitive diagnosis, but also can be referred for one of the many forms of treatment. Undiagnosed or untreated OSA has a 37% higher five-year morbidity and mortality; therefore, finding the most effective form of treatment is critical to aid long-term survival (Marti et al., 2002).

TREATMENT PROTOCOLS

Several forms of treatment are available to the OSA patient (Sanders et al., 1983; Smith et al., 1985; Prinsell, 2002; Fleisher and Krieger, 2007; Thompson et al., 2007). Ideally, a comprehensive approach should be used to determine the most effective form of treatment for each patient. In addition to evaluation by a sleep medicine specialist, a comprehensive approach to OSA treatment may include input from otolaryngology, a medical and/or surgical weight-loss specialist, an oral and maxillofacial surgeon, as well as a dentist and/or dental specialist.

The gold standard treatment is positive air pressure (PAP) therapy that delivers pressurized air into the upper airway to maintain patency and to assure proper ventilation (Sanders et al., 1983). PAP therapy can be delivered *via* nasal pillow, nasal, or oro-nasal masks. The particular mask is selected based on patient comfort and minimizing leak to increase adherence to therapy. For patients with significant facial imbalances including severe midface deficiency (Class III), mandibular deficiency (Class II) or dramatic asymmetry, custom masks may be required to prevent air leakage. However, custom-made masks are not available routinely in clinical practice.

Attended PAP titration in a sleep laboratory is the gold standard method to determine the therapeutic pressure setting and assures that the minimum effective air pressure is delivered for treatment. For selected patients, auto-titrating PAP devices may be considered. All PAP users should have ongoing follow-up to ensure patient comfort, adherence to therapy and clinical response to treatment.

While PAP is a highly successful treatment (Sanders et al., 1983), it is not without risks including—but not limited to—upper respiratory tract infections, sinusitis, rhinorrhea, hoarseness, dry throat and gingivitis (Fig. 1; Calik, 2016). PAP also requires patient adherence to therapy and multiple reports demonstrate moderate-to-low adherence (Sanders et al., 1983; Calik, 2016; BaHammam et al., 2017). Interventions that improve PAP adherence include patient education and PAP "de-sensitization," but long-term adherence still may drop to as low as 40% (Gulati et al., 2017), further highlighting the needed for ongoing clinical follow-up.

For this reason, other forms of treatment must be considered, including both soft- and hard-tissue surgery (Chen et al., 2014). The goal of soft tissue surgery generally is to excise or re-contour exuberant or redundant tissues (e.g., tonsils, adenoids, tongue base and soft palate). While effective in the short term, the soft tissue surgery procedures (other than tonsillectomy and adenoidectomy in children) generally are considered ineffective (Coleman, 1998).

Among the hard tissue surgery options, advancement genioplasty (Riley et al., 1986), mandibular advancement (Bear and Priest, 1980) or combined maxillomandibular advancement (Prinsell, 2002) are the most common with varying degrees of success. Early reports recommended a staged approach where a genioplasty was performed (Riley et al., 1986). If the patient's symptoms were unresolved or resolved only partially, further jaw surgery was performed (Hochban et al., 1994). As greater numbers of patients received care, most reports demonstrated that a single surgical stage incorporating orthognathic jaw surgery with genioplasty is the more successful and preferred approach.

For large advancements (e.g., telegnathic surgery, meaning 10 mm or larger advancement of the maxilla, mandible or both), surgeons may

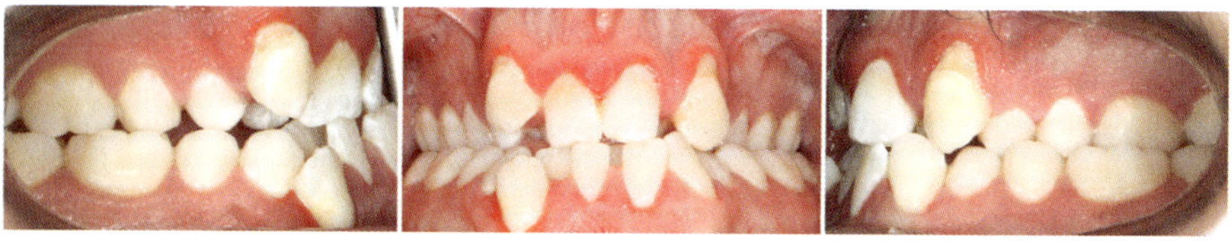

Figure 1. Gingivitis is a common side effect resulting from PAP therapy. The tissues are dried continually by the PAP therapy, particularly in the anterior portion of the maxillary and mandibular dental arches.

perform distraction osteogenesis (Thompson et al., 2007) in order to titrate the advancement to the desired PSG result. Despite the numerous reports of high success with telegnathic surgery, it is not without risk including paresthesia, dysesthesia (an unpleasant, abnormal sense of touch), infection, malocclusion and negative facial esthetics (Fig. 2). To minimize irreversible complications, other forms of treatment—including oral appliances—are used for treatment of OSA.

ORAL APPLIANCE (OA) THERAPY

OA therapy for OSA patients requires several critical steps to assure that appropriate care and appropriate diagnosis are made. While dentists and dental specialists are equipped to screen for and recognize signs/symptoms of OSA, they must refer the patient for definitive diagnosis by a physician. Dentists who fail to utilize this approach not only potentially place the patient at risk of under/over treatment by treating a condition that mimics OSA, but also place themselves at risk of operating outside their scope of care.

Following the physician's diagnosis and either failure or noncompliance of PAP therapy, patients may receive a referral to the dentist or dental specialist for OA care. The ideal patient has mild-to-moderate OSA, will be compliant, non-obese and with normal skeletal jaw position (Lowe et al., 1996). OA therapy also may be considered for patients with moderate-to-severe OSA who are unwilling to unable to use CPAP. The goal of OA therapy is to position the patient's lower jaw downward and forward to place the oropharyngeal muscles under tension to make airway collapse more difficult (Chan et al., 2007; Sutherland et al., 2014). Where to position the lower jaw has been the subject of several investigations; the consensus is approximately 50-66% of the patient's

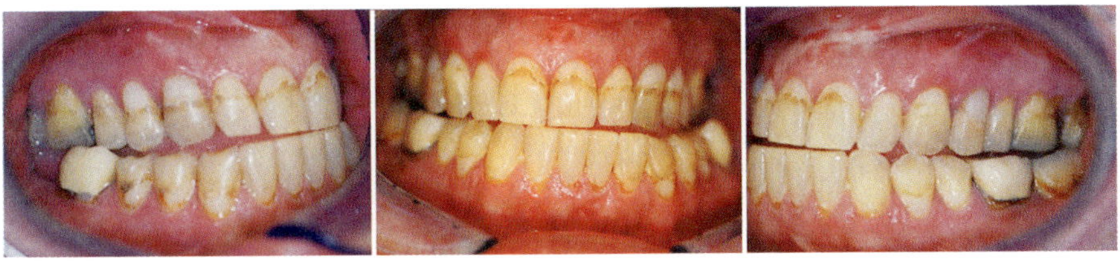

Figure 2. Malocclusion is a known risk factor following orthognathic or telegnathic jaw surgery for the treatment of OSA. This patient's OSA was resolved, but the patient developed a lateral shift to the bite. Orthodontic therapy to optimize and coordinate the dental arches prior to surgery can minimize the risk of malocclusion, but due to altered sensation and large surgical moves, malocclusion still can result.

maximum protrusion (Chan et al., 2007). For most patients, this mandibular repositioning will result in 5 to 7 mm of lower jaw advancement.

To replicate this bite registration for appliance fabrication, the same Projet bite recorder (Great Lakes Orthodontic Products, Tonawanda, NY) used in taking a bite registration for mandibular functional appliances can be used (Fig. 3). Several lab- or custom in-house-fabricated appliances exist, with the final selection based on patient comfort (Fig. 4). No single appliance is effective or ineffective universally. After fabrication, the appliance first should be adjusted and then be titrated. In this context, the term *adjust* refers to manipulating the appliance to assure optimal patient comfort, while *titrate* refers to manipulating the amount of mandibular protrusion or bite opening to obtain the maximum treatment effect. Both are essential: patients will not wear appliances that are uncomfortable; similarly, they do not receive treatment benefit from wearing appliances that are ineffective. To assure efficacy, the patient must wear the OA during a follow-up sleep study.

The ADA has developed paired statements both on the role of dentistry in the management of sleep-related breathing disorders and an evidence brief for oral appliances (OAs) for treatment of sleep-related breathing disorders. Both of these are available online with the links provided at the conclusion of the chapter (ADA Evidence Brief, 2017; ADA Policy Statement, 2017). Many of the specialty groups (AAOMS and AAO) have provided white papers or comments in support of the ADA's recommendations.

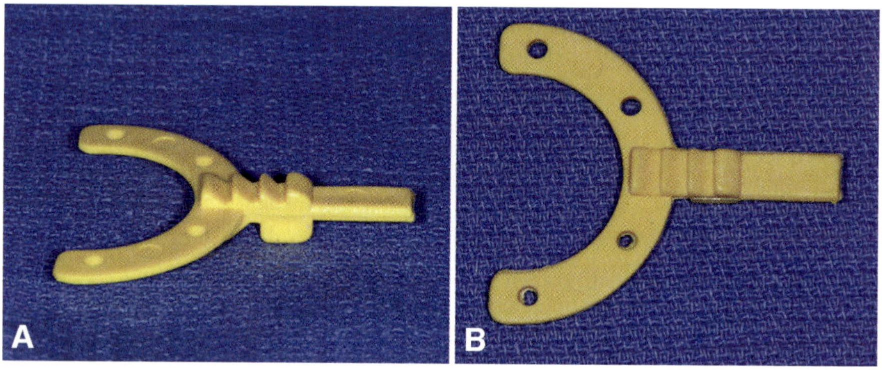

Figure 3. The Projet™ bite fork or bite recorder is common to orthodontists for use with mandibular functional appliances in the growing child. This bite fork can be used in a similar way to help position the patient in a stable, repeatable protruded position. *A:* Viewed from the sagittal perspective, it often is helpful to position the bite fork upside down so that the patient can bite only in one position. *B:* Occlusal view. An injectable PVS material can be placed around the arch and into the holes so it does not separate from the bite fork.

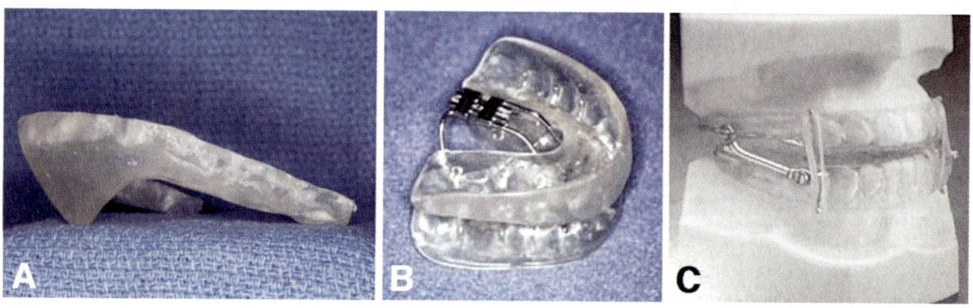

Figure 4. Many forms of mandibular advancement devices (MADs) are available. *A:* An in-office appliance fabricated by the dentist or dental specialist. *B:* Kleerway appliance. *C:* Removable Herbst appliance.

EFFICACY

The current level of evidence for OA therapy includes case reports, retrospective case series, prospective case series, prospective randomized trials and a smaller number of systematic reviews and meta-analyses. Much of the outcome-based assessments examine the effect of a single oral appliance; others studies compare two types of OAs and a third group of studies examines oral appliance therapy *versus* the gold standard PAP therapy. Accurate efficacy results are difficult to obtain due to limited patient groups, high patient dropout rates, lack of controls and generally short study periods.

The term *evidence* must be discussed critically. Most of the early evidence was available in the form of case reports for patients who were treated successfully. It is well known, however, that case reports generally are low on the evidence pyramid and that the treatment effect observed may or may not be generalizable to the overall population of patients. The outcome of a successful case report may exceed the typical benefit observed in an average patient. In other words, the case report may document a response two or more standard deviations beyond the normal treatment effect (Fig. 5).

Subsequent papers described subjective evaluation of patients treated with one or more oral appliance. Further, work included systematic reviews of the subjective and patient-reported treatment effects (Ahrens et al., 2010). Using 14 of a possible 1,475 studies that met their initial search criteria, Ahrens and colleagues (2011) evaluated different OAs (one or two piece) *versus* both a control appliance and each other. They concluded that mandibular advancing appliances (MADs), which represent one specific type of oral appliance, performed better than controls with 66% of treated patients demonstrating an improvement in their AHI index score. MAD appliances are the most common type of OAs. Other classes of OAs include tongue-retaining devices, maxillary and/or

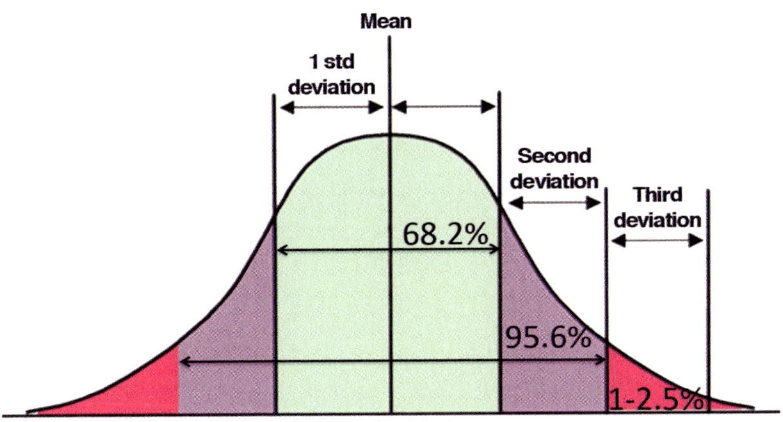

Figure 5. Case reports often are the first level of evidence available in the literature for new treatment modalities. One must interpret and apply these findings with caution given the tremendous range in treatment effect that typically is observed in population studies. Once prospective randomized clinical trials begin, it often becomes clear that the initial case reports were "outliers," with treatment effects two or more standard deviations beyond the average.

mandibular expansion devices. No one device is superior to another; practitioners who use OAs should be familiar with all of the types available and adjust according to individual patient preferences and needs. There was no difference among one-piece MAD designs, no difference for 50% or 75% maximum protrusion and no single appliance type (one- or two-piece design) clearly was superior. While systematic reviews are among the highest level of treatment evidence, one must consider the question being asked. Later reports have demonstrated that patients who feel better may not receive a treatment benefit and that subjective results are not correlated with objective (e.g., PSG) findings. Such evidence has resulted in the current standards of care requiring objective assessment.

Based on the deficiencies of subjective assessments, objective evidence is required and more recently, higher levels of evidence using PSG in prospective randomized control studies have emerged, several of which were of short duration (Tsuiki et al., 2004; Blanco et al., 2005; Lawton et al., 2005; Ng et al., 2006; Hoekema et al., 2007; Petri et al., 2008; Deane et al., 2009). The longer duration studies ranged from six months (Gauthier et al., 2009) to as many as 120 months (Jauhar et al., 2008).

Few papers included control groups (Blanco et al., 2005; Petri et al., 2008). Most papers compared two different appliances without a control group (Tsuiki et al., 2004; de Almeida et al., 2005; Lawton et al., 2005; Ng et al., 2006; Hoekema et al., 2007; Marklund and Franklin, 2007; Jauhar et al., 2008; Zeng et al., 2008; Deane et al., 2009; Gauthier et al., 2009; Ghazal et al., 2009; Giannasi et al., 2009). When examined as a pool, most studies demonstrated that a significant but varying proportion of patients will benefit from OA therapy, but it is difficult to conclude precisely who, how many of the study cohort and to what degree patients will benefit.

To date, few papers performed the most rigorous level of investigation to determine not only the efficacy of OA therapy, but also to compare OA therapy to the gold standard of PAP. The study by Okuno and colleagues (2014) demonstrated that OAs improved AHI more than control appliances, although less than CPAP. Contrary to previous investigators (Hoekema et al., 2007; Lam et al., 2007), the study group of Okuno and coworkers (2014) demonstrated similar compliance rates with OAs or CPAP. Compliance adherence varies among different papers, but generally is based upon patient's self-reports. With CPAP, many (but not all) models will have digital display of the amount of time, leak(s) and pressure. Unfortunately, there is no way to measure adherence with OAs other than by self-report.

Phillips and coworkers (2013) evaluated the short-term prospective randomized crossover study to compare the results of CPAP and a MAD. With over 100 patients completing both arms of the study, the MAD achieved complete resolution in 40% and partial resolution in another 25%. By comparison, 75% of CPAP patients obtained complete resolution and with an additional 15% of CPAP patients obtained partial resolution. Because MAD patients received full or partial resolution in only 65% of cases compared to 90% of PAP patients, MAD patients were 50% as likely to receive full effect demonstrating that PAP provided a superior form of treatment. However, patients preferred the MAD to CPAP by a 2:1 margin. The reported compliance was 6.5 ± 1.3 hours for the MAD *versus* 5.2 ± 2 hours/night for CPAP. For such situations, the clinician and patient must weigh partial benefit with adherent oral appliance therapy *versus* no benefit from non-adherent PAP therapy.

For long-term results, Ghazal and associates (2009) compared two OAs over several years: a modified Herbst appliance (Intranasal Snoring Therapy Appliance; IST) and a prosthodontic appliance (Thornton Anterior Positioning Appliance; TAP). The study utilized 103 consecutively enrolled and randomly assigned middle-aged adults. At six months, both appliances improved the AHI, with the TAP having a higher percentage of success. By the end of the study (42 months), both groups showed similar results. Caution must be taken with these conclusions, as there was significant patient dropout and loss to follow up, leaving less than 50% of the original study population. Of note, this group of investigators was among the first to examine not only the AHI, but also the effects of oral appliance therapy on blood pressure, an important consideration given the observation that controlling blood pressure may be more important than AHI in reducing the adverse health effects of OSA (Marin et al., 2005; Yaggi et al., 2005).

With the increasing number of prospective randomized studies and systematic reviews, meta-analyses now have been performed examining different aspects of treatment. Using seven separate studies with a combined pool of nearly 400 patients, Iftikhar and colleagues (2013) evaluated OAs and their effect on blood pressure demonstrating a modest decrease in systolic, diastolic and mean arterial pressure, although there was no correlation between reduced blood pressure and decreased AHI.

Finally, Li and coworkers (2013) performed a systematic review using 14 prospective randomized trials comparing the gold standard CPAP with oral appliance therapy. The results indicated CPAP was significantly more effective in reducing AHI and apnea index and increasing the minimum oxygen saturation (SpO_2) than oral appliances, and there was no compliance difference between the two treatment approaches. Their conclusion was that while CPAP is better, OAs are appropriate to prescribe to patients who are unable or unwilling to wear CPAP.

TREATMENT LIMITATIONS AND SIDE EFFECTS

One must consider both the treatment limitations (i.e., some patients do not respond to either CPAP or OA) and potential side effects of any prescribed treatment (Otsuka et al., 2006*a*; Hoekema et al., 2007). In addition, concern has arisen with OAs regarding possible dental changes (Fig. 6; Almeida et al., 2006*a,b*). In the study by Almeida and associates (2006*a*) over a seven-year period, 14% of oral appliance patients showed no dental change, 41% experienced favorable change and 44% experienced unfavorable bite changes. Favorable change was described as patients with Class II who improved; unfavorable change was observed in Class I patients who became Class III.

A more recent two-year prospective randomized study by Doff and associates (2013) evaluated potential dental changes in both CPAP and oral appliances. The oral appliance group demonstrated a 1.1 mm decrease in overbite, a 1.5 mm decrease in overjet and a reduced number of posterior contacts; CPAP demonstrated smaller occlusal changes (albeit smaller and not statistically significant due to patient dropout). CPAP also demonstrates a higher number of moderate-to-severe side effects (e.g., nasal congestion, rhinorrhea, eye irritation and sense of suffocation) that must be considered: moderate side effects included dental movement and bite change; serious side effects included sinus infection, pneumonia, cardia arrhythmia and death.

In an attempt to reduce dental side effects through skeletal anchorage and to treat patients better with excessive numbers of missing teeth, a novel micro-implant-retained device was attempted in a small number (ten) of patients (Ngiam and Kyung, 2012). All patients' AHI

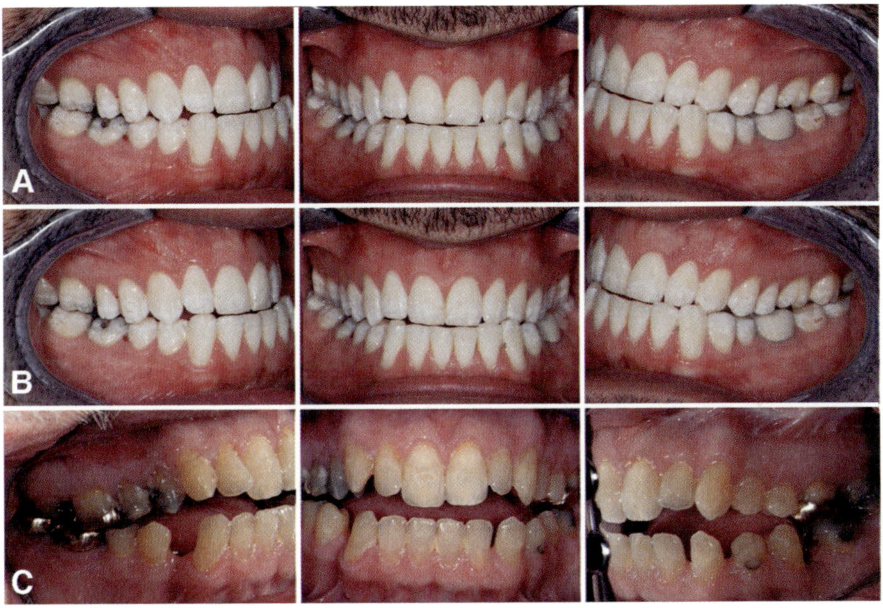

Figure 6. OAs can have side effects just as orthognathic and telegnathic surgery can. Side effects range from mild (*A*), to moderate (*B*) and ultimately to severe (*C*) where the patient loses his/her stable occlusal function. Many of the OAs suggest the use of morning bite repositioning appliances that are fabricated to the patient's pre-treatment bite position. Morning repositioning appliances remain controversial and their effect remains unclear.

improved over the six-month study period with 80% of temporary anchorage devices (TADs) remaining stable and "no dental side effects were seen."

PREDICTION

The improved compliance of OA therapy over PAP has led investigators to attempt to predict patients who will and will not respond favorably to OA care (Lowe et al., 1995; Hoekema et al., 2007). Based on failure analysis of early OA studies, the conclusion reached retrospectively suggested that obesity, age and adjusted neck circumference would be the best predictors (Otsuka et al., 2006*a*). Therefore, younger, non-obese patients were recommended more commonly for OA care.

Based on both the failure of younger non-obese patients and the desire of older obese patients who were unable to comply with PAP therapy attempts to refine predictive criteria were re-examined. Multiple OA studies attempted by Remmers and associates were performed (Tsai et al., 2004; Chan et al., 2007; Remmers et al., 2013). The initial study prospectively examined 23 patients with an respiratory disturbance index (RDI) of $\geq$ 15 (Tsai et al., 2004). Following the use of a remote controlled mandibular positioning appliance in a prospective and blinded study of 23 patients, the investigators determined that they were able to predict positively that OA therapy would be a success. The test also was able to predict correctly those who were successful 90% of the time while demonstrating a "false negative" rate of 11%. Limitations of the initial study included a liberal criteria for success (RDI < 15) and only 50% of the patients achieved success.

A follow-up study used a similar method to protrude the mandible remotely and incrementally, but altered the criteria for success to be the absence of an obstructive event (apnea or hypopnea) for at least five minutes in the supine position (Dort et al., 2006). The investigators were able to recruit a larger study population (44 subjects) and a greater number of patients (33) completed the study. The relatively high dropout rate (25%) made interpreting the results more challenging, but the investigators reported that 80% of the patients who were predicted to succeed actually were successful and 78% of patients who were predicted to fail OA therapy were left with untreated sleep apnea.

In a final attempt to assess predictive capability, a larger study population of 67 consecutive patients diagnosed by a sleep center was recruited (Remmers et al., 2013). Based on this larger population and more stringent success criteria (e.g., reducing the AHI to 10 or less as opposed to the more flexible criteria of 20 or less), the authors reported an 86% sensitivity and 92% specificity. The predictive range was somewhat broad (83-94%), which reinforced the continued need to assess patients following the adjustment and titration period.

Each of the predictive papers has one or more study limitations. Most notably, patients were re-tested with a home sleep apnea test, not a full PSG, which indicates that some respiratory events either may have been missed or reported inaccurately. Additionally, both false positives (patients predicted to succeed with OA treatment, but did not) and false negatives (patients predicted to fail with OA, but improved) occurred. Despite these limitations, the studies successfully expanded the scope of OA care. Clinicians now may consider using a remote controlled mandibular positioning appliance for patients with higher BMI and older age that previously may not have been considered OA therapy.

An alternate method that attempted to predict OA care was reported by Chan and coworkers (2007), who examined 35 patients using awake and upright inspiratory and expiratory flow rates. Their underlying hypothesis was that reductions both in inspiratory and expiratory flow rates might be able to predict collapsibility of the airway and, therefore, predict the response to OA treatment. Based on the criteria they established, the investigators documented a sensitivity of 35% and a specificity of 80%. The study team concluded that the flow-volume curves were not sufficient to predict response to OA treatment accurately.

ADA PROPOSED GUIDELINES

Within the past year, the American Dental Association (ADA) circulated a pair of white papers: the first regarded proposed policies regarding the role of dentistry in the treatment of sleep-related breathing disorders; the second detailed the current evidence for oral appliance treatment for OSA. The ADA recommended that general dentists and dental specialists assure that they are knowledgeable of the official position relating to scope of care, referral patterns and other medicolegal aspects that enable individual practitioners to determine whether or not they wish to provide care of patients with sleep-related breathing disorders.

Among the ADA's nine official recommendations, the first states that dentists should screen patients for OSA using one or more of the methods presented earlier in this chapter. Based on the outcome of the screening tools, dentists should refer patients to the appropriate physician(s) for diagnosis. The authors of the brief also included dental and dental specialty-related treatment options for OSA. Specifically included within the brief is the recommendation that informed consent should be obtained appropriately prior to delivery of care.

Given the life-threatening implications of failed treatment, the term "should" may be considered too weak and the term "must" has been suggested instead. For most forms of consent, the term "appropriately " generally is understood to be written and documented within the chart so it can be demonstrated later in the same manner as every other form of informed consent within dentistry and its associated specialties.

The ADA also has provided the authority of dentists to assess OA therapy through the use of portable unattended monitors. Because both Type 3 and 4 monitors are portable and placed by the patient, they measure far fewer items. A Type 3 monitor typically will measure at least four items including both heart rate and oxygen saturation; some models include equipment to measure body position and/or leg movement. Type 4 monitors are far less sophisticated and typically measure only oxygen saturation without the option for body position, leg movement or respiratory effort. Given the medically-based diagnosis of OSA, this recommendation must be viewed with caution. Dentists generally are not trained suitably to read and interpret Type 3 or 4 portable monitors.

While it is true that many patients either are unable or unwilling to obtain a follow-up PSG, the full-scope test provides the most accurate assessment of treatment efficacy and also provides the patient with the highest standard of care. Because untreated OSA patients are at risk of premature death, dentists should be cautious accepting a level of responsibility for assessment they may not be qualified to perform. Additional recommendations provided by the ADA white statement pertain not only to the dentist's ability to recognize and manage side effects, but also to remain current in their continuing education. Finally, the policy statement recommends that the dentist should maintain regular communications with the referring physician.

Just as each practitioner decides what s/he will and will not treat specifically related to dentistry, every practitioner must decide for her/himself whether one wishes to provide medically related treatment for patients with suspected sleep-related breathing disorders. If a clinician wishes to expand her/his scope of care to include sleep-related breathing disorders, it is recommended that s/he not only be aware of, but also follow the officially recommended policies published by the ADA.

CONCLUSIONS

OSA is a complex and multi-level disease that has a spectrum of therapeutic options across multiple disciplines of medicine and dentistry, ranging from conservative to highly invasive. While PAP remains the gold standard, oral appliance therapy can be a highly successful treatment modality for select patients. Both the general dentist and the dental specialist should engage their local sleep medicine teams actively to expand the range of treatment options for patients with OSA. When delivering OA care, dentists must remain cognizant of their role on the team, their legal scope of care and the ongoing treatment assessment needs. When they do, dentists and dental specialists can provide treatment to a growing segment of the population.

REFERENCES

ADA evidence brief: Oral appliances for sleep-related breathing disorders. http://www.ada.org/~/media/ADA/Member%20Center/FIles/ADA_SCI_OralAppl_SRBD_Brief_Final_15.pdf?la=en. Accessed May 20, 2017.

ADA proposed policy statement on the role of dentistry in the treatment of sleep-related breathing disorders. http://www.ada.org/~/media/ADA/Member%20Center/FIles/Role_of_Dentistry_in_the_Treatment_of_Sleep.pdf?la=en. Accessed May 20, 2017.

Ahrens A, McGrath C, Hägg U. A systematic review of the efficacy of oral appliance design in the management of obstructive sleep apnoea. Eur J Orthod 2011;33(3):318-324.

Ahrens A, McGrath C, Hägg U. Subjective efficacy of oral appliance design features in the management of obstructive sleep apnea: A systematic review. Am J Orthod Dentofacial Orthop 2010;138(5):559-576.

Almeida FR, Lowe AA, Otsuka R, Fastlicht S, Farbood M, Tsuiki S. Long-term sequellae of oral appliance therapy in obstructive sleep apnea patients: Part 2. Study-model analysis. Am J Orthod Dentofacial Orthop 2006*a*;129(2):205-213.

Almeida FR, Lowe AA, Sung JO, Tsuiki S, Otsuka R. Long-term sequellae of oral appliance therapy in obstructive sleep apnea patients: Part 1. Cephalometric analysis. Am J Orthod Dentofacial Orthop 2006*b*;129 (2):195-204.

BaHammam AS, Singh T, George S, Acosta KL, Barataman K, Gacuan DE. Choosing the right interface for positive airway pressure therapy in patients with obstructive sleep apnea. Sleep Breath 2017;21:569.

Bear SE, Priest JH. Sleep apnea syndrome: Correction with surgical advancement of the mandible. J Oral Surg 1980;38(7):543-549.

Bixler EO, Vgontzas AN, Lin HM, Ten Have T, Rein J, Vela-Bueno A, Kales A. Prevalence of sleep-disordered breathing in women: Effects of gender. Am J Respir Crit Care Med 2001;163(3 Pt 1):608-613.

Blanco J, Zamarrón C, Abeleira Pazos MT, Lamela C, Suarez Quintanilla D. Prospective evaluation of an oral appliance in the treatment of obstructive sleep apnea syndrome. Sleep Breath 2005;9(1):20-25.

Calik MW. Treatments for obstructive sleep apnea. J Clin Outcomes Manag 2016;23(4):181-192.

Chan AS, Lee RW, Cistulli PA. Dental appliance treatment for obstructive sleep apnea. Chest 2007;132(2):693-699.

Chen S, Shi S, Xia Y, Liu F, Chen D, Zhu M, Li M, Zheng H. Changes in sleep characteristics and airway obstruction in OSAHS patients with multilevel obstruction following simple UPPP, UPPP-GA, or UPPP-TBA: A prospective, single-center, parallel group study. ORL J Otorhino-laryngol Relat Spec 2014;76(4):179-188.

Coleman JA Jr. Laser-assisted uvulopalatoplasty: Long-term results with a treatment for snoring. Ear Nose Throat J 1998;77(1):22-24, 26-29, 32-24.

Dawson A, Loving RT, Gordon RM, Abel SL, Loewy D, Kripke DF, Kline LE. Type III home sleep testing *versus* pulse oximetry: Is the respiratory disturbance index better than the oxygen desaturation index to predict the apnoea-hypopnoea index measured during laboratory polysomnography? BMJ Open 2015;5(6):e007956.

de Almeida FR, Lowe AA, Tsuiki S, Otsuka R, Wong M, Fastlicht S, Ryan F. Long-term compliance and side effects of oral appliances used for the treatment of snoring and obstructive sleep apnea syndrome. J Clin Sleep Med 2005;1(2):143-152.

Deane SA, Cistulli PA, Ng AT, Zeng B, Petocz P, Darendeliler M. Comparison of mandibular advancement splint and tongue stabilizing device in obstructive sleep apnea: A randomized controlled trial. Sleep 2009;32(5):648-653.

Doff MH, Finnema KJ, Hoekema A, Wijkstra PJ, de Bont LG, Stegenga B. Long-term oral appliance therapy in obstructive sleep apnea syndrome: A controlled study on dental side effects. Clin Oral Investig 2013;17(2):475-482.

Donovan LM, Kapur VK. Prevalence and characteristics of central compared to obstructive sleep apnea: Analyses from the Sleep Heart Health Study Cohort. Sleep 2016;39(7):1353-1359.

Dort LC, Hadjuk E, Remmers JE. Mandibular advancement and obstructive sleep apnoea: A method for determining effective mandibular protrusion. Eur Respir J 2006;27(5):1003-1009.

Escourrou P, Grote L, Penzel T, McNicholas WT, Verbraecken J, Tkacova R, Riha RL, Hedner J; ESADA Study Group. The diagnostic method has a strong influence on classification of obstructive sleep apnea. J Sleep Res 2015;24(6):730-738.

Fleisher KE, Krieger AC. Current trends in the treatment of obstructive sleep apnea. J Oral Maxillofac Surg 2007;65(10):2056-2068.

Gauthier L, Laberge L, Beaudry M, Laforte M, Rompré PH, Lavigne GJ. Efficacy of two mandibular advancement appliances in the management of snoring and mild-moderate sleep apnea: A cross-over randomized study. Sleep Med 2009;10(3):329-336.

Ghazal A, Sorichter S, Jonas I, Rose EC. A randomized prospective long-term study of two oral appliances for sleep apnoea treatment. J Sleep Res 2009;18(3):321-328.

Giannasi LC, Almeida FR, Magini M, Costa MS, de Oliveira CS, de Oliveira JC, Kalil Bussadori S, de Oliveira LV. Systematic assessment of the impact of oral appliance therapy on the temporomandibular joint during treatment of obstructive sleep apnea: Long-term evaluation. Sleep Breath 2009;13(4):375-381.

Gulati A, Ali M, Davies M, Quinnell T, Smith I. A prospective observational study to evaluate the effect of social and personality factors on continuous positive airway pressure (CPAP) compliance in obstructive sleep apnoea syndrome. BMC Pulm Med 2017;17(1):56.

Hochban W, Brandenburg U, Peter JH. Surgical treatment of obstructive sleep apnea by maxillomandibular advancement. Sleep 1994;17(7): 624-629.

Hoekema A, Doff MH, de Bont LG, van der Hoeven JH, Wijkstra PJ, Pasma HR, Stegenga B. Predictors of obstructive sleep apnea-hypopnea treatment outcome. J Dent Res 2007;86(12):1181-1186.

Iftikhar IH, Hays ER, Iverson MA, Magalang UJ, Maas AK. Effect of oral appliances on blood pressure in obstructive sleep apnea: A systematic review and meta-analysis. J Clin Sleep Med 2013;9(2):165-174.

Jauhar S, Lyons MF, Banham SW, Cameron DA, Orchardson R. Ten-year follow-up of mandibular advancement devices for the management of snoring and sleep apnea. J Prosthet Dent 2008;99(4):314-321.

Kapur VK, Auckley DH, Chowdhuri S, Kuhlmann DC, Mehra R, Ramar K, Harrod CG. Clinical practice guideline for diagnostic testing for adult obstructive sleep apnea: An American Academy of Sleep Medicine Clinical Practice guideline. J Clin Sleep Med 2017;13(3):479-504.

Koo BB. Restless leg syndrome across the globe: Epidemiology of the restless legs syndrome/Willis-Ekbom disease. Sleep Med Clin 2015; 10(3):189-205.

Lam B, Sam K, Mok WY, Cheung MT, Fong DY, Lam JC, Lam DC, Yam LY, Ip MS. Randomised study of three non-surgical treatments in mild to moderate obstructive sleep apnoea. Thorax 2007;62(4):354-359.

Lawton HM, Battagel JM, Kotecha B. A comparison of the Twin Block and Herbst mandibular advancement splints in the treatment of patients with obstructive sleep apnoea: A prospective study. Eur J Orthod 2005; 27(1):82-90.

Li W, Xiao L, Hu J. The comparison of CPAP and oral appliances in treatment of patients with OSA: A systematic review and meta-analysis. Respir Care 2013;58(7):1184-1195.

Lowe AA, Fleetham JA, Adachi S, Ryan CF. Cephalometric and computed tomographic predictors of obstructive sleep apnea severity. Am J Orthod Dentofacial Orthop 1995;107(6):589-595.

Lowe AA, Ono T, Ferguson KA, Pae EK, Ryan CF, Fleetham JA. Cephalometric comparisons of craniofacial and upper airway structure by skeletal subtype and gender in patients with obstructive sleep apnea. Am J Orthod Dentofacial Orthop 1996;110(6):653-664.

Marin JM, Carrizo SJ, Vicente E, Agusti AG. Long-term cardiovascular outcomes in men with obstructive sleep apnoea-hypopnoea with or without treatment with continuous positive airway pressure: An observational study. Lancet 2005;365(9464):1046-1053.

Marklund M, Franklin KA. Long-term effects of mandibular repositioning appliances on symptoms of sleep apnoea. J Sleep Res 2007;16(4):414-420.

Marti S, Sampol G, Muñoz X, Torres F, Roca A, Lloberes P, Sagalés T, Quesada P, Morell F. Mortality in severe sleep apnoea/hypopnoea syndrome patients: Impact of treatment. Eur Respir J 2002;20(6): 1511-1518.

Ng AT, Qian J, Cistulli PA. Oropharyngeal collapse predicts treatment response with oral appliance therapy in obstructive sleep apnea. Sleep 2006;29(5):666-671.

Ngiam J, Kyung HM. Microimplant-based mandibular advancement therapy for the treatment of snoring and obstructive sleep apnea: A prospective study. Angle Orthod 2012;82(6):978-984.

Ohayon MM, Priest RG, Zulley J, Smirne S, Paiva T. Prevalence of narcolepsy symptomatology and diagnosis in the European general population. Neurology 2002;58(12):1826-1833.

Okuno K, Sato K, Arisaka T, Hosohama K, Gotoh M, Taga H, Sasao Y, Hamada S. The effect of oral appliances that advanced the mandible forward and limited mouth opening in patients with obstructive sleep apnea: A systematic review and meta-analysis of randomised controlled trials. J Oral Rehabil 2014;41(7):542-554.

Otsuka R, Almeida FR, Lowe AA, Ryan F. A comparison of responders and nonresponders to oral appliance therapy for the treatment of obstructive sleep apnea. Am J Orthod Dentofacial Orthop 2006*a*; 129(2):222-229.

Otsuka R, Ribeiro de Almeida F, Lowe AA, Linden W, Ryan F. The effect of oral appliance therapy on blood pressure in patients with obstructive sleep apnea. Sleep Breath 2006*b*;10(1):29-36.

Petri N, Svanholt P, Solow B, Wildschiødtz G, Winkel P. Mandibular advancement appliance for obstructive sleep apnoea: Results of a randomised placebo controlled trial using parallel group design. J Sleep Res 2008;17(2):221-229.

Phillips CL, Grunstein RR, Darendeliler MA, Mihailidou AS, Srinivasan VK, Yee BJ, Marks GB, Cistulli PA. Health outcomes of continuous positive airway pressure *versus* oral appliance treatment for obstructive sleep apnea: A randomized controlled trial. Am J Respir Crit Care Med 2013; 187(8):879-887.

Prinsell JR. Maxillomandibular advancement surgery for obstructive sleep apnea syndrome. J Am Dent Assoc 2002;133(11):1489-1497.

Remmers J, Charkhandeh S, Grosse J, Topor Z, Brant R, Santosham P, Bruehlmann S. Remotely controlled mandibular protrusion during sleep predicts therapeutic success with oral appliances in patients with obstructive sleep apnea. Sleep 2013;36(10):1517-1525, 1525A.

Riley RW, Powell NB, Guilleminault C, Nino-Murcia G. Maxillary, mandibular, and hyoid advancement: An alternative to tracheostomy in obstructive sleep apnea syndrome. Otolaryngol Head Neck Surg 1986;94(5):584-588.

Rosenthal LD, Dolan DC. The Epworth sleepiness scale in the identification of obstructive sleep apnea. J Nerv Ment Dis 2008;196(5):429-431.

Sanders MH, Moore SE, Eveslage J. CPAP *via* nasal mask: A treatment for occlusive sleep apnea. Chest 1983;83(1):144-145.

Smith PL, Gold AR, Meyers DA, Haponik EF, Bleecker ER. Weight loss in mildly to moderately obese patients with obstructive sleep apnea. Ann Intern Med 1985;103(6 Pt 1):850-855.

Strollo PJ Jr, Rogers RM. Obstructive sleep apnea. N Engl J Med 1996; 334(2):99-104.

Sutherland K. Vanderveken OM, Tsuda H, Marklund M. Gagnadoux F, Kushida CA, Cistulli PA. Oral appliance treatment for obstructive sleep apnea: An update. J Clin Sleep Med 2014;10(2):215-227.

Thompson SH, Quinn M, Helman JI, Baur DA. Maxillomandibular distraction osteogenesis advancement for the treatment of obstructive sleep apnea. J Oral Maxillofac Surg 2007;65(7):1427-1429.

Tsai WH, Vazquez JC, Oshima T, Dort L, Roycroft B, Lowe AA, Hajduk E, Remmers JE. Remotely controlled mandibular positioner predicts efficacy of oral appliances in sleep apnea. Am J Respir Crit Care Med 2004;170(4):366-370.

Tsuiki S, Lowe AA, Almeida FR, Fleetham JA. Effects of an anteriorly titrated mandibular position on awake airway and obstructive sleep apnea severity. Am J Orthod Dentofacial Orthop 2004;125(5):548-555.

Woods CE, Usher K, Maguire GP. Obstructive sleep apnoea in adult indigenous populations in high-income countries: An integrative review. Sleep Breath 2015;19(1):45-53.

Yaggi HK, Concato J, Kernan WN, Lichtman JG, Brass LM, Mohsenin V. Obstructive sleep apnea as a risk factor for stroke and death. N Engl J Med 2005;353(19):2034-2041.

Young T, Palta M, Dempsey J, Skatrud J, Weber S, Badr S. The occurrence of sleep-disordered breathing among middle-aged adults. N Engl J Med 1993;328(17):1230-1235.

Zeng B, Ng AT, Qian J, Petocz P, Darendeliler MA, Cistulli PA. Influence of nasal resistance on oral appliance treatment outcome in obstructive sleep apnea. Sleep 2008;31(4):543-547.

SURGICAL MANAGEMENT OF OBSTRUCTIVE SLEEP APNEA: FROM CHILDHOOD TO ADULTHOOD

Sharon Aronovich and Joseph I. Helman

ABSTRACT

Obstructive sleep apnea (OSA) is a chronic multi-factorial condition that has far-reaching effects on a person's health and psychological wellbeing. In the management of patients with OSA who are unable to tolerate positive airway pressure (PAP), several non-surgical and surgical treatment options may be considered. Non-PAP treatments should be tailored individually based on severity of OSA, severity of patient symptoms, cardiovascular health, medical co-morbidities, psychological co-morbidities and anatomical factors. Surgical interventions (e.g., skeletal expansion *via* maxillomandibular advancement) may be highly effective in carefully selected individuals. However, patients with morbid obesity, central sleep apnea along with OSA, craniofacial syndromes and hypotonicity can be expected to require multi-modal treatment to manage OSA adequately.

This chapter takes a case-based approach to discuss potential ramifications of OSA and surgical treatment options that are available. Herein we refer to a hypothetical patient, a boy named Chris, who faces challenges associated with OSA symptoms. As we discuss various aspects of the clinical scenario, we will cite the literature that guides medical decision-making for similar patients, to exemplify the ramifications of this chronic condition and the surgical treatment options that are available.

KEY WORDS: obstructive sleep apnea (OSA), maxillomandibular advancement (MMA), sleep-disordered breathing (SDB), airway, surgical treatment

---

Case Description: Part One

Chris works hard and wants to do well in school, but he struggles to remember things or to concentrate long enough to comprehend concepts he is trying to learn. As a young child, he had signs of excessive hyperactivity and difficulty staying on task, which frequently were dismissed as normal or as "just a phase." When his teacher mentioned he was the only child regularly disrupting class and acting aggressively at times, his parents decided to see his pediatrician. On exam, Chris' physician notes that he is obese, has enlargement of the tonsils and adenoids, a thick neck and a high palatal vault. The pediatrician noted that Chris should have a comprehensive evaluation by a pediatric neuropsychologist and that sleep-disordered breathing (SDB) should be considered as part of the differential diagnosis.

Scientific Correlation

Sleep-related breathing disorders (e.g., obstructive sleep apnea [OSA]) frequently disrupt normal sleep architecture, leading to frequent nocturnal awakenings (fragmented sleep) and nocturnal hypoxemia. As a result, affected children may present with symptoms of hyperactivity, impulsivity, decreased cognitive function, emotional instability, irritability and various other behavioral disorders (Chervin et al., 2007). Other symptoms and complications of childhood OSA include failure to thrive (growth retardation), enuresis, headaches, snoring and excessive daytime sleepiness (Rosen, 1996). In the pediatric population, the incidence of snoring and OSA is 3 to 12% and 1 to 10%, respectively. Moreover, studies reveal a strong association between mild OSA and ADHD (Schecter, 2002). In school-aged children, those with OSA have higher odds of being in the lower 25% of the class due to learning problems (Schecter, 2002).

Case Description: Part Two

Chris underwent an in-lab polysomnogram (PSG) that revealed an Apnea-Hypopnea Index (AHI) of 14/hour. This finding is consistent with severe childhood OSA (AHI > 10, or more than ten apneas and/or hypopneas/hour). Additionally, a careful review of Chris' family history indicated that his father and grandfather have OSA. Could this contribute to why he has OSA?

Scientific Correlation

Studies on the heritability of OSA have found a genetic influence in 21 to 84% of cases. Methodological bias may exist in these studies, as the sample included tends to be families of OSA patients or those referred for a sleep study. Moreover, obesity is a significant confounding factor that also is heritable.

To reduce the effects of these biases, de Paula and colleagues (2016) conducted a study to assess the inheritance pattern of OSA in a rural part of southern Brazil. They examined a randomly selected population without obesity (median BMI = 25) or a known family history of OSA. With nearly 600 participants from 91 families, the rate of OSA (AHI > 5) was 19% and heritability was found at a rate of 25%.

Case Description: Part Three

What is the first-line treatment option available for Chris' OSA?

Scientific Correlation

As in adults, the cause of sleep-related airway obstruction in children may be multi-factorial. If adenotonsillar hypertrophy is present, adenotonsillectomy (AT) is recommended as first-line treatment for childhood OSA by the American Academy of Pediatrics (Mitchell, 2007). A prospective study on the outcomes of AT in 79 patients found a large reduction in the AHI—from 27.5 to 3.5 events/hour; however, OSA was not resolved completely in a significant number of subjects (Mitchell, 2007).

Another meta-analysis on the outcomes of AT in obese children with OSA analyzed the outcomes of four studies (N = 110 children). While there was a substantial reduction in the mean AHI from 29.4 to 10.3 events/hour, only 12% of patients achieved resolution of pediatric OSA (AHI < 1) and 49% met criteria for mild OSA (AHI from 1 to 4.9). Costa and Mitchell (2009) concluded that AT does not resolve OSA in the majority of obese children.

In a three-year prospective longitudinal study at Chang Gung Memorial Hospital in Taiwan, 88 children between the ages of six and twelve years underwent AT for OSA. AHI initially decreased from a mean of 13.5 +/- 7.2 to 3.5 +/- 8.4 events/hour at the six-month follow-up PSG. However, at the 36-month follow-up PSG, the mean AHI was 6.5 +/- 5.6 per hour and 68% of patients continued to have at least mild OSA (AHI > 1). This study found that the overall success rate decreases over time from 47% at six months to 32% at 36 months.

It can be concluded that OSA does improve with AT, but it does not remit spontaneously in the majority of patients who undergo AT. This study also highlights the need for long-term follow-up in pediatric patients with OSA. It is notable that the risk of recurrence or persistent OSA was highest in those with an elevated body mass index (BMI > 95th percentile on the growth chart for same sex and age = obesity) and severe OSA at baseline (Huang et al., 2014).

Case Description: Part Four

Chris had enlarged tonsils and adenoids and underwent AT. He felt better after surgery , snored less and had substantial improvement on follow-up—specifically, his teachers and parents noticed an improvement in several areas including staying on task, concentration and academic performance. His PSG demonstrated an AHI of 6/hour with a minimum blood oxygen saturation of 86%. While the improvement from AHI at 14/hour (severe OSA) to AHI at 6/hour (moderate OSA) is associated with symptomatic benefit for Chris, the residual OSA seen post-operatively merits follow-up by a pediatric sleep medicine specialist. The pediatric sleep medicine specialist reviewed a number of potential treatment options, starting with discussion of non-invasive treatments.

Scientific Correlation

Occasionally, non-invasive modalities can be used in this situation including a trial of corticosteroid nasal spray to manage obstructive nasal breathing, an oxygen supplementation trial or a continuous positive airway pressure (CPAP) trial. Response to treatment should be assessed longitudinally, as the severity of OSA may worsen over time. For many patients, the etiology of this worsening is multi-factorial. Presence of a thick neck, macroglossia, significant retrognathia and a narrow maxilla are anatomical risk factors for OSA. The Center for Disease Control data demonstrates a rising rate of obesity in both the adult and pediatric populations. Based on data from 2011-2012, the prevalence of obesity was 21% in children (ages six to eleven years) and 18% in adolescents (ages 12 to 19 years).

The effects of obesity on the upper airway have been demonstrated by way of volumetric MRI studies. Obesity leads to reversible thickening of the lateral pharyngeal walls, an increase in the size of parapharyngeal fat pads and significant fat accumulation of adipose tissue in the base of the tongue with associated macroglossia (Welch et al., 2002; Kim et al., 2014). Attempts to modify dietary choices, eating habits and physical activity are recommended (Siegfried et al., 1999; Verhulst et al.,

2009). However, achieving significant short- and long-term weight control with medical supervision remains a challenge for many patients.

Case Description: Part Five

Chris and his parents would like to learn more about other non-surgical treatments for OSA in children.

Scientific Correlation

Between the ages of eight and thirteen, maxillary expansion is another viable non-surgical treatment option for appropriately selected children with OSA. In a group of 31 non-obese children without adenotonsillar hypertrophy and clinical evidence of maxillary constriction, a mean rapid maxillary expansion (RME) of 4.3 mm was found to reduce the AHI from 12.2 ± 2.6 to 0.4 ± 1.1, eliminating the occurrence of nocturnal hypoxemia (Pirelli et al., 2004).

It must be noted that the maxillary expander must be removed prior to obtaining a follow-up PSG, in that the presence of an expander has been shown to be associated with persistent moderate to severe OSA. Presumably, the bulk of the maxillary expander may lead to a downward and posterior tongue posture that narrows the airway. A twelve-year follow-up of 23 subjects from this expansion group revealed no recurrent of OSA (Pirelli et al., 2015).

When managing OSA in either the deciduous and mixed dentition, a bonded maxillary expander or a bone-borne expander may represent viable options as well (Ashok et al., 2014). Alleviation of nasal obstruction with volumetric enlargement of the nasal cavity and nasal valve area are well-known benefits of RME (De Fellippe et al., 2008; McNamara et al., 2015). However, more recent studies have demonstrated significant increases (99.4^2 mm or 60%) in the retropalatal cross-sectional area, a significant increase in the pharyngeal airway volume and a raised tongue posture (Chang et al., 2013; Iwasaki et al., 2013).

In children with findings of adenotonsillar hypertrophy and maxillary constriction, there is evidence that both expansion and AT will be required to treat OSA adequately for most children. The sequence of treatment may vary, however, depending on patient age, tonsil grade and patient or parent preference (Guilleminault et al., 2008). It must be emphasized that these results apply to non-obese children with maxillary constriction. There are no current data on the use of maxillary expansion in children with a normal transverse maxillary dimension. Such children remain a group to explore as treatment with RME is non-invasive and typically is tolerated well. Finally, clinicians must be cautious to avoid promising resolution of OSA with RME treatment. This technique currently is supported by only a small number of studies with relatively limited sample sizes.

Case Description: Part Six

Chris has already undergone adenotonsillectomy. Are there other surgical options to consider for treatment of pediatric OSA?

Scientific Correlation

Several other surgical procedures are available for children with OSA. These procedures include lingual tonsillectomy, epiglottoplasty (trimming of a long epiglottis or tightening a ptotic epiglottis), supraglottoplasty (in cases of significant laryngomalacia, a congenital softening of the soft tissue of the larynx above the vocal cords), hyoid suspension, mandibular distraction osteogenesis (MDO) or mandibular advancement with an inverted L-osteotomy. Many of the above procedures lack rigorous outcome assessment research and their use remains at the discretion of the treating providers. There are data on MDO in infants and toddlers, but given the invasiveness of mandibular advancement procedures, early use of these options is limited to patients with severe mandibular hypoplasia and severe OSA.

Several minimally invasive treatment modalities (palatal implants, laser-assisted uvuloplasty [LAUP], sclerotherapy, cautery-assisted palatal stiffening and somnoplasty, the latter using heat energy to modify the tissues of the uvula and soft palate) have been advocated for the management of OSA; however, their success rates have been dismal (Table 1).

Case Description: Part Seven

Chris and his parents wonder if they should wait on treatment until he is older. Are there are other options are available to address multi-level airway narrowing experienced by many adults with severe OSA? Is there value in stage 1 (multi-level) surgery (e.g., uvulopalatopharyngoplasty [UPPP]) and hyoid suspension with genioglossus advancement (GAHS)?

Table 1. Success rate of maxillomandibular advancement (MMA) surgery compared to operations targeting an elongated soft palate, low upper airway tone, macroglossia and complete bypass of the upper airway. UPPP = uvulopalatopharyngoplasty; LAUP = laser-assisted uvuloplasty; RFA = radiofrequency ablation; BOT TORS = base of tongue trans-oral robotic surgery.

	MMA	UPPP	LAUP	RFA	Upper airway stimulation (Inspire)	BOT TORS	Tracheostomy
Studies	9	15	2	8	2	5	4
N	234	950	34	175	69	68	159
% reduction in AHI success	87% (80-92) 89%	33% (23-42) low	18% low	34% low	68% 66%	50% 50% (25-80%)	Most studies lacked mean AHI data 96.2%
AHI change (events/hour)	54.4-7.7	40.3-29.8	18.6-14.7	23.4-14.2	29.3 → 9.0	59.4 → 29.6	N/A

Scientific Correlation

These procedures attempt to alleviate airway obstruction at the retro-palatal, retroglossal and hypopharyngeal areas by removing redundant soft palatal tissue, reducing collapsibility at the base of the tongue, repositioning the tongue anteriorly and advancing the hyoid bone.

In subjects with severe OSA, however, the PSG results of UPPP and GAHS reveal a high incidence of residual OSA requiring further PAP treatment post-operatively. To be clear, while many studies demonstrate a significant reduction in AHI and improvement in nocturnal hypoxemia, the post-operative mean AHI or RDI frequently is greater than 20/hour (Vilaseca et al., 2002; Neruntarat, 2003; Miller et al., 2004; Bowden et al., 2005; Stuck et al., 2005; Baisch et al., 2006). To date, no study has determined significant predictors of treatment success. However, it is possible that the use of multi-level airway surgery may be of greatest benefit in younger, non-obese patients who do not have macroglossia or hypotonicity and present for the treatment of mild-to-moderate OSA.

Trans-oral robotic surgery (TORS) for base of tongue (BOT) reduction is another alternative or adjunctive treatment modality. A meta-analysis including 353 patients found that TORS leads to a 68% success rate (defined as AHI < 20 and AHI reduction > 50%) with mean AHI reduction from 44.3 ± 22.4 to 17.8 ± 16.5, but a cure rate (AHI < 5) of only 24% (Miller et al., 2017).

Finally, the latest addition to the armamentarium of procedures for OSA involves hypoglossal nerve stimulation or upper airway stimulation (Strollo et al., 2014). The results showed some improvement in the patients treated. The selection criteria implemented was very strict, which resulted in an enrollment of only 126 (14%) from a total of 929 patients initially screened. Among those who had the nerve stimulator implanted, 66% had an adequate response rate (defined as AHI < 20 and AHI reduction > 50%), with an overall 51% reduction in mean AHI reduction from 32.0 ± 11.8 to 15.3 ± 16.1 (Strollo et al., 2014).

Thus far, this modality is recommended only for use in patients with AHI between 15 and 65, BMI < 32 and those without large tonsils (grade 3 or 4) or concentric airway collapse on drug-induced sleep endoscopy. Some authors have questioned the generalizability of these results to a broader OSA population and have expressed concern about the cost-effectiveness of this treatment modality (Pengo and Steier, 2015). The cost of the device and surgery ranges from $30,000-$40,000 and the replacement of the battery can cost another $17,000—a high cost for a very selective patient population. More data may clarify the cost-benefit ratio of the hypoglossal nerve stimulation, as well as the success rates for broader group of patients affected by OSA.

Case Description: Part Eight

Chris's parents read online about the option of MMA surgery and ask for more information about this procedure.

Scientific Correlation

MMA involves a large (typically 10 mm) advancement of the maxilla and mandible *via* Le Fort 1 maxillary osteotomy and bilateral sagittal split osteotomy. The latter surgery effectively expands the facial skeleton anteriorly and leads to multi-level airway enlargement at the hypopharynx, oropharynx and velopharynx.

Using CBCT scans with three-dimensional (3D) airway analysis, studies have found clinically and statistically significant changes in airway volume and minimal cross-sectional area (Abramson et al., 2011; Makovey et al., 2017). Moreover, assessment of the airway *via* nasopharyngoscopy demonstrates reduced collapsibility of the lateral pharyngeal walls (Kasey et al., 2002). A meta-analysis of eight retrospective studies (Table 2) determined that patients undergoing MMA experienced a mean success rate (AHI < 10) of 89.9% (range = 65.2 to 100%; Waite et al., 1989; Conradt et al., 1997; Prinsell, 1999; Bettega et al., 2000; Goh and Lim, 2003; Caples et al., 2010).

In a retrospective study of 29 patients treated by MMA between 1996 and 2002 (Magliocca and Helman, 2007), a success rate of 78% was realized. The authors investigated the subgroups at risk of failure and found that patients with a BMI of more than 32 or an AHI of more than 70 had an overall success rate of 60% regardless of previous airway surgery. Patients with a BMI below 32 and an AHI below 70 had more than a 90% chance of success.

It seems that the severity of the disease and co-morbid obesity had a significant impact on the success of traditional MMA of 10 mm. To overcome the low success rate in the extremely severe and morbidly obese sleep apnea patients, the authors of the study decided to manage this subgroup of patients with a 25 mm bimaxillary advancement through distraction osteogenesis (DO) at a rate of 1 mm/day. The pre-operative data of the nine patients subjected to DO showed an average BMI of 34.3

Table 2. Success of MMA surgery based on the percentage of patients who achieved a post-operative AHI < 10. Adapted from Caples et al., 2010.

AUTHOR	N (SAMPLE SIZE)	FOLLOW-UP (IN MONTHS)	AHI BEFORE	AHI AFTER	SUCCESS RATE (%)
Waite, 1989	23	1.5	62.8	15.2	65.2
Riley, 1990	30	6	72	8.8	96.7
Hochban, 1997	38	2	44.4	2.5	97.4
Prinsell, 1999	50	5.2	59.2	4.7	100
Li, 2000	40	50.7	69.6	8.9	90
Bettega, 2000	20	6	59.3	11.1	75
Goh, 2003	11	7.7	70.7	11.4	81.8
Dattilo, 2004	15	1.5	76.2	12.6	86.7
All	234	12.4	54.4	7.7	89.9

and an AHI of 91.0. After the advancement of 25 mm by DO, the average AHI was 6.1 with a 100% success rate (Magliocca and Helman, 2007).

MMA appears to have high success rates in carefully selected patients and is the second most successful surgical procedure after a tracheostomy (Table 2). MMA surgery also was found to improve quality of life as measured using the functional outcomes of sleep questionnaire (FOSQ). In Lye and colleagues' study (2008), patients who underwent MMA surgery reported significant improvements in general productivity, activity level, vigilance and social life.

Patients who have upper airway narrowing and obstructions because of maxillomandibular retrognathia are ideal candidates for MMA. In such cases, facial esthetics typically are enhanced by the procedure and importantly, the odds of meeting success criteria improve (Makovey et al., 2017). Many patients with moderate-to-severe OSA, however, present with risk factors for unfavorable facial alterations. Those include patients who present with a straight or protrusive profiles (SNA > 84°) at baseline, thin soft tissues (females > males), upturned nasal tip, wide alar base and short anterior face height (Cohen-Levy et al., 2013).

With advancement of the maxillomandibular complex, expected facial alterations include nasolabial fullness and bimaxillary protrusion (simian), upper lip protrusion, widening of the alar bases and nostrils, flattening and upturning of the nasal tip and effacement of the nasolabial fold. Facial shortening from excessive impaction of the maxillofacial complex also increases the risk of jowling (neck/cheek ptosis) and premature aging of the face.

In light of these expected changes, patients who are considering MMA surgery must be willing to have significant facial changes occur. Excessive maxillary advancement is the most common cause of unsatisfactory facial changes. To minimize this outcome, counterclockwise rotations of the occlusal plane should be utilized to help maximize advancement at Point B and pogonion while minimizing changes at ANS and Point A (Figs. 1-2). For patients with compensated Class II malocclusions, it is preferable to use surgical orthodontics to retract the lower incisors and facilitate a greater mandibular advancement while limiting the extent of maxillary advancement required (Goncalves et al., 2006; Wei et al., 2017).

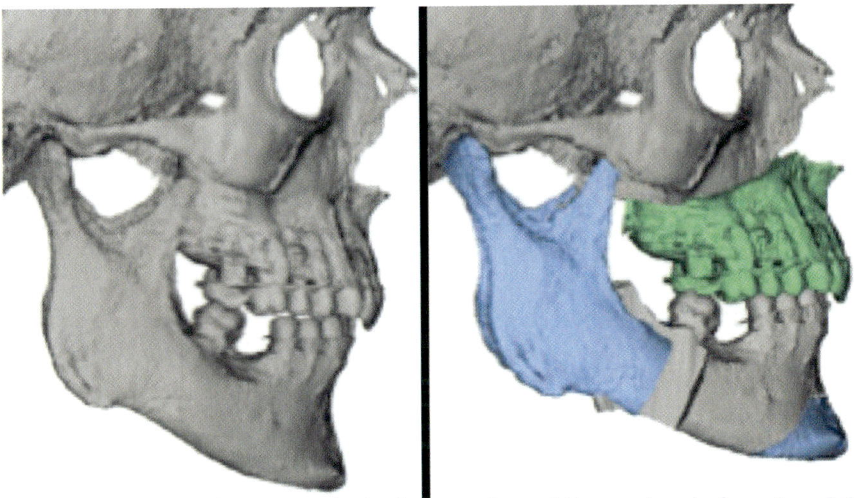

Figure 1. MMA with counterclockwise rotation of the occlusal plane to minimize the advancement at Point A and anterior nasal spine while maximizing the mandibular advancement.

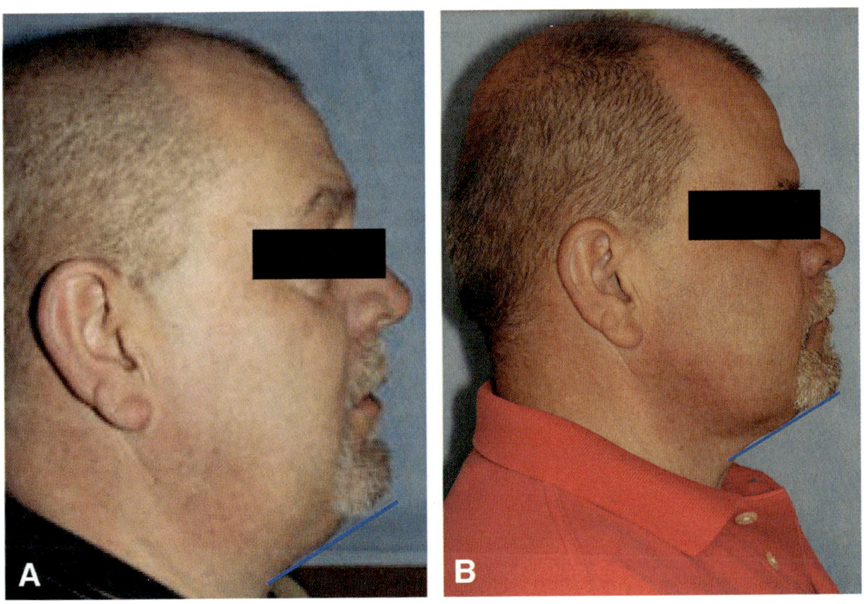

Figure 2. *A:* A patient with severe OSA (AHI = 89) pre-operatively. *B:* Facial profile changes with MMA and counterclockwise rotation, as planned in Fig. 1.

Case Description: Part Nine

Chris and his parents are intrigued by the potential benefit of MMA, but are concerned about pursuing such an aggressive intervention. Is there a way to predict who will do well with MMA for treatment of OSA?

Scientific Correlation

While MMA surgery is a valuable tool in the management of OSA, careful patient selection remains the key to good outcomes. Compared to a younger orthognathic patient population, adult patients who undergo MMA frequently are obese and have significantly greater number of co-morbid cardiovascular, pulmonary, musculoskeletal and psychological conditions. In a comparative study on complications of OSA, the OSA cohort demonstrated a 3.04 higher relative risk for complications compared to adolescents who present for a similarly extensive orthognathic procedure (Table 3). Patients in the OSA group tend to have more difficult intubations, need more vigilant peri-operative airway monitoring, have longer hospital stays and overall recovery, experience more pain, need more analgesics, have a higher rate of infections and hardware failures, and have a greater number of returns to the operating room (Passeri et al., 2016).

Table 3. Comparison of the complication rate in 28 patients with OSA undergoing MMA to 26 patients with dentofacial deformities undergoing the same extent of surgery. RR = relative risk; DFD = dentofacial deformity. Adapted from Passeri et al., 2016.

	OSA	DFD
Post-op complications	28	19
Dysesthesia (abnormal sense of touch)	6 (common early)	0
Infection	18 (25 instances)	6
Hardware removal	11 (14 instances)	0
Non-union	2	0
Re-operation	9 (10 instances)	0
Total number of post-op complications (POC)	108	33
Absolute risk of POC	3.86	1.27
RR of POC in OSA *versus* DFD	3.04	

Ultimately, some patients who undergo MMA will have persistent OSA post-operatively. Despite the high overall success rate of MMA, an incomplete response to surgical treatment with MMA has been reported in up to 45% of patients (Waite et al., 1989; Susarala et al., 2011; Makovey et al., 2017). Studies comparing successful and unsuccessful cases could not find any significant difference between the amount of skeletal advancement or the extent of airway enlargement in those groups. This experience highlights the need to view OSA as a multi-factorial condition that may require multi-modal treatment. Options for managing residual OSA are outlined in Figure 3.

In certain cases, maxillomandibular expansion (MME) may be considered as an alternative to MMA surgery (Fig. 4). A recent meta-analysis assessed aggregate data on 36 adult patients who underwent maxillary expansion *via* surgically-assisted rapid palatal expansion (SARPE) and three patients who underwent MME *via* SARPE and symphyseal distraction. Maxillary expansion and MME reduced the mean AHI by 59% (AHI = 24.3 ± 27.5 pre-op -> 9.9 ± 13.7 post-op) and 78% (AHI = 47.5 ± 29.8 pre-op -> 10.7 ± 3.2 post-op), respectively.

While a very small sample, this treatment appears to be a promising approach that requires further investigation with prospective studies and larger sample sizes. The advantages may include short surgical time, decreased surgical complexity, decreased blood loss, decreased incidence of neurosensory disturbances and limited facial changes. The cost of orthodontics for bimaxillary expansion and full-fixed appliances is a major obstacle for many patients and highlights the need for medical coverage of orthodontic treatment in the management of OSA (Abdullatif et al., 2016).

Case Description: Part Ten

Chris and his parents are grateful to have received a thorough overview of surgical treatment options for OSA. They plan to review this information in more detail and schedule a follow-up visit when they decide how they would like to proceed.

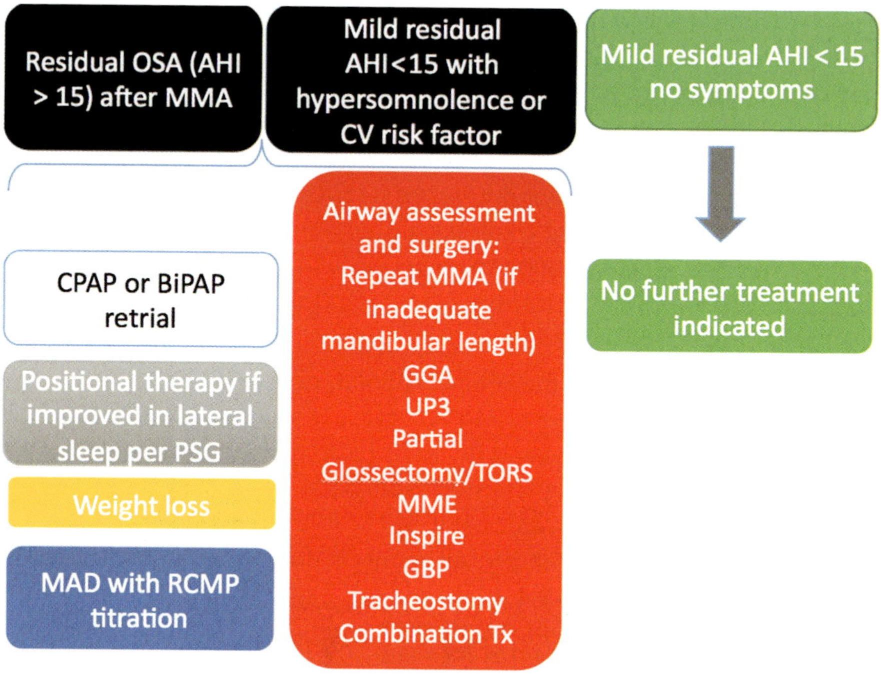

Figure 3. Flow chart on the management of residual OSA after MMA. On the far left are non-surgical treatment options for patients with residual OSA that requires treatment. The center column lists surgical procedures that can be performed to manage residual OSA. On the far right, for patients with mild residual OSA but without hypersomnolence or cardiovascular co-morbidities, no further treatment is indicated. CPAP = continuous positive airway pressure; BiPAP = bi-level positive airway pressure; PSG = polysomnogram; MAD = mandibular advancement device; RCMP = remote controlled mandibular protrusion; MMA = maxillomandibular advancement; GGA = genioglossus advancement surgery; UP3 = uvulopalatopharyngoplasty; TORS = trans-oral robotic surgery; MME = maxillomandibular expansion; GBP = gastric bypass; Combination Tx = a combination of above surgical and non-surgical treatments.

CONCLUSIONS

OSA is a chronic multi-factorial condition that has far-reaching effects on a person's health and psychological wellbeing. Both children and adults should undergo routine screening for signs and symptoms of OSA. Patients who screen positive—or those with anatomical, hereditary or other risk factors—should be referred to a sleep specialist for an evaluation and PSG.

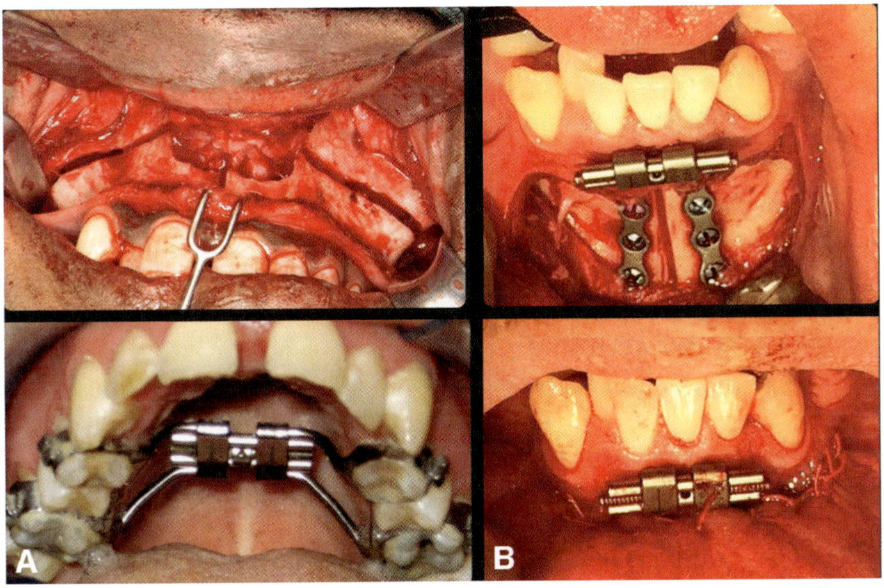

Figure 4. *A:* MME with typical maxillary osteotomies and Hyrax expander for a SARPE. *B:* Symphyseal distraction using a bone-borne distractor. The fixation plates are tunneled down through small incisions in the mucosa above the plane of dissection to allow complete wound closure and minimize risk of dehiscence.

In the management of patients with OSA who are unable to tolerate PAP, there are several non-surgical and surgical treatment options to be considered. The latter should be tailored individually based on severity of OSA, severity of patient symptoms, cardiovascular health, medical and psychological co-morbidities, and anatomical factors. Surgical interventions (e.g., skeletal expansion *via* MMA) may be highly effective in carefully selected individuals. However, patients with morbid obesity, complex central and OSA, craniofacial syndromes and hypotonicity can be expected to require multi-modal treatment to manage OSA adequately.

REFERENCES

Abdullatif J, Certal V, Zaghi S, Song SA, Chang ET, Gillespie MB, Camacho M. Maxillary expansion and maxillomandibular expansion for adult OSA: A systematic review and meta-analysis. J Craniomaxillofac Surg 2016;44(5):574-578.

Abramson Z, Susarala SM, Lawler M, Bouchard C, Troulis M, Kaban LB. Three-dimensional computed tomographic airway analysis of patients with obstructive sleep apnea treated by maxillomandibular advancement. J Oral Maxillofac Surg 2011;69(3):677-686.

Ashok N, Varma NK, Ajith VV, Gopinath S. Effect of rapid maxillary expansion on sleep characteristics in children. Contemp Clin Dent 2014;5(4): 489-494.

Baisch A, Maurer JT, Hörmann K. The effect of hyoid suspension in a multilevel surgery concept for obstructive sleep apnea. Otolaryngol Head Neck Surg 2006;134(5):856-861.

Bettega G, Pépin JL, Veale D, Deschaux C, Raphaël B, Lévy P. Obstructive sleep apnea syndrome: Fifty-one consecutive patients treated by maxillofacial surgery. Am J Respir Crit Care Med 2000;162(2 Pt 1):641-649.

Bowden MT, Kezirian EJ, Utley D, Goode RL. Outcomes of hyoid suspension for the treatment of obstructive sleep apnea. Arch Otolaryngol Head Neck Surg 2005;131(5):440-445.

Caples SM, Rowley JA, Prinsell JR, Pallanch JF, Elamin MB, Katz SG, Harwick JD. Surgical modifications of the upper airway for obstructive sleep apnea in adults: A systematic review and meta-analysis. Sleep 2010;33(10):1396-1407.

Chang Y, Koenig LJ, Pruszynski JE, Bradley TG, Bosio JA, Liu D. Dimensional changes of upper airway after rapid maxillary expansion: A prospective cone-beam computed tomography study. Am J Orthod Dentofacial Orthop 2013;143(4):462-470.

Chervin RD, Weatherly RA, Garetz SL, Ruzicka DL, Giordani BJ, Hodges EK, Dillon, JE, Guire KE. Pediatric sleep questionnaire: Prediction of sleep apnea and outcomes. Arch Otolaryngol Head Neck Surg 2007;133(3): 216-222.

Cohen-Levy J, Petelle B, Vieille E, Dumitrache M, Fleury B. Changes in facial profile after maxillomandibular advancement surgery for obstructive sleep apnea syndrome. Int Orthod 2013;11(1):71-92. [In English and French.]

Conradt R, Hochban W, Brandenburg U, Heitmann J, Peter JH. Long-term follow-up after surgical treatment of obstructive sleep apnoea by maxillomandibular advancement. Eur Respir J 1997;10(1):123-128.

Costa DJ, Mitchell R. Adenotonsillectomy for obstructive sleep apnea in obese children: A meta-analysis. Otolaryngol Head Neck Surg 2009; 140(4):455-460.

de Paula LK, Alvim RO, Pedrosa RP, Horimoto AR, Krieger JE, Oliveira CM, Pereira AC, Lorenzi Filho G. Heritability of OSA in a rural population. Chest 2016;149(1):92-97.

Goh YH, Lim KA. Modified maxillomandibular advancement for the treatment of obstructive sleep apnea: A preliminary report. Laryngoscope 2003;113(9):1577-1582.

Goncalves JR, Buschang PH, Goncalves DG, Wolford LM. Surgical stability of oropharyngeal airway changes following counter-clockwise maxilla-mandibular advancement surgery. J Oral Maxillofac Surg 2006;64(5): 755-762.

Guilleminault C, Quo S, Huynh NT, Li K. Orthodontic expansion treatment and adenotonsillectomy in the treatment of obstructive sleep apnea in prepubertal children. Sleep 2008;31(7):953-957.

Huang YS, Guilleminault C, Lee LA, Lin CH, Hwang FM. Treatment outcomes of adenotonsillectomy for children with obstructive sleep apnea: A prospective longitudinal study. Sleep 2014;37(1):71-76.

Iwasaki T, Saitoh I, Takemoto Y, Inada E, Kakuno E, Kanomi R, Hayasaki H, Yamasaki Y. Tongue posture improvement and pharyngeal airway enlargement as secondary effect of rapid maxillary expansion: A cone-beam computed tomography study. Am J Orthod Dentofacial Orthop 2013;143(2):235-245.

Kim AM, Keenan BT, Jackson N, Chan EL, Staley B, Poptani H, Torigian DA, Pack AI, Schwab RJ. Tongue fat and its relationship to obstructive sleep apnea. Sleep 2014;37(10);1639-1648.

Li KK, Guilleminault C, Riley RW, Powell NB. Obstructive sleep apnea and maxillomandibular advancement: An assessment of airway changes using radiographic and nasopharyngoscopic examinations. J Oral Maxillofac Surg 2002;60(5):526-530.

Lye KW, Waite PD, Meara D, Wang D. Quality of life evaluation of maxillomandibular advancement surgery for treatment of obstructive sleep apnea. J Oral Maxillofac Surg 2008;66(5):968-972.

Magliocca K, Helman JI. Distraction osteogenesis in the management of obstructive sleep apnea syndrome. Bell WH, Guerrero CA, eds. In: *Distraction Osteogenesis of the Facial Skeleton*. Hamilton, Ontario: BC Decker 2007;431-436.

Makovey I, Shelgikar AV, Stanley JJ, Robinson A, Aronovich S. Maxillomandibular advancement surgery for patients who are refractory to continuous positive airway pressure: Are there predictors of success? J Oral Maxillofac Surg 2017;75(2):363-370.

McNamara JA Jr, Lione R, Franchi L, Angelieri F, Cevidanes LH, Darendeliler MA, Cozza P. The role of rapid maxillary expansion in the promotion of oral and general health. Prog Orthod 2015;16:33.

Miller FR, Watson D, Boseley M. The role of the genial bone advancement trephine system in conjunction with uvulopalatopharyngoplasty in the multilevel management of obstructive sleep apnea. Otolaryngol Head Neck Surg 2004;130(1):73-79.

Miller SC, Nguyen SA, Ong AA, Gillespie MB. Transoral robotic base of tongue reduction for obstructive sleep apnea: A systematic review and meta-analysis. Laryngoscope 2017;127(1):258-265.

Mitchell RB. Adenotonsillectomy for obstructive sleep apnea in children: Outcome evaluated by pre- and postoperative polysomnography. Laryngoscope 2007;117(10):1844-1854.

Neruntarat C. Genioglossus advancement and hyoid myotomy under local anesthesia. Otolaryngol Head Neck Surg 2003;129(1):85-91.

Oliveira De Felippe NL, Da Silveira AC, Viana G, Kusnoto B, Smith B, Evans CA. Relationship between rapid maxillary expansion and nasal cavity size and airway resistance: Short and long-term effect. Am J Orthod Dentofacial Orthop 2008;134(3):370-382.

Passeri LA, Choi JG, Kaban LB, Lahey ET III. Morbidity and mortality rates after maxillomandibular advancement for the treatment of obstructive sleep apnea. J Oral Maxillofac Surg 2016;74(10):2033-2043.

Pengo MF, Steier J. Emerging technology: Electrical stimulation in obstructive sleep apnoea. J Thorac Dis 2015;7(8):1286-1297.

Pirelli P, Saponara M, Guilleminault C. Rapid maxillary expansion in children with obstructive sleep apnea syndrome. Sleep 2004;27(4):761-766.

Pirelli P, Saponara M, Guilleminault C. Rapid maxillary expansion (RME) for pediatric obstructive sleep apnea: A 12-year follow-up. Sleep Med 2015;16(8):933-935.

Prinsell JR. Maxillomandibular advancement surgery in a site-specific treatment approach for obstructive sleep apnea in 50 consecutive patients. Chest 1999;116(6):1519-1529.

Rosen CL. Obstructive sleep apnea syndrome (OSAS) in children: Diagnostic challenges. Sleep 1996:19(10 Suppl):S274-S277.

Schecter MS. Section on pediatric pulmonology, subcommittee on obstructive sleep apnea syndrome: Technical report. Diagnosis and management of childhood obstructive sleep apnea syndrome. Pediatrics 2002;109(4):369.

Siegfried W, Siegfried A, Rabenbauer M, Hebebrand J. Snoring and sleep apnea in obese adolescents: Effect of long-term weight loss-rehabilitation. Sleep Breath 1999;3(3):83-88.

Strollo PJ Jr, Soose RJ, Maurer JT, de Vries N, Cornelius J, Froymovich O, Hanson RD, Padhya TA, Steward DL, Gillespie MB, Woodson BT, Van de Heyning PH, Goetting MG, Vanderveken OM, Feldman N, Knaack L, Strohl KP. Upper-airway stimulation for obstructive sleep apnea. N Eng J Med 2014;370(2):139-149.

Stuck BA, Neff W, Hörmann K, Verse T, Bran G, Baisch A, Düber C, Maurer JT. Anatomic changes after hyoid suspension for obstructive sleep apnea: An MRI study. Otolaryngol Head Neck Surg 2005;133(3):397-402.

Susarala SM, Abramson ZR, Dodson TB, Kaban LB. Upper airway length decreases after maxillomandibular advancement in patients with obstructive sleep apnea. J Oral Maxillofac Surg 2011;69(11):2872-2878.

Verhulst SL, Franckx H, Van Gaal L, De Backer W, Desager K. The effect of weight loss on sleep-disordered breathing in obese teenagers. Obesity (Silver Spring) 2009;17(6):1178-1183.

Vilaseca I, Morelló A, Monserrat JM, Santamaria J, Iranzo A. Usefulness of uvulopalatophyaryngoplasty with genioglossus and hyoid advancement in the treatment of obstructive sleep apnea. Arch Otolaryngol Head Neck Surg 2002;128(4):435-440.

Waite PD, Wooten V, Lachner J, Guyette RF. Maxillomandibular advancement surgery in 23 patients with obstructive sleep apnea syndrome. J Oral Maxillofac Surg 1989;47(12):1256-1261.

Wei S, Zhang Y, Guo X, Yu W, Wang M, Wao K, Sun H, Zhang H, Lu X. Counterclockwise maxillomandibular advancement: A choice for Chinese patients with severe obstructive sleep apnea. Sleep Breath 2017; 21(4):853-860.

Welch KC, Foster GD, Ritter CT, Wadden TA, Arens R, Maislin G, Schwab RJ. A novel volumetric magnetic resonance imaging paradigm to study upper airway anatomy. Sleep 2002;25(5):532-542.

OBSTRUCTIVE SLEEP APNEA: A PATIENT'S PERSPECTIVE

Louis Chmura

ABSTRACT

Obstructive sleep apnea (OSA) is epidemic and the comorbidities that are associated with this condition include cardiac arrest, high blood pressure and stroke, to name a few. The solutions presented to patients are limited (e.g., continuous positive airway pressure [CPAP] and oral appliance therapy [OAT]) and require compliance to be effective. As a sufferer of OSA, I can attest that the OSA-associated reduction in daily quality-of-life-degradation of memory, general malaise, irritability, loss of friends and intimacy is particularly concerning.

Orthodontists are positioned well to make permanent positive changes in dental and facial anatomy that can reduce or eliminate OSA with less need for compliance. Orthodontics plays a role in a number of potential OSA treatment options (e.g., jaw surgery or increasing tongue space) and thus is becoming an effective tool in treating OSA in selected patients.

Schendel and colleagues (2012) demonstrated that the airway lumen increases until about 20 years of age and then decreases. Orthodontic goals should revolve around maximizing the airway space. Removing enlarged tonsils and adenoids often is a first step for children, but following with a CPAP can be counterproductive, negatively affecting growth for some. Managed properly, expanding the maxilla, protracting the mandible forward and normalizing facial growth have shown good promise in providing long-term improvement in OSA.

The majority of patients who do not use CPAP will require more than one modality of OSA treatment: adenotonsillectomy and expansion; orthodontic treatment and orthognathic surgery; OAT; or some other combination. Practitioners of different specialties must work together to provide optimal solutions, based on the specific needs of the patient.

KEY WORDS: obstructive sleep apnea, orthodontics, airway

INTRODUCTION

Being asked to speak at the Moyers Symposium is a singular honor for me. Dr. Robert Moyers was my mentor and he is the reason I am an orthodontist today. He took me under his wing when I was a dental student, hired me to work a summer at the Center for Human Growth and Development (an interdisciplinary research unit on the Ann Arbor campus of The University of Michigan) and later allowed me to help in his practice. He taught me orthodontics, ethics, integrity and forgiveness. To this day, his picture stands in my office to remind me how to conduct myself.

Dr. James McNamara taught us histology in dental school. I remember marveling at his ability to teach such a complex subject in a way that was understandable. Jim also was my clinical director in the orthodontic program and he still is one of the top lecturers of whom I have had the pleasure to listen. As a clinician, it is a true honor to share the stage with so many accomplished academics. I hope to share with you what it is like to have sleep apnea from a patient's perspective.

My introduction to sleep apnea occurred in 1980 on the 6-Foster Wing of Sparrow Hospital in Lansing, MI. 6-Foster was the neuro floor at Sparrow, housing all the patients in comas, along with those with paralysis (diplegia or quadriplegia), dementia and delirium tremors. I was an orderly on the midnight shift whose primary responsibility was to turn those who could not turn themselves every two hours to prevent bedsores. At one point, I sat at the bedside of a severely obese man in a coma. He periodically stopped breathing, I would poke him, he would curse and begin breathing again. We repeated this sequence all night long. At the time, I had no idea this condition was called sleep apnea or how deeply sleep apnea would affect my life in the future.

Another result of working at Sparrow was meeting my wife, Penny, whom I married in 1980. I completed dental school in 1985 and my orthodontic residency in 1987. Penny completed her masters in nursing in 1988.

By 1988, I had put on a few pounds, but was still in reasonably good shape. I could run a five-minute mile and loved the stop/start sports (e.g., basketball, racquetball) that helped keep me in shape. I sometimes snored, but thought that was just normal. I started private practice first in Centreville, VI, then in Marshall, MI, where I have been practicing since 1991.

In 2002, I tore my Achilles tendon playing basketball, which spelled the end of the quick stop/start sports that I loved. By 2012, I was a heart attack waiting to happen (Fig. 1). Despite our advanced medical training, neither Penny nor I saw this coming. We sought help only when Penny started staying up at night, poking me when I stopped breathing. A series of evaluations, an overnight polysomnogram (PSG) and an obstructive sleep apnea (OSA) diagnosis resulted in me using a continuous positive airway pressure (CPAP) machine.

A PSG is an overnight sleep study. I have had several, both in a hospital-based sleep clinic and in a converted hotel room. Numerous electronic leads are attached—fourteen channels in all—including electroencephlography (EEG), electrocardiography (EKG), electromyography (EMG), abdominal and thoracic effort belts (to quantify respiratory effort), a nasal pressure transducer and a nasal-oral thermistor (Fig. 2). The patient is videotaped and observed during sleep. While it may be difficult to imagine sleeping under these conditions, when suffering from sleep apnea, insomnia often is not a problem.

The baseline PSG is a diagnostic study done to assess the presence and severity of sleep-disordered breathing. A technologist analyzes

Figure 1. A heart attack waiting to happen (2012).

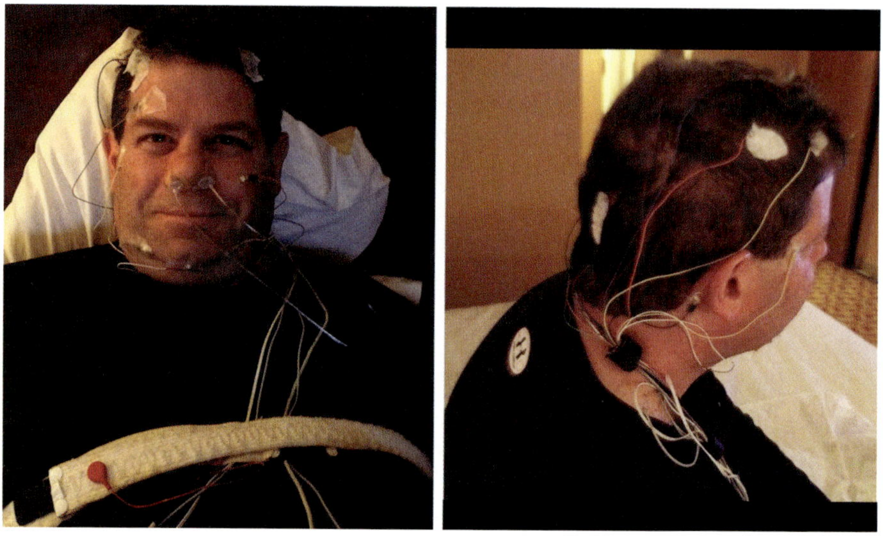

Figure 2. Polysomnogram (overnight sleep study; PSG) with all the leads for the various channels. I jokingly call it my "wire charity mullet."

the raw data, counts the number of apneas and hypopneas and averages them over the time spent asleep. The Apnea-Hypopnea Index (AHI) is the average number of apneas and hypopneas per hour of sleep. A board-certified sleep medicine physician reviews the data, including sleep staging and respiratory scoring, to confirm if OSA or some other sleep-disordered breathing if present. When the diagnosis is OSA, the patient typically is scheduled for a second sleep study, called a CPAP titration study, to determine the most effective treatment setting.

The most common first-line treatment for sleep apnea is CPAP. Essentially, this treatment acts as a pneumatic splint to prevent collapse of the upper airway. Positive airway pressure is delivered *via* a mask, which may cover just the nose, rest in the nostrils (i.e., nasal pillows), or cover both the nose and mouth. Patients may have difficulties getting used to CPAP for a host of reasons, often categorized as mask intolerance (the mask is uncomfortable, does not fit correctly, or the patient feels claustrophobic) or pressure intolerance (the patient finds the positive airway pressure to be uncomfortable).

For me, a full-face mask was uncomfortable and I chose the nasal pillows (Fig. 3), a name that speaks to marketing more than reality. My prescribed pressure setting was so high that air would blow into my nose

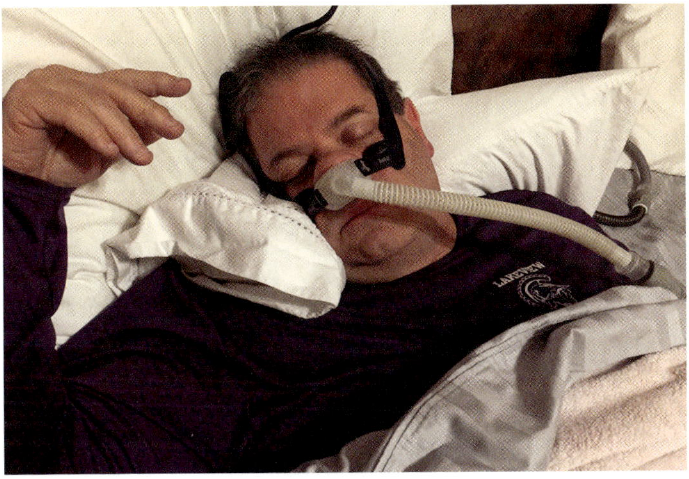

Figure 3. CPAP with a nasal pillow mask.

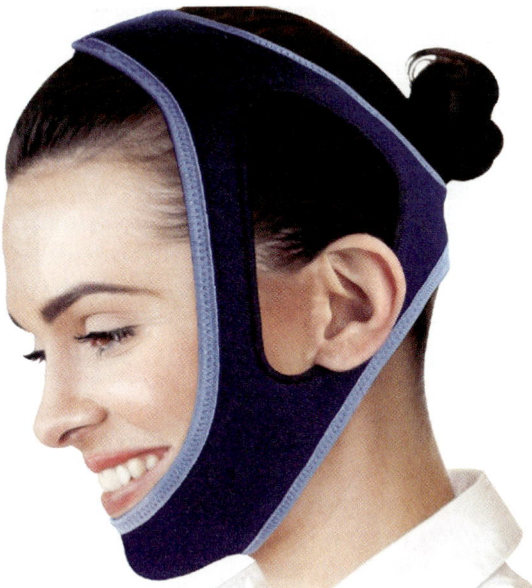

Figure 4 . Head mask to hold mouth closed.

and out of my mouth. It is possible to wear a head strap to keep your mouth closed (Fig. 4), but I chose to tape my mouth closed (Fig. 5). You may imagine how this changed our bedtime habits—Penny and I no longer discussed the day's events as we drifted off to sleep.

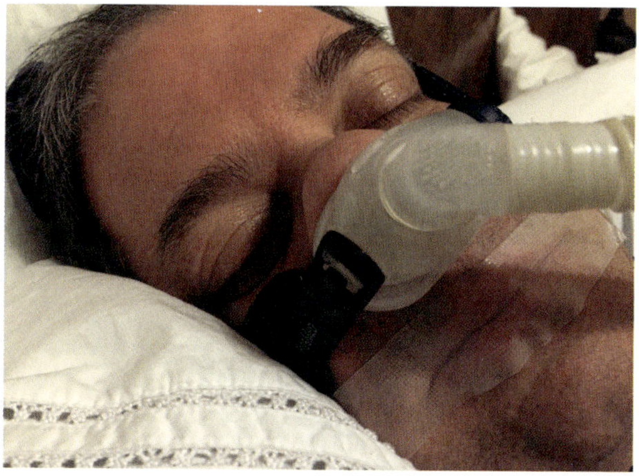

Figure 5. Mouth taped shut to keep the air from escaping out of my mouth.

It is important to note that OSA crept up on us. Though both highly trained healthcare professionals, the changes occurred so gradually that we both missed them. I liken it to the old tale about a frog in a pan of water: place a frog in a pan of boiling water and it will sense the heat and jump out. Place the frog in a pan of room temperature water and slowly bring it to a boil, and the frog will not notice the gradual heating until it is too late. Similarly, both and Penny and I missed the signs until I had signs and symptoms of severe OSA.

It also is important to note that patients do not have an accurate frame of reference. Symptoms appear gradually; the day-to-day changes are so subtle that they do not trigger patients to seek help for themselves or loved ones. It is easy to miss big changes when they progress gradually over time.

This point was driven home when Penny and I took a trip to Atlanta and I had been wearing a CPAP every time I went to sleep for six months. Unfortunately, I had packed my CPAP in a checked bag and our luggage was lost. I was shocked at how awful I felt after only two days without CPAP use! I was reminded of how I felt at the end of a difficult finals week: absolutely exhausted after staying up all night, completing last-minute assignments and studying under pressure.

Now imagine that you never recover—that last day of finals week goes on and on and never stops. The devastating effects of this constant state of tension and pressure have terrible long-term effects—increases in cardiac arrest, stroke, high blood pressure, diabetes mellitus type II and motor vehicle accidents (Brooks et al., 1994; Kryger et al., 2002; Young et al., 2002, 2008; Fritscher et al., 2007; NIH, 2010; Javaheri, 2013). These are devastating long-term effects, but in my experience, the degradation of daily quality of life is even more disturbing. Here are just a few of the negative effects I experienced due to untreated OSA:

1. Stamina: Where before I had boundless energy, I now had to force myself to do even the things I love.
2. Memory/brainpower: Where before I could juggle numerous complex and disparate thoughts and ideas in my head, I now found myself writing things down. Even simple things like a grocery list got written down. Big projects, like writing this chapter, seemed overwhelming. Remembering names was a challenge.
3. Friends: One of the side effects of chronic exhaustion is irritability and reduction in the ability to filter. I lost long-term friends—I do not completely know why—during the time when we were trying to determine what was wrong with me. I am convinced it is because I said something that was inappropriate for the situation.
4. Self-confidence: When I first developed OSA, I believed I unknowingly offended referring dentists, the staff working for me and some patients and/or their parents. Members of my team began questioning my judgment in almost every area of the practice and I eventually began questioning myself. As time went by, I found I was not quite as "out of it" as they suggested, but it affected my confidence and caused me to question everything I did and was doing at the time.
5. Food/weight: Two effects of OSA are a reduction in the level of leptin, the hormone that suppresses appetite and an increase in ghrelin, the hormone that increases appetite for high carb, "fast-energy" foods (Harsch et al., 2003). These hormonal changes, coupled with generally feeling tired due to lack of restful sleep, makes exercise difficult. Together, this situation

leads to weight gain which in turn exacerbates the likelihood of having OSA—a downward spiral.

As a consequence of OSA, I developed diabetes mellitus type II (DM-II), which means I had to be conscious of what I ate. If I cheated and ate foods high in carbohydrates, I paid for it with exhaustion. I am blessed to be able to attend numerous conferences each year, but the breakfast and lunch spreads at most hotels are not designed for diabetics. I found myself being purposeful in choosing what to eat, making sure that I was able to present appropriately to my audience.

6. Hydration: One July 4th, I was hospitalized with a massive headache. After running multiple tests over eight hours, it appeared that I was simply dehydrated. Why? Dehydration resulted from a combination of DM-II and blowing air up my nose all night. I had to aware acutely of how much water I was consuming each day. I avoided all caffeinated drinks and found myself paying close attention when I drank any alcohol.
7. Intimacy: Some couples spend the last few minutes before drifting off to sleep sharing their thoughts on how the day went or saying sweet things to each other. Not so when you tape your mouth shut and wear a CPAP. In addition, testosterone is released during deep sleep. I am blessed by having been married to a saint of a woman for the past 36 years, but sleep apnea certainly has changed the way we relate.
8. Joy: The greatest joy in my life is helping others see possibilities where they did not see them—whether it is teaching, mentoring or serving as an example—possibilities are my passion. When I first contracted sleep apnea, I did not realize that much of the joy I had experienced previously was lost. Later, when I began to get restful sleep, I realized that I was getting hopeful again, but I found myself distracted by all the

"stuff" I now had to pay attention to (e.g., hydration, food, sleep).

I have come to grips with the fact that all the above simply are part of my life and I am regaining the joy I once had. It has been difficult, however, not to grieve for what might have been had I not been saddled with this disease and all its side effects. I wear my CPAP faithfully, but I could not avoid thinking there must be a better way to treat OSA.

Once diagnosed, I decided to learn as much as I could about sleep apnea. I attended introductory courses with the American Academy of Dental Sleep Medicine and numerous others hosted by sleep appliance companies. I read hundreds of articles and several books on the subject.

In nearly every course I took, oral appliance therapy (OAT) was touted as *the* alternative to CPAP for mild-to-moderate sleep apnea and for patients who could not tolerate CPAP therapy (Clete and Kushida, 2006). Note, the criteria for success in the sleep appliance literature was *different* than for the CPAP literature. Success was determined either by an AHI < 5/hour or by cutting the AHI in half (Ferguson et al., 2006). So, a patient with an AHI of 29 (moderate) was termed a success if the AHI dropped to 14.5 (just short of moderate); a patient with an AHI of 65 was a success, even though s/he still had severe apnea!

I fabricated several sleep appliances for myself and my father and I learned that it often is easier to wear a sleep appliance than a CPAP. My father was the primary caregiver when my mother was failing, so he needed to get up every time she did. With the CPAP, he explained that it took 10 to 20 minutes each time to adjust the CPAP mask when they went back to bed. With the oral appliance, he could drink and talk with it in, so it was a more practical solution.

As a long-term treatment strategy, compliance is key for both CPAP and OAT. CPAP therapy is exceptionally effective when the CPAP is worn, but there is a high incidence of non-compliance (Ferguson et al., 2006). OAT, on the other hand, was less effective, with 33% or more not reducing AHI to low levels—but compliance was better (Ferguson et al., 2006). There also are several potential side effects associated with OAT (e.g., tooth movement, muscle soreness, dry mouth). It seemed to me that orthodontists, as architects of the oral cavity, potentially could offer a longer-lasting treatment for OSA sufferers who did not require long-term compliance.

Numerous surgeries have been suggested as solutions for sleep apnea and snoring. The clear majority, however, either were severely invasive (tracheotomy) and were less than 50% effective long-term (uvulopalatopharyngoplasty [UPPP]; Khan et al., 2009), or had more severe dental side effects than OAT (e.g., pulsing the tongue). The most predictable results were associated with moving the upper and lower jaws forward surgically, particularly when accompanied by counterclockwise rotation of the mandible (Nehra et al., 2001). Lye and associates (2008) showed a 93% improvement in quality of life with bimaxillary advancement. I decided to pursue this option.

I had first heard of the counterclockwise surgical approach from Drs. William Arnett and Michael Gunson in Santa Barbara, CA, and I arranged a remote consultation with Dr. Gunson. I wanted as precise a result as possible, so we used a digital treatment solution (Dolphin 3D, Dolphin Imaging, Chatsworth, CA) to develop my treatment plan. We planned to down-fracture the maxilla, more in the posterior than the anterior, to normalize the palatal plane and increase incisor display (Fig. 6). This approach allowed rotating the mandible superiorly in the anterior and inferiorly in the posterior, which is considered a "counterclockwise" movement and resulted in a predicted 10 mm mandibular advancement without moving Point A too far forward and avoiding the concurrent deleterious esthetic effects on the nose. Now I was ready to move teeth appropriately.

To maximize precision, I used SureSmile™ to position my teeth to within 0.1 mm and 1°, then had a robot bend a precise set of archwires to accomplish those exact movements prior to surgery (Fig. 7). I also was able to perform virtual surgery to match the prediction and we set it up for a three-piece maxillary surgery (to adjust for width) with segments from upper canine to upper canine and bilaterally from first premolar to second molar. In that I already work closely with both companies, we made sure the same references were used in both SureSmile™ and Dolphin 3D.

Although my original plan was to go to Santa Barbara for the surgery, I was discussing a different patient's surgery (planned for a counterclockwise approach) with a local surgeon. When the surgeon agreed that

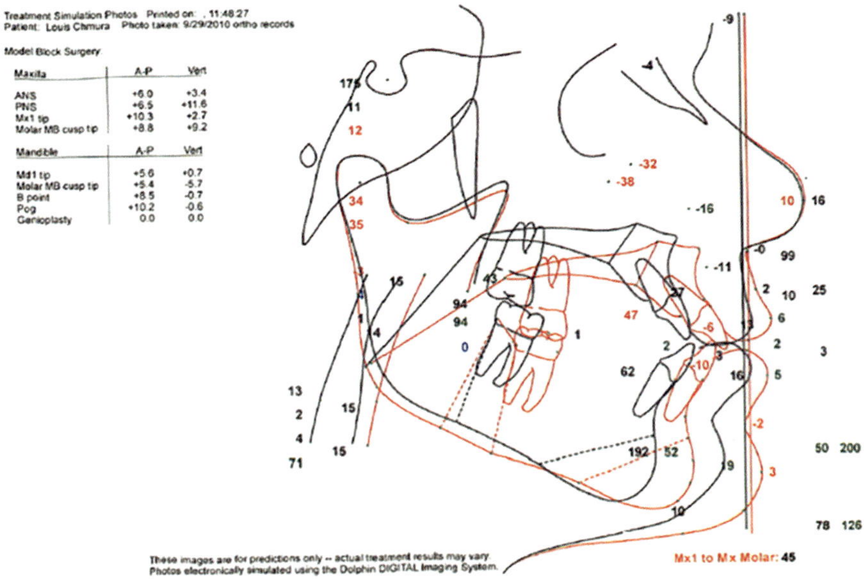

Figure 6. Surgical treatment plan with normalization of palatal plane and counterclockwise rotation of mandible.

this approach was feasible, I discussed my case with him in more detail. I shared the prediction tracings (Fig. 6), explained the approach I wanted; his response was "no problem." I completed tooth movements, took records and the surgeon and I worked my case up with Simplant Medical Modeling (Golden, CO), a third high-tech solution (Fig. 8). Unfortunately, Medical Modeling did not have the same references as SureSmile™ or Dolphin 3D, which led to complications.

I had my first orthognathic surgery on May 31, 2011 at 9:00 a.m. and went home the afternoon of the next day. I went in expecting a horrible recovery, but it really was not that bad. I took nothing stronger than Advil and Tylenol after leaving the hospital and my worst symptoms were severe nasal congestion, initial high blood sugars (384 because they were feeding me apple juice in the hospital), facial swelling for three days after the steroids stopped and difficulty cleaning around the surgical splint, which I wore for a full six weeks. I had to sleep upright in a recliner chair.

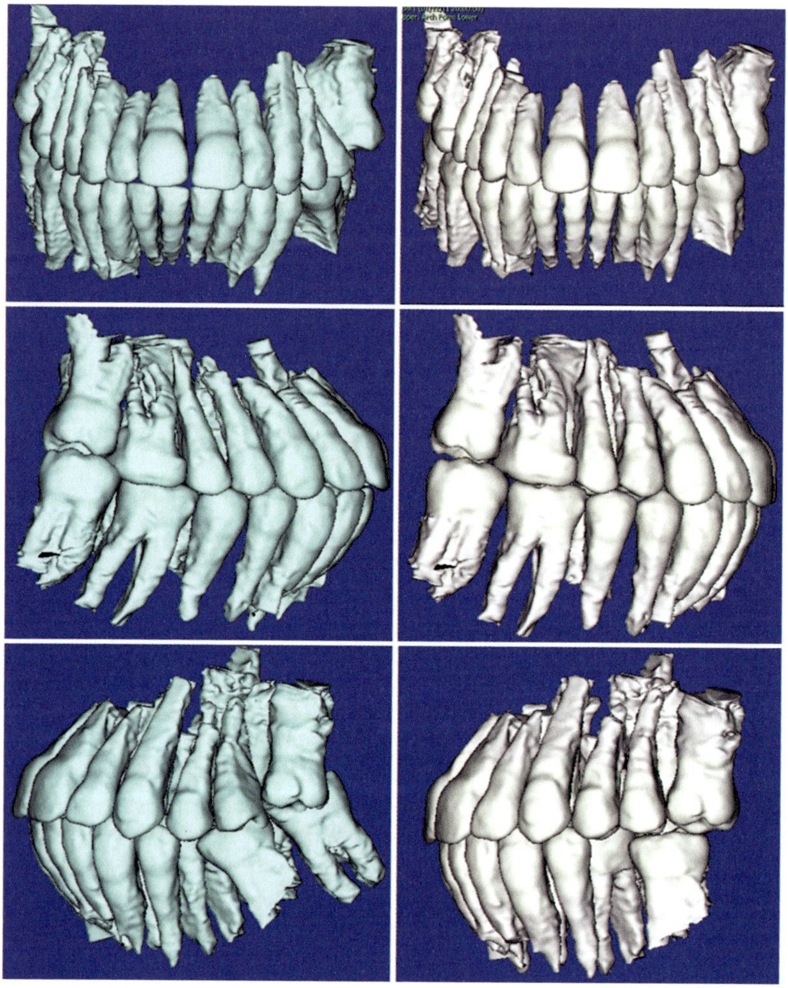

Figure 7. SureSmile™ setup. Before (*left panels*). After (*right panels*).

Week 1 (Fig. 9*A*): I slept about two hours at a time and found it difficult to ingest sufficient liquids. This situation was solved by replacing the syringe that they sent home with a plastic "catsup" bottle, which allowed me to squirt a swallow of water into my cheek, then swallow and repeat.

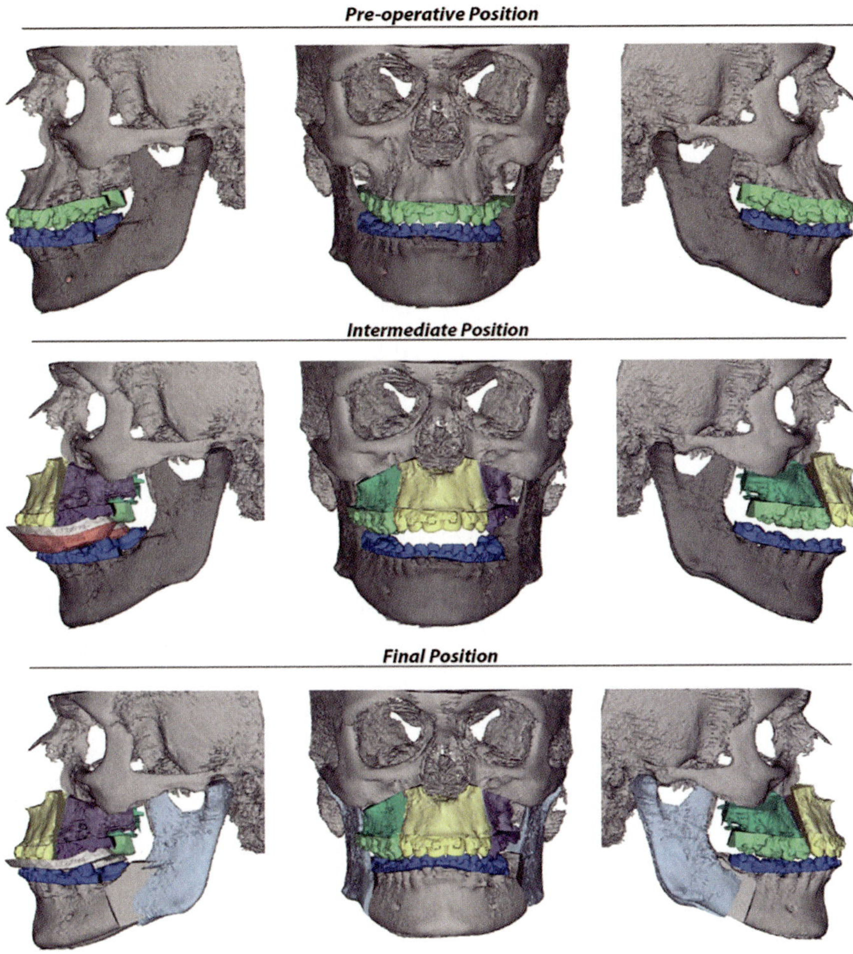

Figure 8. Prediction using medical modeling.

Week 2 (Fig. 9*B*): Nosebleeds and congestion were reduced, sleeping was better and I walked 30 minutes on a treadmill. I returned to my office after 11 days and saw patients in the mornings.

Week 3 (Fig. 9*C*): I slept great, but got lightheaded at the driving range. At the end of the third week, I could complete a four-hour drive in my automobile.

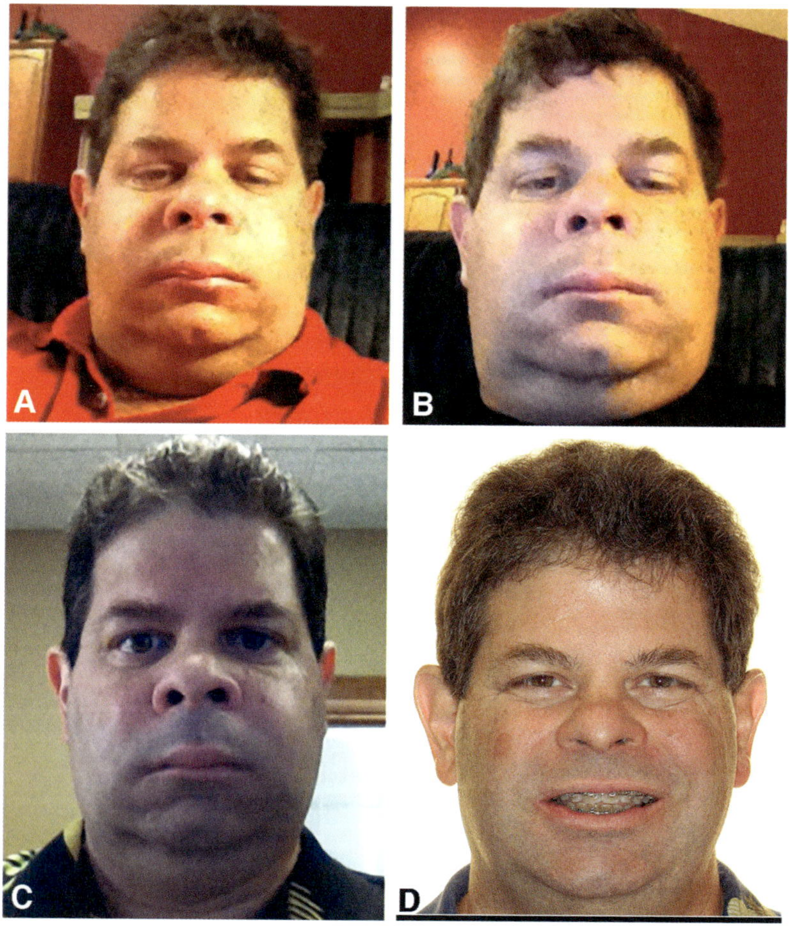

Figure 9. *A:* Week 1 after surgery. *B:* Week 2 after surgery. *C:* Week 3 after surgery. *D:* Week 6 after surgery.

Week 4: I started eating blended soups and was completely off all meds—no Advil, Metformin or Diovan. My blood pressure was 116/68 and I felt great walking miles outside and even playing golf.

Week 6 (Fig. 9*D*): off meds, playing golf and long walks. I felt better than I had in 20 years, I was overwhelmed with the amount of oxygen and energy I had, and I was singing the praises of orthognathic surgery for sleep apnea. Six months later, I was devastated when a home sleep test showed residual apnea (AHI = 14/hour) and I was back on an auto-titrating CPAP device.

In reviewing the before and after cephalometric tracings, the first surgery missed our goals (Fig. 10). Instead of moving the maxilla forward and down, with more posterior inferior movement than anterior, the maxilla was impacted, which allowed little counterclockwise movement and only 3.6 mm of mandibular anterior movement. The severe nasal bleeding and congestion was due to grinding down Point A to avoid negative nasal esthetics. I looked 20 years older, had an anterior open bite and still had OSA. I spent the next year looking for an orthodontic solution.

At this point, my close friend Dr. Ed Lin stepped in and asked me, "When are we going to redo the surgery, Lou? I can't stand to see you trying to eat." I still wanted to solve the apnea, so I had my second orthognathic surgery on March 25, 2013 in Green Bay, WI, with Dr. Lin's surgeon. This recovery was better than the first in many ways.

First, the surgery was performed with no splint, which allowed me to clean my teeth properly. Second, we stayed in Ed and Siri's condo, which had no TV. When I was up, I read. When I was tired, I slept. I also had the benefit of Siri's culinary skills and incredibly tasty clear liquid broths. This time, I took pain meds for a day and a half before switching to Advil and Tylenol. We stayed two weeks, primarily to satisfy the surgeon. I was ready to go home much earlier and hopeful that I finally would have gotten rid of this dreaded disease, but it was not to be.

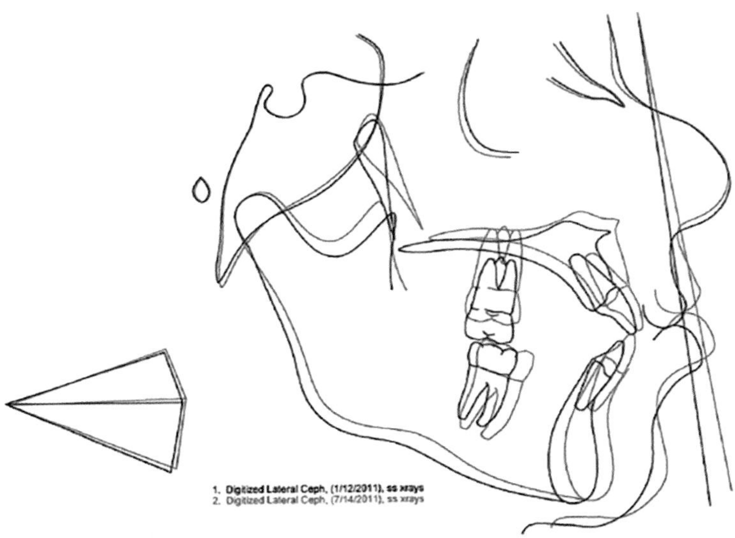

Figure 10. Actual before/after superimposition from my first surgery.

The second surgeon could not overcome the scar tissue from the first surgery. My maxilla was moved down, which improved my smile arc (helping me look more than ten years younger) and my bite improved such that I could touch my front teeth together. I still had residual OSA, however, and went back on PAP therapy.

Given my experience, one would expect me to be sour on surgery. Absolutely not—bimaxillary orthognathic surgery still is the most predictable long-lasting solution available for appropriately selected adults with OSA and while the recovery is difficult, it is not overwhelming. When successful, the surgical approach alters one's anatomy and obviates wearing something every night and having to nap for the rest of one's life. The surgery must be planned correctly, often with a counterclockwise mandibular movement. It is imperative that the same references are used across technologies. In retrospect, I should have taken a more questioning approach when talking with Medical Modeling. In any case, my interest in providing better solutions continued.

In November 2012, I was invited to help establish a task force focused on finding orthodontic solutions for OSA. Our original group of nine orthodontists evolved over the next five years: some members dropped out, other members joined and added different expertise to the group. We reviewed the sleep literature and attended numerous sleep apnea meetings. I personally went to six sleep apnea courses. We designed a two-day course created specifically for orthodontists, which gave a complete background in sleep and OAT, but also showed the unique position that orthodontists have in altering anatomy non-surgically and how this can help patients with sleep apnea and airway limitations.

Our findings: adults and children are different. Not only are the symptoms of sleep apnea different (e.g., daytime sleepiness *versus* hyperactivity), but the long-term consequences also are different. As a result, we developed different protocols for adults and children.

ADULT SLEEP APNEA AND TREATMENT

Screening

As orthodontists, I believe we are positioned to screen patients for sleep apnea. Whether the patient is an adult or a child, by understanding the signs and symptoms of sleep apnea, we can identify patients who we potentially can help. For adults, the Epworth Sleepiness Scale or STOP BANG questionnaire should be available for every adult patient (and parents of younger patients). Observe clinical signs like head posture, midface hypoplasia, restrictive nares, or retrognathic mandibles. If necessary, follow up with an intra-oral examination to observe the size and position of and space available the tongue and any evidence of acid reflux, any or all of which may the patient's likelihood of having OSA.

The next step is to confirm the diagnosis of OSA with a baseline PSG. Certain patients may be candidates for a less detailed home sleep apnea test, which is done in conjunction with a board-certified sleep medicine physician. OSA treatment is recommended with a diagnosis of OSA with: 1) an AHI > 5-15/hour <u>and</u> presence of excessive daytime sleepiness or hypertension; or 2) an AHI above 15/hour regardless of whether the patient has symptoms of daytime sleepiness.

Continuous Positive Airway Pressure (CPAP)

CPAP is the gold standard treatment for adult OSA. CPAP is effective in treating sleep apnea, as it can treat multiple areas of obstruction—the physician can increase pressure generated by the CPAP apparatus until the airway is opened. The problem with CPAP is compliance. Most insurance companies require four hours/day for 70% of days in a month, before they will reimburse patients for CPAP equipment costs. Most importantly, CPAP does not address OSA when it is not worn, so this option, while considered the "gold standard," has limitations.

Oral Appliance Therapy (OAT)

Bringing the mandible forward also brings the tongue and hyoid apparatus forward, which can open the airway. OAT often is considered for mild-to-moderate OSA and for CPAP-intolerant patients. These appliances certainly are within the skillset of dentists and orthodontists and are relatively simple to make. The difficulties lie in managing/avoiding side effects, learning to work together with physicians and medical billing. In addition, it does matter where the obstruction is; if the obstruction is not associated with the tongue or the muscles attached to the hyoid, OAT will not address the problem. Thus, approximately a third of patients achieve an AHI < 5/hour, another third will have a 50% reduction in AHI and the remaining third will have even less effectiveness (Sutherland et al., 2017).

Orthognathic Surgery

Orthognathic surgery nearly always is preceded by orthodontic alignment. Despite this predictability, UPPP and less effective surgeries frequently are recommended first—often due to a commonly held belief (patients and providers) that bimaxillary advancement is too invasive. My understanding is that recovery from double jaw surgery is relatively easy compared to that for UPPP. Other surgical approaches (e.g., somnoplasty, radiofrequency ablation, glossectomy, pillar procedure and even tracheotomy) either have severe negative associations or have shown less than 50% long-term efficacy.

Orthodontic Treatment

In most circles, orthodontic treatment is not part of the discussion for adult OSA. In our task force, however, Dr. David Paquette presented the following treatment of a family member. When orthodontic treatment began, both upper and lower arches of the adult patient had collapsed lingually and developed crowding (Fig. 11). Simply creating a new arch form orthodontically (Fig. 12) dropped the patient's AHI from 52 (severe) to 14 (mild). Figure 13 shows the amount of expansion required for this result.

Retention for post-orthodontic patients who exhibit an AHI of 14/hour or lower can incorporate OAT. In addition, it is easier to fabricate an oral appliance now that the teeth have been aligned. When malalignment is severe, it is difficult to fabricate a well-fitting oral appliance—relieving the undercuts due to malocclusion often compromises retention. In this patient, the oral appliance dropped his AHI to 5.8/hour (Fig. 14).

TREATMENT FOR CHILDHOOD SLEEP APNEA AND COMPROMISED AIRWAY

Childhood Sleep Apnea, Airway and Treatments

The comorbidities of sleep apnea in adults are severe (e.g., high blood pressure, stroke, cardiac arrest, Type II diabetes; Brooks et al., 1994; Young et al., 2002, 2008; Fritscher et al., 2007; NIH, 2010; Javaheri, 2013), but the consequences are even more dire for children. Deprivation of oxygen and/or restful sleep while children are growing, maturing and

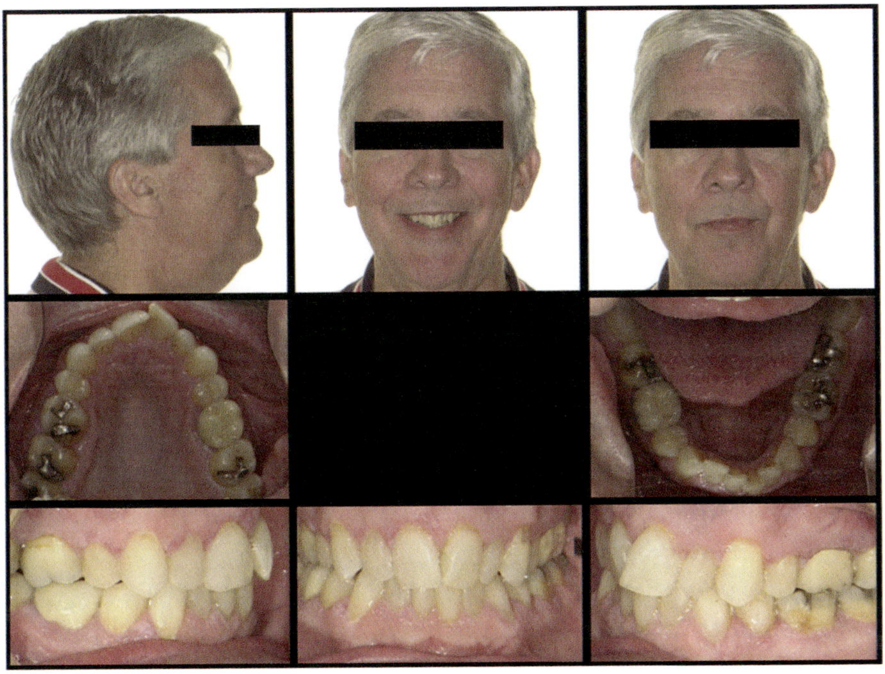

Figure 11. Prior to orthodontic treatment. AHI = 52.

learning is associated with behavioral problems, ADD, ADHD (Miano et al., 2008), difficulties in school (Hallblower et al., 2006; Kheirandish and Gozal, 2006) and even a drop in IQ (Hallblower et al., 2006). In addition, many of these same comorbidities are associated with airway resistance (e.g., snoring or Upper Airway Resistance Syndrome).

3D Cone-beam Computed Tomography (CBCT)

We now have the capability of using 3D CBCT in an orthodontic office with very low exposures (less than a panoramic film, 100x less than a hospital CAT scan). The data revealed with CBCT have given us insights into the development of childhood airway problems.

Schendel and coworkers (2012) studied 1,300 children of various ages and plotted the average minimum cross-sectional area (MCA) at various ages (Fig. 15). At first glance, it appears that the MCA is 10x the patient's age. Looking closer, however, the standard deviations are large, suggesting these data are not the ideal for establishing norms.

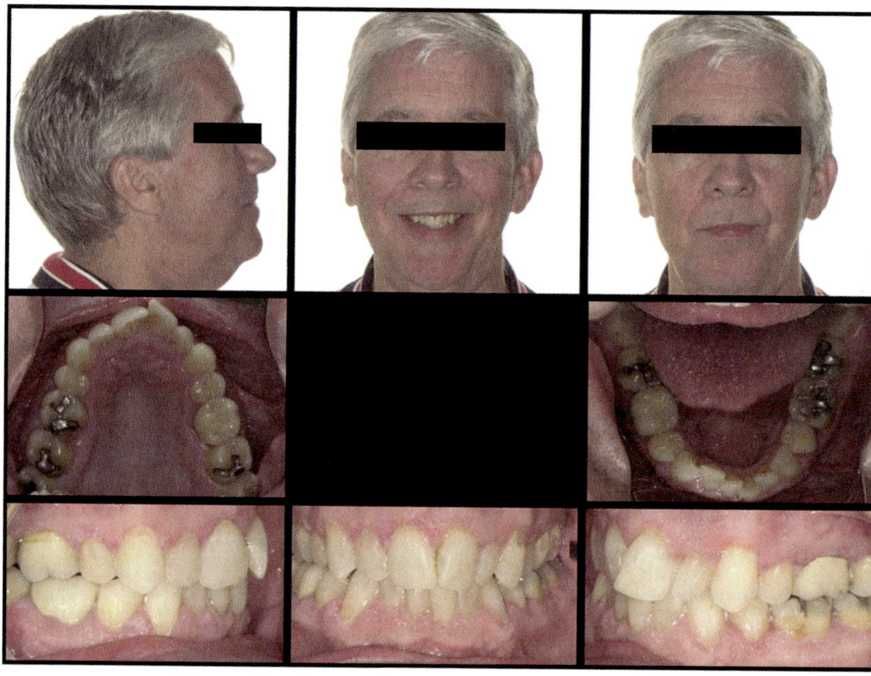

Figure 12. After orthodontic alignment. AHI = 14.

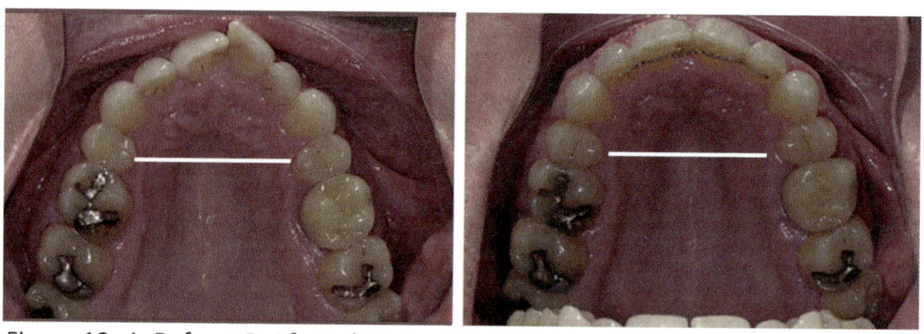

Figure 13. *A:* Before. *B:* After, showing the amount of expansion associated with orthodontic alignment.

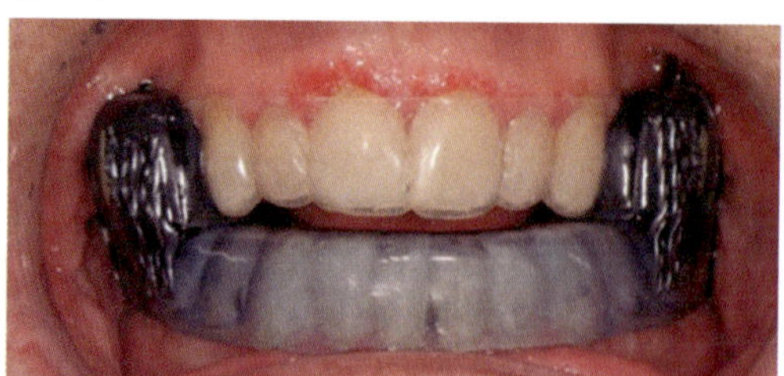

Figure 14. Oral appliance used as a retainer. AHI = 5.8.

Smallest Airway Area	Age Group (yr)													
	6-8	9-11	12-14	15-17	18-20	21-25	26-30	31-35	36-40	41-45	46-50	51-55	≥56	Average	SD
Average	77.70	89.89	128.64	169.13	171.55	160.97	172.18	159.22	157.31	149.26	144.39	143.18	121.83	141.94	30.15
Average SD	48.78	47.77	66.31	86.19	113.98	80.97	81.28	81.68	84.54	120.83	75.44	81.66	82.15	80.89	20.63
Female															
Average	70.92	95.38	138.21	150.16	185.97	184.49	177.77	168.70	171.63	137.98	150.29	146.19	107.28	145.00	35.48
Average SD	44.29	49.10	70.00	72.98	140.55	94.27	84.86	80.33	86.31	75.44	74.45	79.61	65.99	78.32	23.32
Male															
Average	84.21	83.47	120.80	188.87	155.92	137.53	159.15	143.06	132.92	161.48	134.34	137.33	136.99	136.62	29.03
Average SD	52.35	45.85	62.68	94.84	73.82	56.89	71.90	82.53	76.51	155.84	77.08	86.44	94.46	79.32	27.56

Schendel et al. 3D Analysis of Airway Growth. J Oral Maxillofac Surg 2012.

Figure 15. Average smallest airway area by age group (Chervin et al., 2000). Reprinted with permission from Elsevier.

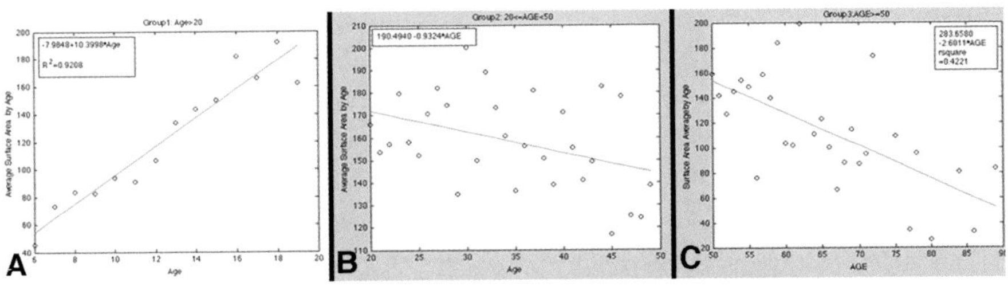

Figure 16. Linear regression best fit line of smallest airway area *versus* age. *A:* Ages 0-20. *B:* Ages 20-50. *C:* Ages 50+. Reprinted with permission from Elsevier.

When plotted over time, however, the trends are telling (Fig. 16). The MCA increases from ages 6 to 20, decreases from ages 20 to 50 and decreases at a greater rate after age 50.

There also is evidence to show that nasal breathing is associated with more normal facial growth. Oropharynx and nasopharynx volumes and MCA were lower significantly in Class II subjects (El and Palomo, 2011) and nasal breathers had significantly greater airway volumes and MCA (Alves et al., 2011). As orthodontists, our treatment planning not only should align teeth, but also encourage nasal breathing and maximize the airway.

Screening the Pediatric Patient

The American Academy of Pediatrics has recommended that all children should be screened for snoring and, if present, should be seen by a sleep specialist and undergo a PSG (Marcus et al., 2012). Similar to adults, with children we start with a questionnaire—in this case the Pediatric Sleep Questionnaire (PSQ), which has sensitivity of 87% and specificity of 87% (Chervin et al., 2000). A positive answer on snoring or labored breathing is a positive indicator for an airway problem.

We then follow with the same clinical exam indicators and, if necessary, recommend a home sleep test; as of this writing, there are two Food and Drug Administration (FDA)-approved tests for use with children. It is not always necessary to obtain a diagnosis, as treating in an airway-friendly manner always is appropriate.

Treatment for Childhood Sleep Apnea or Airway Issues

OSA in children has negative consequences on growth, development and socialization that compels correction as soon as possible. The most direct assessment of OSA involves before and after PSGs, but these studies are expensive and intrusive, and there are relatively few studies of treatment modalities with this measure. Two treatment modalities that have been studied with before and after PSGs are: 1) removal of enlarged tonsils and adenoids (American Academy of Pediatrics, 2002; Sheldon et al., 2005); and 2) rapid palatal expansion (RPE; Guilleminault et al., 2008, 2011).

OSA is associated with enlarged tonsils and/or adenoids, narrow maxillae and reduced nasal breathing, but causation is unclear. Most children require both expansion and tonsillectomy/adenoidectomy for resolution of symptoms (Guilleminault et al., 2008). Numerous studies show the efficacy of RPE on childhood OSA (Hershey et al., 1976; Cistulli et al., 1996, 1998; Seto et al., 2001; Doruk et al., 2004; Johal and Conaghan, 2004; Pirelli et al., 2004; Villa et al., 2007, 2011; Guilleminault et al., 2008, 2011; Ramires et al., 2008; De Felippe et al., 2009; Miano et al., 2009; Altug-Atac et al., 2010; Sökücü et al., 2010; Baratieri et al., 2011; Gorgulu et al., 2011; Aloufi et al., 2012; Marino et al., 2012; Muge et al., 2012; Smith et al., 2012) with effects including reduced nasal resistance (Hershey et al., 1976; Cistulli et al., 1996; Doruk et al., 2004), increased nasal airway volume (Seto et al., 2001; Johal and Conaghan, 2004; Ramires et al., 2008; De Felippe et al., 2009; Altug-Atac et al., 2010; Sökücü et al., 2010; Baratieri et al., 2011; Görgülü et al., 2011; Aksu et al., 2012; Aloufi et al., 2012; Smith et al., 2012), increased pharyngeal volume (Aloufi et al., 2012) and reduction of symptoms (Cistulli et al., 1998; Pirelli et al., 2004; Villa et al., 2007, 2011; Miano et al., 2009; Marino et al., 2012).

There are a number of studies comparing "semi-rapid" and/ or "slow" expansion to RPE (Akkaya et al., 1999; Işeri and Ozsoy, 2004; Machado Júnior and Crespo, 2006; Lima Filho and Ruellas, 2007; Cao et al., 2009; Huynh et al., 2009; Shetty et al., 2009; Kilic and Oktay, 2010; Corbridge et al., 2011; Wong et al., 2011; Martina et al., 2012). "Slow expansion" encompasses a wide range of expansion modalities (one week of RPE turning 2x daily, followed by turning 3x weekly; Quad Helix, Minne-Expander) and show similar results to RPE for many linear measurements (inter-molar distance, inter-canine width, inter-molar width and arch circumference). However, none of these studies utilize before and after PSGs, so while they may increase the size of the tongue space, it is not clear the effect on relieving OSA or on the nasal airway.

The airway is enlarged both with mandibular advancement (Schütz et al., 2011) and with surgical procedures that increase the vertical dimension, so long as the lips still can close comfortably (Step Back Technique, 2013). In essence, to increase airway in growing individuals, we want to expand arches transversely, anteroposteriorly and vertically, while maintaining lip closure (Fig. 17).

MULTI-FACTORIAL PROBLEM, MULTI-SPECIALTY SOLUTION

During the Moyers Symposium, I was surprised at how many presenters felt that because their solution did not completely treat OSA, it should be discontinued. OSA is a multi-factorial disease and in my experience, it often takes a combination of treatments to solve it.

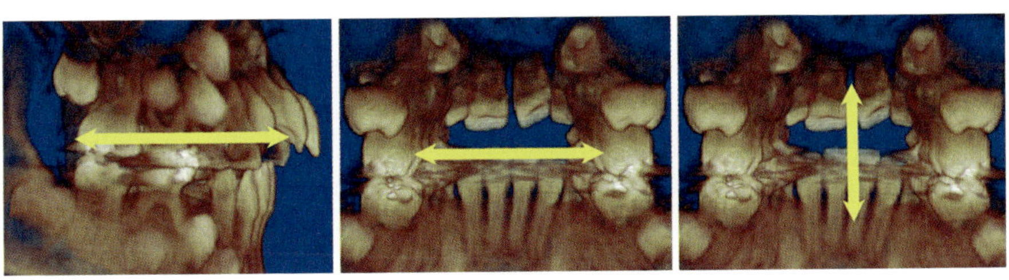

Figure 17. Directions of preferred movements to increase airway.

Such approaches may include rapid maxillary expansion combined with adenotonsillectomy, or an oral appliance (to allow a better seal and lower pressures) combined with CPAP. Another combined approach is orthodontic alignment combined with orthognathic advancement or even alignment of teeth to improve retention and reduce protrusion and then fabricating an oral appliance. We still need to define optimal protocols, but OSA patients suffer marked degradation in quality of life and even if it takes more than one modality to improve quality of life, it is a worthy endeavor.

Case Example: Miles

I will illustrate the type of bite and quality-of-life improvement "airway friendly" orthodontics can have with a patient example. Miles was under the treatment of a physician for sleep apnea. In June 2014, his PSG showed an AHI of 1.6/hour (threshold for children is < 1/hour) and an AHI during rapid eye movement (REM) sleep of 4.3/hour. He was referred for removal of tonsils and adenoids in September 2014; a follow-up PSG showed an AHI of 5.1/hour and the AHI during REM increased to 10.7/hour. The patient subsequently was given a CPAP to wear each night.

Miles was nine years, eleven months old when referred to me in September 2015 (Fig. 18). Our assessment showed an initial airway of 127 mm^2 (Fig. 19), with the lower molars tipped mildly lingually (Fig. 20). We performed RPE with a Hyrax expander to bands on the upper first molars combined with a lower removable expander to upright the lower posterior teeth. Over the course of twelve months, his minimal cross-sectional area (MCA) increased to 310 mm^2.

With 3D CBCT, we now are able to superimpose T1 and T2 volumes on the bones within the anterior cranial base, allowing us to show the actual results of treatment (Fig. 21). Obviously, we had expansion (Fig. 22), but we also had anterior positioning of both the maxilla and mandible (Fig. 23) and in effect, a counterclockwise movement of the mandible.

→ Figure 19. Image of airway. *A:* Before orthodontic treatment; MCA = 127 mm^2. *B:* After orthodontic treatment; MCA = 310 mm^2.

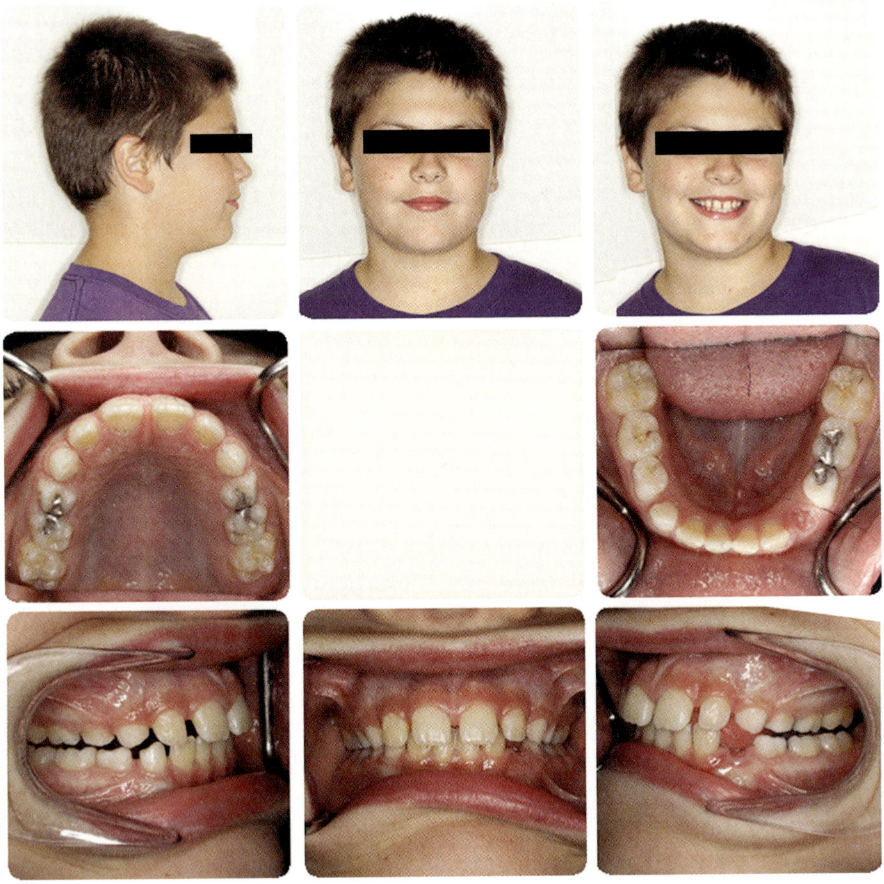

Figure 18. Miles upon referral.

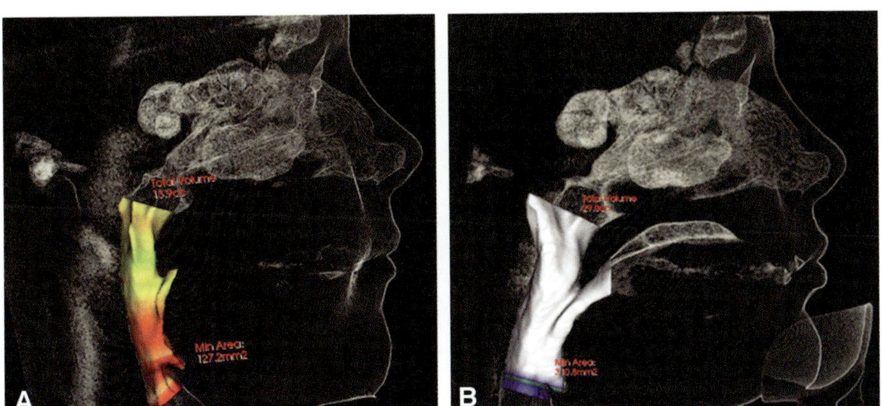

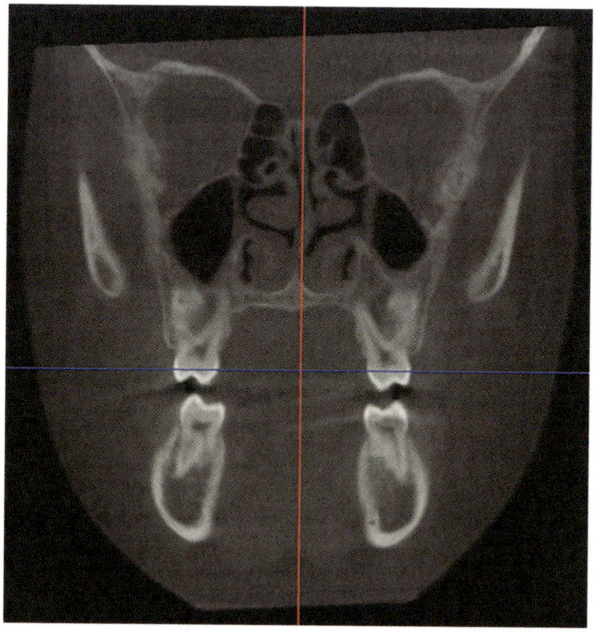

Figure 20. Cross-section at level of first molars showing lower posteriors tipped lingually.

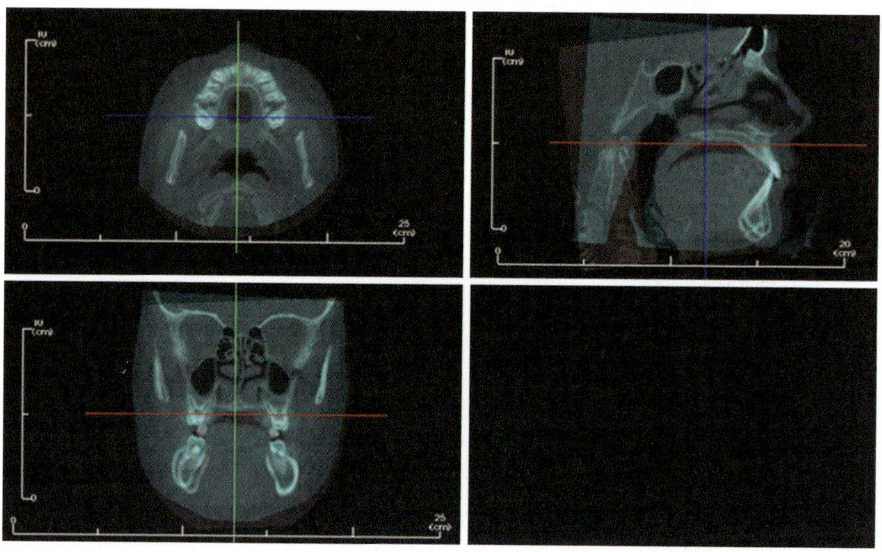

Figure 21. Miles: superimposition of T1 and T2 CBCT volumes on anterior cranial base.

→ Figure 23. Superimposition showing anterior growth of both maxilla and mandible.

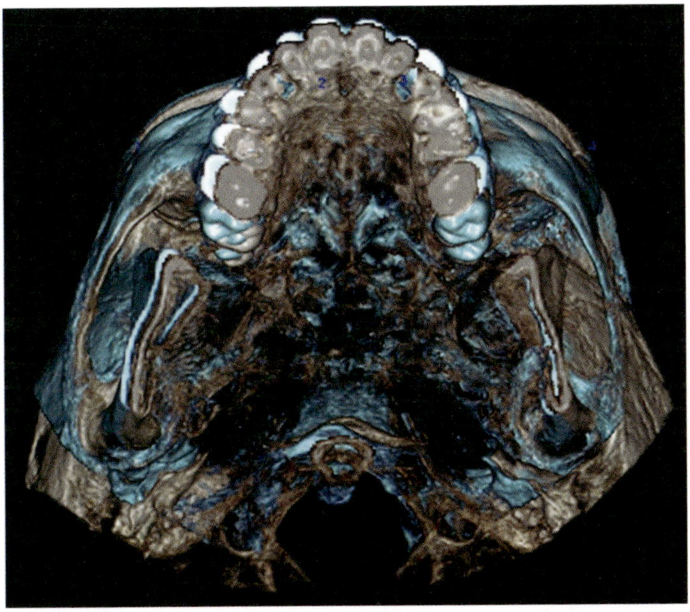

Figure 22. Superimposition showing maxillary expansion. White = T1; blue = T2.

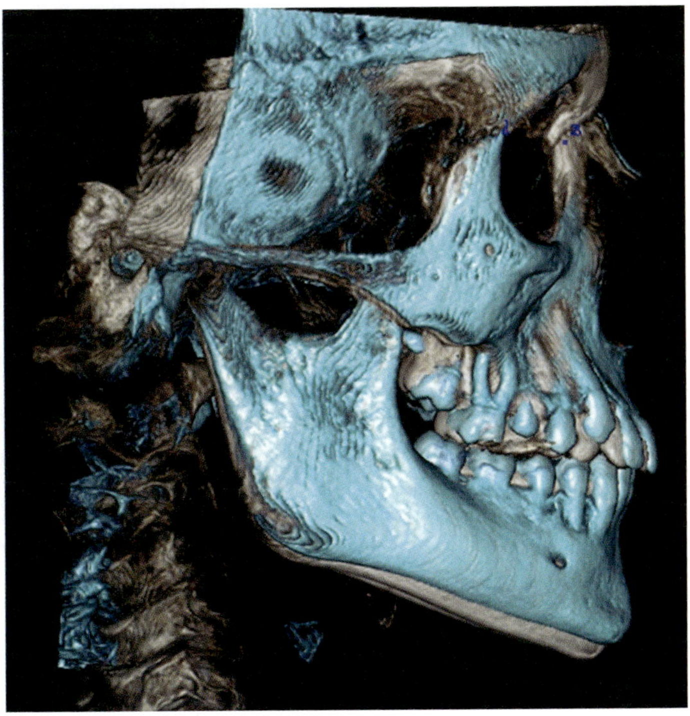

Although his AHI still was 1.5/hour and his REM AHI 3.4/hour, he was symptom free. Moreover, he no longer had to wear the CPAP, which was life changing for him and his family. I interviewed his mother—below is the transcript of that conversation:

> *When we really noticed Miles was having some sort of difficulty, we really didn't know what it was. It was probably in kindergarten. He would fall asleep a lot during class, was not very focused or attentive during class when he was awake—school was a struggle in general. The sleep doctor did prescribe Miles to have a CPAP machine because of the severity of his sleep apnea. He didn't like it—it rashed around his nose. He was uncomfortable with it. We didn't like looking at him with a mask on (Fig. 24).*
>
> *Once the sleep doctor referred us to Dr. Chmura's office, we had a lot of questions. When we got here, we were welcomed. We got to see what the office was all about. In Miles' preliminary exam, Dr. Chmura's office did a 3D x-ray. When the doctor pulled that up on the screen and asked us to look at it (Fig. 19). He explained that Miles' airway was not big enough, so removing the tonsils and adenoids helped some, but with a small airway he was never going to get enough oxygen to sleep well enough to stay awake during school or to keep the focus that he needed to keep during the day.*
>
> *Dr. Chmura gave us an option to help with sleep apnea and that was the expander. We trusted him through that process; it cured Miles' sleep apnea. Once the expansion process was done, we were absolutely amazed at what a new kid we had (Figs. 25-26). We went back for another sleep study—no more CPAP. He didn't have to haul it out to the camper when we went camping and he didn't have to take it to a friend's house when he went overnight. He was able to sleep, finally sleep. It was well worth it and it did cure the sleep apnea.*

Obviously, this case example is just one of many. I developed a good relationship with our local board-certified sleep physician, so he is aware of an alternative solution to CPAP with the potential to reduce downward and forward facial growth. We increased Miles' airway significantly, which improved both his and his family's quality of life, as well as the amount and direction of growth.

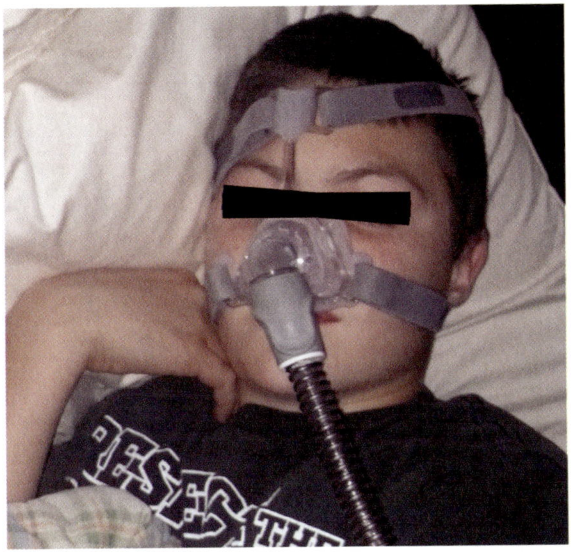

Figure 24. Miles wearing CPAP.

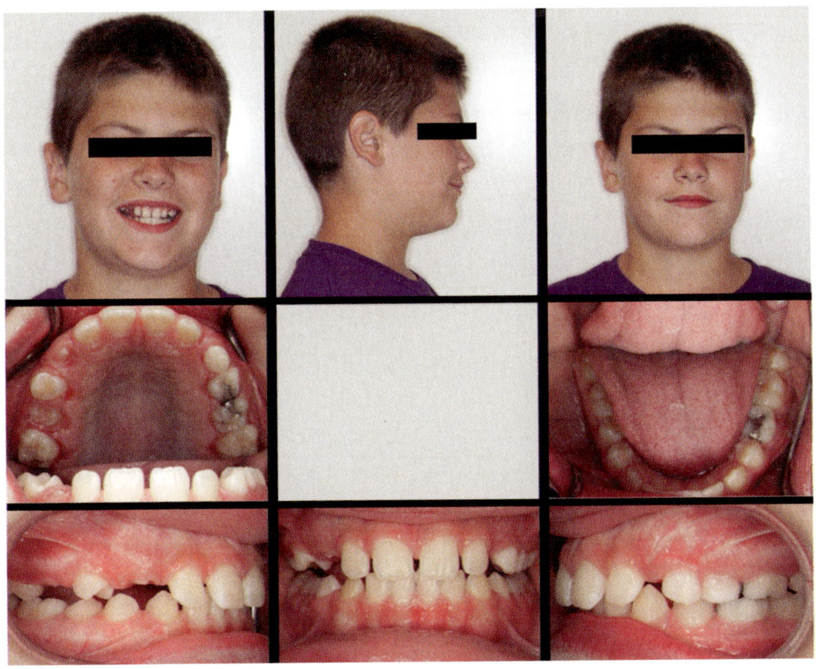

Figure 25. Miles after Phase I orthodontic treatment.

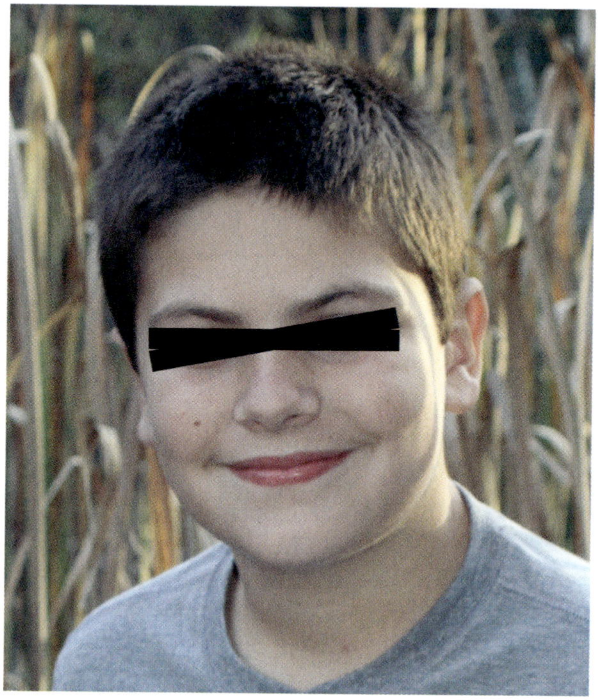

Figure 26. Miles after Phase I RPE.

As we move forward, we will have more before and after CBCT images to see exactly what happened with various therapies. Meanwhile, orthodontists—as the architects of the dental and facial anatomy—must consider airway in treatment decisions to educate our colleagues about what we can do to help OSA suffers.

CONCLUSIONS

1. OSA and sleep-disordered breathing strikes men, women and children; up to 80% of these are undiagnosed.
2. People with OSA are suffering not just long-term complications, but the day-to-day loss of function, reduced ability to think and degradation of relationships that are side effects of untreated OSA.

3. OSA develops gradually and these patients/parents often do not have a frame of reference to understand the extent of their loss.
4. CPAP and OAT both require compliance and neither addresses the underlying anatomical problems causing apnea. We need to establish which anatomical features are solved with OAT to reduce the number of unsuccessful outcomes.
5. Orthodontists can alter anatomy to reduce and potentially prevent sleep apnea and should be considered in treatment planning for sleep apnea, often as the first option. We have directions—increasing tongue space, nasal airflow and MCA—and need to establish magnitudes of change needed to normalize AHI.
6. Bimaxillary orthognathic advancement, particularly utilizing counterclockwise mandibular movement, has the greatest predictability in curing sleep apnea by advancing the mandible without degrading facial esthetics. Recovery from orthognathic surgery is less traumatic than expected and this mode of treatment should be considered as a first surgical option.
7. If we document our cases and show the true superimpositions, we should be able to establish guidelines on which orthodontic solution is best for which particular anatomy.
8. In the vast majority of cases, it takes more than one modality to address sleep apnea (e.g., adenotonsillectomy and RPE, orthodontics and OAT, orthodontics and orthognathic surgery, even OAT and CPAP). The different specialties must collaborate to establish optimal protocols, realizing it may take more than one, perhaps working together in a manner similar to a cleft palate team.

REFERENCES

Akkaya S, Lorenzon S, Uçem TT. A comparison of sagittal and vertical effects between bonded rapid and slow maxillary expansion procedures. Eur J Orthod 1999;21(2):175-180.

Aksu M, Taner T, Sahin-Veske P, Kocadereli I, Konas E, Mavili ME. Pharyngeal airway changes associated with maxillary distraction osteogenesis in adult cleft lip and palate patients. J Oral Maxillofac Surg 2012; 70(2):133-140.

Aloufi F, Preston CB, Zawawi KH. Changes in the upper and lower pharyngeal airway spaces associated with rapid maxillary expansion. ISRN Dent 2012:2012:290964.

Altug-Atac AT, Atac MS, Kurt G, Karasud HA. Changes in nasal structures following orthopaedic and surgically assisted rapid maxillary expansion. Int J Oral Maxillofac Surg 2010;39(2):129-135.

Alves M Jr, Baratieri C, Nojima LI, Nojima MC, Ruellas AC. Three-dimensional assessment of pharyngeal airway in nasal- and mouth-breathing children. Int J Pediatr Otorhinolaryngol 2011;75(9):1195-1199.

American Academy of Pediatrics. Section on Pediatric Pulmonology, Subcommittee on Obstructive Sleep Apnea Syndrome. Clinical practice guideline: Diagnosis and management of childhood obstructive sleep apnea syndrome. Pediatrics 2002;109(4):704-712.

Baratieri C, Alves M Jr, de Souza MM, de Souza Araújo MT, Maia LC. Does rapid maxillary expansion have long-term effects on airway dimensions and breathing? Am J Orthod Dentofacial Orthop 2011;140(2): 146-156.

Brooks B, Cistulli PA, Borkman M, Ross G, McGhee S, Grunstein RR, Sullivan CE, Yue DK. Obstructive sleep apnea in obese noninsulin-dependent diabetic patients: Effect of continuous positive airway pressure treatment on insulin responsiveness. J Clin Endocrinol Metab 1994; 79(6):1681-1685.

Cao Y, Zhou Y, Song Y, Vanarsdall RL Jr. Cephalometric study of slow maxillary expansion in adults. Am J Orthod Dentofacial Orthop 2009; 136(3):348-354.

Chervin RD, Hedger K, Dillon JE, Pituch KJ. Pediatric sleep questionnaire (PSQ): Validity and reliability of scales for sleep-disordered breathing, snoring, sleepiness, and behavioral problems. Sleep Med 2000;1(1): 21-32.

Cistulli PA, Palmisano RG, Poole MD. Treatment of obstructive sleep apnea syndrome by rapid maxillary expansion. Sleep 1998;21(8):831-835.

Cistulli PA, Richards GN, Palmisano RG, Unger G, Berthon-Jones M, Sullivan CE. Influence of maxillary constriction on nasal resistance and sleep apnea severity in patients with Marfan's syndrome. Chest 1996;110(5):1184-1188.

Corbridge JK, Campbell PM, Taylor R, Ceen RF, Buschang PH. Transverse dentoalveolar changes after slow maxillary expansion. Am J Orthod Dentofacial Orthop 2011;140(3):317-325.

De Felippe NL, Bhushan N, Da Silveira AC, Viana G, Smith B. Long-term effects of orthodontic therapy on the maxillary dental arch and nasal cavity. Am J Orthod Dentofacial Orthop 2009;136(4):490.e1-e8.

Doruk C, Sökücü O, Sezer H, Canbay EI. Evaluation of nasal airway resistance during rapid maxillary expansion using acoustic rhinometry. Eur J Orthod 2004;26(4):397-401.

El H, Palomo JM. Airway volume for different dentofacial skeletal patterns. Am J Orthod Dentofacial Orthop 2011;139(6):e511-e521.

Ferguson KA, Cartwright R, Rogers R, Schmidt-Nowara W. Oral appliances for snoring and obstructive sleep apnea: A review. Sleep 2006;29(2): 244-262.

Findley LJ, Unverzagt ME, Suratt PM. Automobile accidents involving patients with obstructive sleep apnea. Am Rev Respir Dis 1988;138(2): 337-340.

Fritscher LG, Mottin CC, Canani S, Chatkin JM. Obesity and obstructive sleep apnea-hypopnea syndrome: The impact of bariatric surgery. Obes Surg 2007;17(1):95-99.

Görgülü S, Gokce SM, Olmez H, Sagdic D, Ors F. Nasal cavity volume changes after rapid maxillary expansion in adolescents evaluated with 3-dimensional simulation and modeling programs. Am J Orthod Dentofacial Orthop 2011;140(5):633-640.

Guilleminault C, Monteyrol PJ, Huynh NT, Pirelli P, Quo S, Li K. Adenotonsillectomy and rapid maxillary distraction in pre-pubertal children: A pilot study. Sleep Breath 2011;15(2):173-177.

Guilleminault C, Quo S, Huynh NT, Li K. Orthodontic expansion treatment and adenotonsillectomy in the treatment of obstructive sleep apnea in prepubertal children. Sleep 2008;31(7):953-957.

Hallblower AC, Degaonkar M, Barker PB, Earley CJ, Marcus CL, Smith PL, Prahme MC, Mahone EM. Childhood obstructive sleep apnea associates with neuropsychological deficits and neuronal brain injury. PLos Med 2006;3(8):e301.

Harsch IA, Konturek PC, Koebnick C, Kuehnlein PP, Fuchs FS, Pour Schahin S, Wiest GH, Hahn EG, Lohmann T, Ficker JH. Leptin and ghrelin levels in patients with obstructive sleep apnoea: Effect of CPAP treatment. Eur Respir J 2003;22(2):251-257.

Hershey HG, Stewart BL, Warren DW. Changes in nasal airway resistance associated with rapid maxillary expansion. Am J Orthod 1976;69(3): 274-284.

Huynh T, Kennedy DB, Joondeph DR, Bollen AM. Treatment response and stability of slow maxillary expansion using Haas, hyrax, and quad-helix appliances: A retrospective study. Am J Orthod Dentofacial Orthop 2009;136(3):331-339.

Işeri H, Ozsoy S. Semirapid maxillary expansion: A study of long-term transverse effects in older adolescents and adults. Angle Orthod 2004; 74(1):71-78.

Javaheri S. Basics of sleep apnea and heart failure. American College of Cardiology 2013. http://www.acc.org/latest-in-cardiology/articles/2014/07/22/08/25/basics-of-sleep-apnea-and-heart-failure

Johal A, Conaghan C. Maxillary morphology in obstructive sleep apnea: A cephalometric and model study. Angle Orthod 2004;74(5):648-656.

Khan A, Ramar K, Maddirala S, Friedman O, Pallanch JF. Olson EJ. Uvulopalatopharyngoplasty in the management of obstructive sleep apnea: The Mayo Clinic experience. Mayo Clin Proc 2009;84(9):795-800.

Kheirandish L, Gozal D. Neurocognitive dysfunction in children with sleep disorders. Dev Sci 2006;9(4):388-399.

Kilic N, Oktay H. Effects of rapid-slow maxillary expansion on the dentofacial structures. Aust Orthod J 2010;26(2):178-183.

Kryger MH, Mignot E, Orr WC, Ryan D, Walsh JK. Sleep in American Poll, National Sleep Foundation Executive Summary 2002.

Kushida CA, Morgenthaler TI, Littner MR, Alessi CA, Bailey D, Coleman J Jr, Friedman L, Hirshkowitz M, Kapen S, Kramer M, Lee-Chiong T, Owens J, Pancer JP; American Academy of Sleep. Practice parameters for the treatment of snoring and obstructive sleep apnea with oral appliances: An update for 2005. Sleep 2006;29(2):240-243.

Lima Filho RM, Ruellas AC. Long-term anteroposterior and vertical maxillary changes in skeletal Class II patients treated with slow and rapid maxillary expansion. Angle Orthod 2007;77(5):870-874.

Lye, KW, Waite, PD, Meara, D, Wang, D. Quality of life evaluation of maxillomandibular advancement surgery for treatment of obstructive sleep apnea. J Oral Maxillofac Surg 2008;66(5);968-972.

Machado Júnior AJ, Crespo AN. Cephalometric study of alterations induced by maxillary slow expansion in adults. Braz J Otorhinolaryngol 2006;72(2):166-172.

Marcus CL, Brooks LJ, Draper KA, Gozal D, Halbower AC, Jones J, Schechter MS, Ward SD, Sheldon SH, Shiffman RN, Lehmann C, Spruyt K; American Academy of Pediatrics. Diagnosis and management of childhood obstructive sleep apnea syndrome. Pediatrics 2012;130(3):576-584.

Marino A, Ranieri R, Chiarotti F, Villa MP, Malagola C. Rapid maxillary expansion in children with Obstructive Sleep Apnoea Syndrome (OSAS). Eur J Paediatr Dent 2012;13(1):57-63.

Martina R, Cioffi I, Farella M, Leone P, Manzo P, Matarese G, Portelli M, Nucera R, Cordasco G. Transverse changes determined by rapid and slow maxillary expansion: A low-dose CT-based randomized controlled trial. Orthod Craniofac Res 2012;15(3):159-168.

Mehra P, Downie M, Pita MC, Wolford LM. Pharyngeal airway space changes after counterclockwise rotation of the maxillomandibular complex. Am J Orthod Dentofacial Orthop 2001;120(2):154-259.

Miano S, Rizzoli A, Evangelisti M, Bruni O, Ferri R, Pagani J, Villa MP. NREM sleep instability changes following rapid maxillary expansion in children with obstructive apnea sleep syndrome. Sleep Med 2009; 10(4):471-478.

National Institutes of Health (NIH). Sleep apnea tied to increased risk of stroke. https://www.nhlbi.nih.gov/news/press-releases/2010/sleep-apnea-tied-to-increased-risk-of-stroke. April 2010.

Pirelli P, Saponara M, Guilleminault C. Rapid maxillary expansion in children with obstructive sleep apnea syndrome. Sleep 2004;27(4):761-766.

Ramires T, Maia RA, Barone JR. Nasal cavity changes and the respiratory standard after maxillary expansion. Braz J Otorhinolaryngol 2008; 74(5):763-769.

Schendel SA, Jacobson R, Khalessi S. Airway growth and development: A computerized 3-dimensional analysis J Oral Maxillofac Surg 2012; 70(9):2174-2183.

Schütz TC, Dominguez GC, Hallinan MP, Cunha TC, Tifik S. Class II correction improves nocturnal breathing in adolescents. Angle Orthod 2011;81(2):222-228.

Seto BH, Gotsopoulos H, Sims MR, Cistulli PA. Maxillary morphology in obstructive sleep apnoea syndrome. Eur J Orthod 2001;23(6);703-714.

Sheldon SH, Kryger MH, Ferber R, Gozal D. *Principles and Practice of Pediatric Sleep Múedicine.* 1st ed. Philadelphia: Elsevier Saunders Co. 2005:197-267.

Shetty P, Hegde AM, Rai K. Study of stress distribution and displacement of the maxillary complex following application of forces using jack-screw and nitanium palatal expander: 2. A finite element study. J Clin Pediatr Dent 2009;34(1):87-93.

Smith T, Ghoneima A, Stewart K, Liu S, Eckert G, Halum S, Kula K. Three-dimensional computed tomography analysis of airway volume changes after rapid maxillary expansion. Am J Orthod Dentofacial Orthop 2012;141(5):618-626.

Sökücü O, Doruk C, Uysal OI. Comparison of the effects of RME and fan-type RME on nasal airway by using acoustic rhinometry. Angle Orthod 2010;80(5):870-875.

Step Back Technique. Video from Gauge. https://www.youtube.com/watch?v=2lp0P0Q Rqek. May 14, 2013.

Sutherland K, Phillips CL, Cistulli PA. Efficacy *versus* effectiveness in the treatment of obstructive sleep apnea: CPAP and oral appliances. JDSM 2015;2(4):175-181.

Villa MP, Malagola C, Pagani J, Montesano M, Rizzoli A, Guilleminault C, Ronchetti R. Rapid maxillary expansion in children with obstructive sleep apnea syndrome: 12-month follow-up. Sleep Med 2007;8(2): 128-134.

Villa MP, Rizzoli A, Miano S, Malagola C. Efficacy of rapid maxillary expansion in children with obstructive sleep apnea syndrome: 36 months of follow-up. Sleep Breath 2011;15(2):179-184.

Wong CA, Sinclair PM, Keim RG, Kennedy DB. Arch dimension changes from successful slow maxillary expansion of unilateral posterior crossbite. Angel Orthod 2011;81(4):616-623.

Young T, Finn L, Peppard PE, Szklo-Coxe M, Austin D, Nieto FJ, Stubbs R, Hla KM. Sleep disordered breathing and mortality: Eighteen-year follow-up of the Wisconsin sleep cohort. Sleep 2008;31(8):1071-1078.

Young T, Peppard PE, Gottlieb DJ. Epidemiology of obstructive sleep apnea. Am J Respir Crit Care Med 165(9):1217-1239.

GENETIC FACTORS AFFECTING FACIAL MORPHOLOGY ASSOCIATED WITH SLEEP APNEA

James K. Hartsfield, Jr., Lorri Ann Morford, George Jeryn Jacob, G. Thomas Kluemper

ABSTRACT

Obstructive sleep apnea (OSA) is characterized by partial or complete blockage of the upper airway, which leads to a series of apneas (complete cessation of airflow) and/or hypopneas (reduced airflow) that may reduce sleep quality. OSA has a complex etiology that can be influenced by both environmental and genetic factors, including the patient's weight, muscle tone, craniofacial size and/or shape and inherited genetic variations. Early genetic research of OSA focused on examining the role of obesity-related genes on the condition due to the increased prevalence of obesity among OSA patients. More recent evidence, however, suggests that factors that influence craniofacial morphology (i.e., skeletal Class II pattern and facial convexity) also may be important. For example, craniofacial features that reduce airway volume, alone or together with soft tissue abnormalities, may predispose an individual to OSA or contribute to its etiology. The clinical impact of these features may explain why some individuals who are diagnosed with OSA are not obese. With significant reductions in the cost of Next-Generation sequencing of the human genome recently, it now is more feasible than ever to study genetic factors that may influence OSA occurrence. The goal of this chapter is to present a representative patient diagnosed with OSA who otherwise is healthy and exhibits a normal Body Mass Index (BMI). We review known relationships between OSA, BMI and facial convexity, as well as propose several candidate genes for future study in cases of non-obese OSA.

KEY WORDS: non-obese obstructive sleep apnea, Class II malocclusion, facial convexity, body mass index, genetics

---

INTRODUCTION

The "gold standard" for diagnosis of obstructive sleep apnea (OSA) is an overnight polysomnography (PSG; i.e., sleep study). Based on a PSG-determined Apnea-Hypopnea Index (AHI) disease threshold of 15/hour in adults, the prevalence of OSA is estimated at 6-17% within the general population and may be higher in populations of obese and/or individuals of advanced age (Senaratna et al., 2016). While "environmental" factors such as allergic rhinitis and asthma—which may have genetic components—can contribute to or be associated with the development of OSA (Chirakalwasan and Ruxrungtham, 2014; Prasad et al., 2014; Teodorescu et al., 2015), data indicates that OSA also may have a familial (genetic) component.

Although OSA family members often share a similar Body Mass Index (BMI; Young et al., 2002; Kryger et al., 2011), familial aggregation of sleep-disordered breathing symptoms is independent of familial similarities in weight (Redline et al., 2012). The underlying causes of OSA are heterogeneous, with data from clinical and epidemiologic studies indicating that genetic factors influence the expression of OSA (Kryger et al., 2011; Gehrman et al., 2015; Cade et al., 2016). Genetic studies of OSA, however, frequently have resulted in variable and inconsistent results, perhaps due to the complex nature of the disease, genetic heterogeneity, small sample sizes being tested and variation in the phenotypic endpoint being measured (Zhuoqiang et al., 2015).

While obesity often is associated with OSA and may have a common genetic basis (Palmer et al., 2004; Patel, 2005), variations in craniofacial structure also appear to be connected to OSA and may be independent of obesity/high BMI (Young et al., 2002; Miyao et al., 2008; Susarla et al., 2010; Chi et al., 2011; Roedig et al., 2014). When OSA is caused or influenced by a variation in anatomy, the obstruction may originate from soft tissue (including musculature) and/or skeletal components (Chokroverty, 2010).

Multiple studies examining OSA patients have identified key craniofacial features that appear to contribute to the condition (e.g., a decreased antero-posterior dimension of the cranial base, maxillary retrusion, inferior position of the hyoid bone, macroglossia, smaller upper and lower pharyngeal airway, and a retrognathic [convex facial profile] mandible [Battagel et al., 2000; Riha et al., 2005; El and Palomo, 2011]), even in patients under age 18 years (Flores-Mir et al., 2013).

A markedly retruded position of the mandible can reduce the space for the tongue and can affect pharyngeal airway volume, predisposing the patient to obstructive apnea while asleep or awake, as can be seen in infants with Pierre Robin sequence (PRS). This observation has been extended by some to suggest that extraction of teeth also may increase the likelihood of the patient developing OSA. As noted by Larsen and colleagues (2015), some studies found that intra-oral airway space was not decreased by premolar extractions, regardless of the method of space closure (Valiathan et al., 2010; Al Maaitah et al., 2012; Stefanovic et al., 2013), while others found a decrease in the airway volume dependent on the type of space closure (Germec-Cakin et al., 2010; Wang et al., 2011; Chen et al., 2012).

These studies focused on the endpoint of airway volume, however, and not the actual occurrence of OSA with premolar extractions. Using electronic medical and dental health records, Larsen and associates (2015) found that OSA was not greater among subjects with a history of missing premolars in each quadrant (n = 2,792) when compared to matched control subjects without any missing premolars (n = 2,792), while controlling for age, gender and BMI.

Mandibular distraction is advocated commonly in patients with PRS; several studies have documented an improvement in the airway volume post-distraction (Almajed et al., 2017; Khayat et al., 2017). Unfortunately, there are many craniofacial dysmorphology conditions other than PRS, such as Crouzon syndrome (Driessen et al., 2017) and Down syndrome (Skotko et al., 2017), which can present with craniofacial abnormalities and predispose one to OSA (Tan et al., 2016). Up to 15% of children younger than 18 months with craniofacial anomalies presenting with respiratory distress have multiple airway problems (Gungor et al., 2017).

CASE REPORT

The following case illustrates a potential quandary of the nonobese patient with OSA that an orthodontist may encounter in practice. Patient B.P. is a Caucasian male who was seen in consultation at the age of 31 years due to his moderate OSA (AHI of 17.5/hour), which had been diagnosed by PSG at 25 years of age. The patient previously had attempted utilizing Continuous Positive Airway Pressure (CPAP) therapy, Bi-level Positive Airway Pressure therapy (BPAP, a machine that supplies different inspiratory and expiratory pressures) and a mandibular advancement appliance, each with limited to no success. He was using Autotitrating Positive Airway Pressure (AutoPAP) at the time of consultation, but was having trouble tolerating positive airway pressure. His BMI was normal at 21 kg/m^2.

In adolescence, B.P. had undergone camouflage treatment for his mild-to-moderate skeletal Class II malocclusion with headgear and full fixed orthodontic appliances. It was noted at the consultation that the results achieved with this previous orthodontic treatment exhibited solid, naturally maintained retention over the course of 10 to 15 years (Fig. 1), but with posterior facial divergence and a convex profile. His cephalometric findings were consistent with a mild-to-moderate Class II skeletal pattern with an 8° ANB angle and typical dental compensations.

Due to his OSA and craniofacial anatomy, a treatment plan of orthodontics combined with a two-jaw surgical advancement was proposed and accepted. To increase the anterior overjet for a greater Bi-Lateral Sagittal Split Osteotomy (BSSO) advancement, the mandibular first pre-molars were extracted, spaces closed and the mandibular incisors were retracted with maximum posterior anchorage. The maxilla was brought forward and impacted with autorotation of the mandible.

Following this treatment, the patient had a Class I canine and Class III molar occlusion and a straight facial profile; he was free of OSA (Fig. 2). In that he no longer reported OSA symptoms, his insurance company would not pay for post-surgery PSG and the patient did not wish to pay for one out of pocket, so a post-surgery AHI was not available. Interestingly, in hindsight, the patient's father also suffered from OSA and exhibited a Class II skeletal relationship.

OSA, BMI and FACIAL CONVEXITY

This case report raises many questions regarding the occurrence of OSA in subjects with facial convexity who are not obese. Although variations in Class II malocclusion with increased facial convexity have been associated with OSA, this association only recently has been investigated

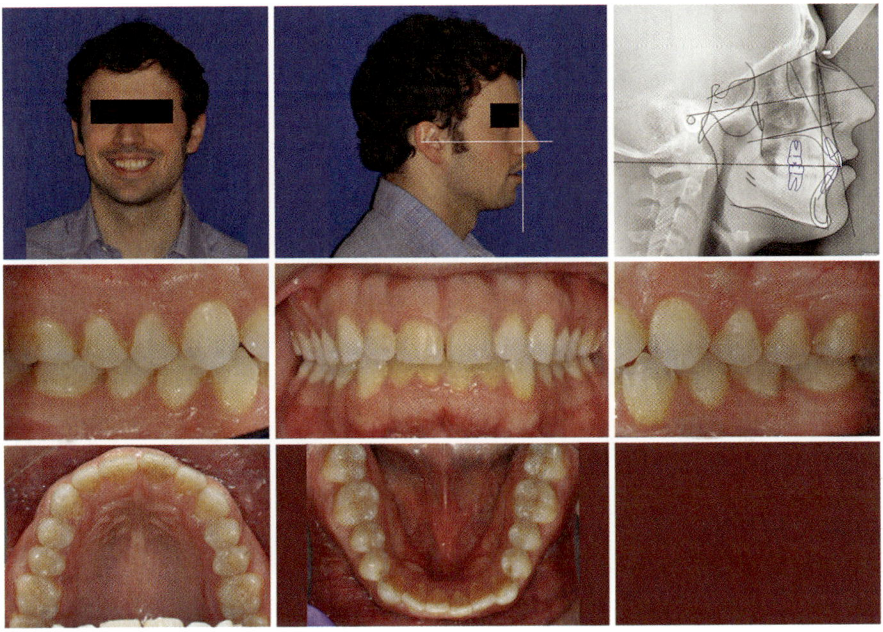

Figure 1. B.P. as he presented to his oral and maxillofacial surgeon, seeking a surgical correction for moderate OSA (AHI 17.5). Notice a mild/moderate skeletal Class II malocclusion with ANB angle of 8°, excess facial convexity and Class I occlusion resulting from orthodontic camouflage correction during adolescence.

along with the BMI of the subjects. OSA patients with skeletal Class II (convex facial profile) have been found to have significantly lower BMIs than the non-Class II OSA patients (Roedig et al., 2014). This finding suggests that the skeletal Class II phenotype may be associated with or directly influence OSA development irrespective of an individual's BMI.

One of the most commonly discussed genetic factors that may have a role in OSA is apolipoprotein-E (*APOE*). The protein produced from this gene has been studied mostly as a carrier of intermediate-density lipoproteins, notably cholesterol and fat soluble vitamins (e.g., vitamin D). Genetic studies of OSA, however, have produced mixed results concerning *APOE,* making the role of this gene in OSA difficult to interpret (Saarelainen et al., 1998; Gottlieb et al., 2004; Larkin et al., 2006; Cosentino et al., 2008; Kalra et al., 2008; Nikodemova et al., 2013; Tisko et al., 2014; Xu et al., 2015; Zhuoqiang et al., 2015). Although the initial study

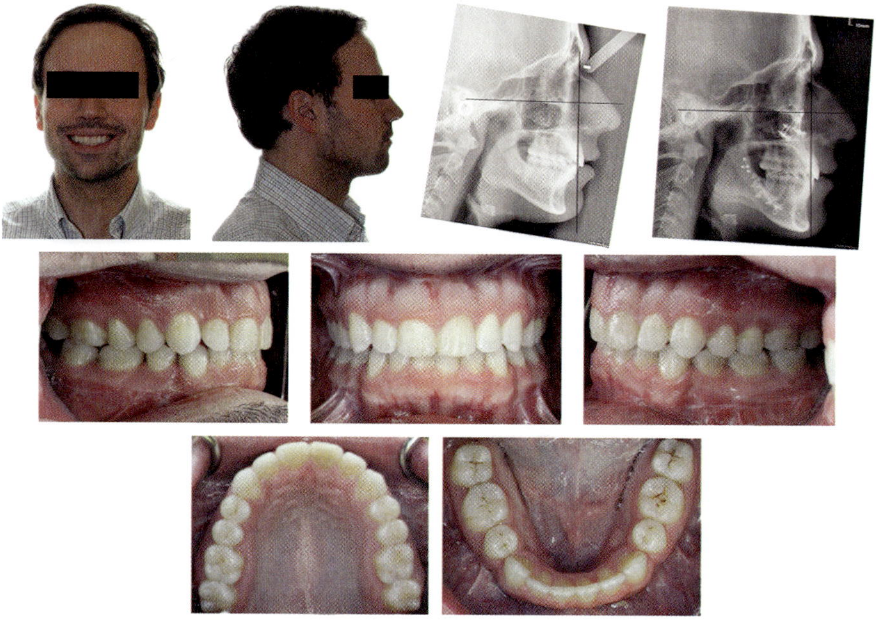

Figure 2. B.P. following combined treatment of orthodontics and two-jaw advancement surgery. Notice reduced facial convexity, overall improved skeletal balance, extraction of mandibular first premolars allowing orthodontic creation of Class II malocclusion and excess overjet prior to surgical advancement. Although B.P. declined a follow-up PSG because it was not covered by insurance, there no longer was a clinical need to use positive airway pressure (PAP) therapy as his daytime sleepiness symptoms resolved post-operatively.

by Roedig and coworkers (2014) found that OSA patients with skeletal Class II had significantly lower BMIs than the non-skeletal Class II subjects, they found no evidence to support an association between the *APOE*-ε4 allele and Class II (*versus* Class I or Class III) in adult OSA subjects. It was noted, however, that the study was underpowered to detect an effect of the *APOE*-ε4 allele by skeletal classification.

Preliminary data with a larger cohort has confirmed that OSA patients with a retrognathic profile and increased facial convexity were more likely to have a lower BMI compared to OSA patients who were non-retrognathic, or who had a decreased facial profile convexity (Wachs et al., unpublished data). Similarly, patients who were retrognathic also were more likely to carry the *APOE*-ε3/4 isoform compared to OSA patients who are not retrognathic, or have an increased facial profile convexity.

Even though an oral appliance was not successful in the case presented, a high percentage of the successful cases using an oral appliance to advance the mandible were young patients with prevalent or exclusive positionality of respiratory events while lying down whom had a low BMI (Maspero et al., 2015).

In addition, craniofacial structure in an Asian sample appears to confer an elevated risk of OSA despite lower rates of obesity. The group had a shorter cranial base, maxilla, midface and mandibular length in the sagittal plane compared to a Caucasian group, which instead had a greater BMI for the same OSA severity (Lee et al., 2010). In Liu and coworkers' study (2000), the length of the soft palate (PNS-P) was increased in Chinese patients with OSA, suggesting that the upper airway soft tissues, along with an increased angulation in the cranial base to occlusal plane angle, may contribute to airway collapse.

FURTHER CONSIDERATIONS

Due to the complex etiology of OSA, it has been proposed that several genetic factors—including those related to obesity and its associated morbidities (e.g., inflammation, neural control of the upper airway muscle tone and craniofacial morphology)—are likely to influence the disease (Patel et al., 2012; Yin et al., 2014; de Paula et al., 2016). Environmental and multiple genetic factors often are crucial to the development of complex traits (Abass and Hartsfield, 2008).

Gene-environment interactions also can predispose a patient to OSA, or make the patient more susceptible to OSA, such as with a hypoxic environment and its impact on the peroxisome proliferator-activated receptor-γ (*PPARG*) gene. In fact, variation in the *PPARG* gene has been linked with obesity (Ek et al., 1999) and has been cited as a risk factor in developing OSA. Because hypoxia can suppress *PPARG* gene transcription (Yun et al., 2002), a patient who is exposed constantly to a hypoxic environment will be at higher risk to influence *PPARG* gene expression and increased the risk for OSA.

The relationship between obesity and OSA has been analyzed in several research studies at the clinical and cellular levels (Tauman and Gozal, 2006; Schwartz et al., 2008; Franklin and Lindberg, 2015; Polak et al., 2016). Obesity can predispose a patient to OSA through changes in the fat distribution within upper airway tissues that may lead to an upper airway collapse, or *via* excess weight on the anterior chest wall that can reduce chest wall compliance and increase efforts during breathing with greater airway resistance.

Twin and family studies have identified genetic factors associated with obesity (Bodurtha et al., 1990; Stunkard et al., 1990). The association of a Class II skeletal pattern and/or increased facial convexity with OSA in individuals who are not obese begs the question of whether there could be a direct connection of the craniofacial morphology to the OSA. Or, do the OSA and retrognathia result as a pleiotropic effect (occuring when one gene influences two or more seemingly unrelated phenotypic traits) in some subjects? It may not be surprising that several genetic risk factors may overlap between these two conditions (Fig. 3).

It would be helpful clinically if we could know which patients with mild-to-moderate retrognathia should be considered for orthognathic surgery, even prior to any onset of OSA, in that not all retrognathic cases go on to develop OSA. However, we are far from being able to assign a risk score for OSA *a priori* based on genetic variations and/or other variables; ultimately, we may not be able to assign reasonable risk estimates to OSA due to the genetic heterogeneity of the condition. If genetic factors could be identified with a high level of predictive confidence, then they could be incorporated into risk stratification and diagnosis of patients who have the characteristic facial morphology with and without OSA. Hence, investigations into the effect that genetic factors play on facial morphology within OSA cohorts are important and should be undertaken.

The identification of gene variations that are associated clearly with OSA may help clinicians to identify individuals at high risk for this condition and allow them to modify the treatment of a malocclusion based on the underlying pathophysiologic and/or anatomic structures and genetic information obtained from the patient. In the section below, we highlight several genes, in addition to the *APOE* gene, that either have been associated with OSA previously in the literature or that may be a reasonable candidate gene for future examination in conjunction with OSA and craniofacial morphology. While severe mutations within many of the genes described below clearly lead to the formation of a craniofacial syndrome, disorder or an extreme phenotype, inheriting one or more less severe variations and/or mutations in these genes in fact may yield a

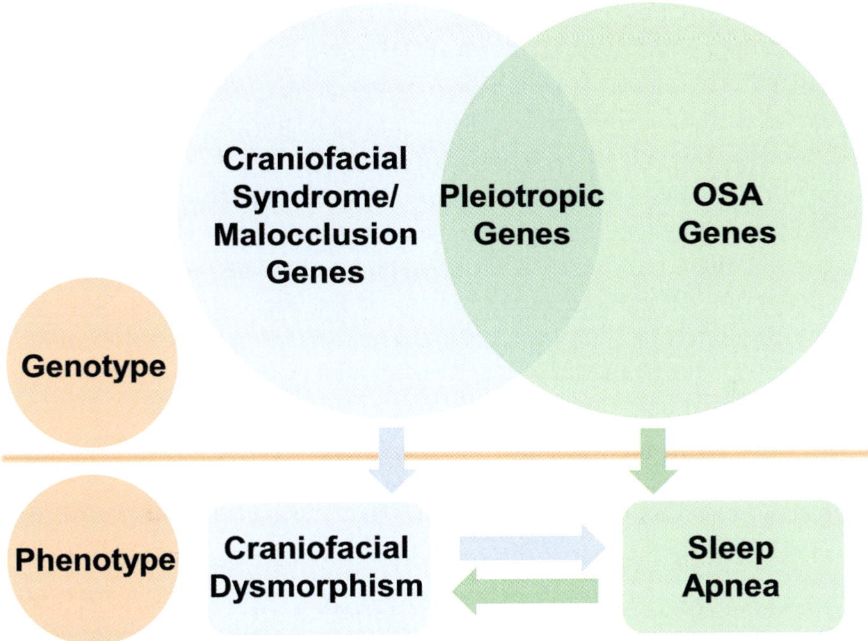

Figure 3. Craniofacial dysmorphism and sleep apnea can be caused by an underlying gene mutation. The pleiotropic effect of a gene mutation can cause both these conditions through an independent mechanism. For example, *PHOX2B* gene mutation has been associated with congenital central hypoventilation syndrome (which affects the central respiratory drive) and also with skeletal Class III malocclusion (maxillary hypoplasia), which can cause OSA from an underlying independent morphology mechanism.

milder craniofacial phenotype that can influence or contribute to the OSA etiology.

Paired-Like Homeobox 2B (PHOX2B) Gene

Association between changes in the *PHOX2B* gene nucleotide sequence and Class III malocclusion (maxillary hypoplasia) in some children with OSA as a part of the autosomal dominant condition Congenital Central Hypoventilation Syndrome (CCHS; OMIM #209880; Lavezzi et al., 2013) also is known as "Ondine's curse" (Deonna et al., 1974). Although most cases of CCHS are caused by a heterozygous mutation of the *PHOX2B* gene located on chromosome 4p13 (Lavezzi et al., 2013), mutations in other genes involved in the development of the nervous system have been identified on rare occasion including RET proto-oncogene

(*RET*), Glial Cell Derived Neurotrophic Factor (*GDNF*), Endothelin 3 (*EDN3*), Brain Derived Neurotrophic Factor (*BDNF*) and the Achaete-Scute Family Basic Helix-Loop-Helix Transcription Factor 1/Mammalian Achaetescute Homolog-1 (*ASCL1/MASH1*).

The soft tissue facial phenotype in children and young adults with *PHOX2B*-determined CCHS generally was shorter and flatter, with decreased upper face height, increased nasal tip protrusion, decreased nasolabial angle and decreased upper lip height. They also had an inferior inflection of the lateral segment of vermilion border on the upper lip compared to matched controls, perhaps reflecting the contribution of neural crest cells to the face structure, as well as the autonomic nervous system (Todd et al., 2006).

CCHS is categorized as a neurocristopathy, which is a diverse class of vertebrate pathologies and disorders that arise due to an abnormal presence of or defects in the migration, differentiation or death of neural crest cells during development. In addition to observed alterations in facial morphology, part of the autonomic nervous system respiratory component is affected in CCHS patients. The *PHOX2B* gene plays an important role in development of the autonomic nervous system and mutations in this gene are causal for CCHS. The type of the *PHOX2B* mutation impacts the disorders severity in terms of hypoventilation during periods of wakefulness and/or sleep (Weese-Mayer et al., 1999). Other neural crest-related anomalies (e.g., Hirschsprung disease, neural crest tumors and congenital heart defects) also may be associated (Lombardo et al., 2017).

Heterozygous mutations in the *PHOX2B* gene (mutations that affect the gene on one, but not both, inherited copies of chromosome 4) are the most common cause of CCHS, particularly polyalanine expansions, which are observed in over 90% of all cases (Berry-Kravis et al., 2006; Rand et al., 2014). Polyalanine expansions occur when there is the abnormal addition of one (or more) tri-nucleotide insertion(s) into the DNA code at a specific location, which forms a new codon(s) and signals the addition of an extra alanine amino acid(s) into the protein. Numerous studies have measured from 15 to 39 extra nucleotides (i.e., 5 to 13 new codons for alanine) contained within the DNA code of a 20-residue polyalanine tract in exon 3 the *PHOX2B* gene for the CCHS-affected individuals, but not control subjects, leading to the addition of 5 to 13 extra alanine amino acids to the *PHOX2B* protein presumably due to non-homologous recombinations during meiosis (Amiel et al., 2003; Weese-Mayer et al., 2003; Berry-Kravis et al., 2006). On occasion in CCHS-affected individuals, when extra nucleotides are not added in multiples of three into the *PHOX2B* third exon, scientists have observed a shift in the "three-nucleotide reading frame" (i.e., codons) downstream of the alteration, producing either completely new amino acids that no longer resemble the *PHOX2B* protein code (missense mutations), truncations of the *PHOX2B* protein (nonsense mutations) after the insertion point and/or deletions in the normal 20-residue alanine tract (Amiel et al., 2003; Berry-Kravis et al., 2006).

Thirty-two percent of 50 Class III patients aged 8 to 14 years with a history of sleep apneic episodes were found to have "silent" point mutations in the *PHOX2B* gene, while 20 age-matched controls did not have the mutations (Lavezzi et al., 2013). Even though a "silent" mutation occurs when the change of a single DNA nucleotide with a protein-coding portion (exon) of a gene does not affect the amino acid sequence of the protein, it still can affect transcription and mRNA splicing resulting in the amount or alteration of the mature protein (Cartegni et al., 2002).

Depending on severity, affected individuals breathe normally while awake and hypoventilate with normal respiratory rates and shallow breathing during sleep, or may hypoventilate both awake and asleep, related to abnormal autonomic control of respiration (Rand et al., 2014). It is noteworthy that although CCHS originally was thought to be congenital—hence the name—it also may occur later (e.g., in mid-life) and be variable. This "later onset" CCHS, or acquired hypoventilation syndrome (AHCS), may have family members with the same *PHOX2B* gene mutation who are asymptomatic (Bygarski et al., 2013). A family history of neurocristopathy along with sleep apnea, particularly but not necessarily congenital, may indicate genetic analysis of the *PHOX2B* gene is warranted.

Lysophosphatidic Acid Receptor 1 (LPAR1) and Endothelin-1 (EDN1) Genes

Lysophosphatidic Acid Receptor 1 (*LPAR1*; Contos et al., 2000) and Endothelin-1 (*EDN1*) gene (Kurihara et al., 1994) knockout mice models have been associated with craniofacial abnormalities. Interestingly, a candidate gene study revealed *LPAR1* associated with AHI among African Americans, suggesting the possibility of a skeletal component for OSA pathogenesis (Patel et al., 2012). The AHI data was combined from two sources for African Americans. The first dataset (n = 647) had an OSA prevalence of 36% and the median (interquartile range) for AHI was 5.9 (1.5, 21.8). The second dataset (n = 459) had an OSA prevalence of 100% with the median (interquartile range) for AHI of 39.2 (5.0, 192.5).

The role of *EDN1* in sleep-disordered breathing among European Americans was suggested in a candidate gene study (Larkin et al., 2010). *EDN1* deficient mice also have demonstrated impaired ventilatory response to hypercapnia and hypoxia (Kuwaki et al., 1996), along with disturbance in pharyngeal arch development (Kurihara et al., 1994), which suggests a possible link between craniofacial abnormalities and ventilatory function.

Vacuolar Protein Sorting 13 Homolog B (VPS13B; alias Cohen Syndrome 1, COH1) Gene

As mentioned previously, various genetic syndromes have been associated with craniofacial phenotypes that may predispose patients to OSA. Cohen Syndrome (OMIM #216550) is a rare autosomal recessive disorder caused either by homozygous mutations in the *COH1* gene located on chromosome 8q22.2, or by inheritance of compound heterozygous mutations. Over 25 different *COH1* gene mutations have been associated with the syndrome, including non-sense mutations, frame shifts, splice site variants, in-frame deletions and a mis-sense mutation (Kolehmainen et al., 1996; Seifert et al., 2006). The *VPS13B* alias *COH1* gene product is involved in glycosylation of proteins and mutations in the gene have been associated with protein N-glycosylation defects in the Golgi and abnormal endosomal-lysosomal trafficking problems (Duplomb et al., 2014). Examples of two proteins that exhibit abnormal glycosylation in these patients include Intercellular Cell Adhesion Molecule 1 (ICAM1) and Lysosomal Associated Membrane Protein 2 (LAMP-2; Duplomb et al., 2014).

Patients diagnosed with Cohen Syndrome can exhibit a variety of clinical features (e.g., facial dysmorphism, truncal obesity with variable BMI, ocular abnormalities, microcephaly, intellectual disability, high nasal bridge, open mouth with prominent lips and/or maxillary central incisors, micrognathia, high narrow palate and/or laryngeal abnormalities that can predispose patients to OSA; Cohen et al., 1973; Balestrazzi et al., 1980; North et al., 1985; Chandler et al., 2003). This syndrome appears to be one in the same with Mirhosseini-Holmes-Walton Syndrome (Norio et al.,

1986; Kondo et al., 1990). While Cohen Syndrome may affect individuals of Finnish or Ashkenazi Jewish descent with a higher frequency than other ethnic groups, reports also have noted in a number of different ethnic backgrounds being affected, including but not limited to Amish, Greek/Mediterranean and Irish (Friedman and Sack, 1981; Chandler et al., 2003; Norio, 2003).

Although Cohen Syndrome typically is described as a condition with autosomal recessive inheritance, approximately 70% of affected individuals have identified mutations in both copies of the gene (alleles), while the other 30% only a mutation in one allele, or no mutation is found (Balikova et al., 2009). The *VPS13B* gene sequence for all of these patients was analyzed due to the features of Cohen Syndrome being present. It is not known whether other mutations/variations in the *VPS13B gene* may exist that do not result in Cohen Syndrome, but still may be associated with some variation in craniofacial growth and development. If *VPS13B* mutations that do not lead to Cohen Syndrome, but that are associated with variation in craniofacial growth and development are found, these mutations may be associated with a phenotype (e.g., OSA).

Fibrillin-1 (FBN1) Gene

Marfan Syndrome is a condition with autosomal dominant inheritance that also is associated with a high prevalence of OSA due to a defect in connective tissue resulting in upper airway tissue laxity and airway collapse (Cistulli and Sullivan, 1995). Marfan Syndrome results from a heterozygous mutation in the *FBN1* located on chromosome 15q21. The role of *FBN1* gene is to produce fibrillin, which provides rigidity to tissues such as bone and muscle. Facial dysmorphism due to maxillary and mandibular retrognathism and narrow palatal vault (Westling et al., 1997), along with upper airway laxity, predisposes these patients to OSA.

Fibroblast Growth Factor Receptor 2 (FGFR2) Gene

Crouzon Syndrome is a form of craniosynostosis with autosomal dominant inheritance due to a heterozygous mutation in the fibroblast growth factor receptor-2 gene (*FGFR2*), which is located on chromosome 10q26.1. Patients with Crouzon Syndrome may have OSA that can be due to several anatomic factors (e.g., maxillary hypoplasia, constricted maxillary arch, high-arched palatal vault, posteriorly displaced tongue and lengthened soft palate; Hui et al., 1998).

Twist Family Basic Helix-Loop-Helix Transcription Factor 1 (TWIST1) Gene

Saethre-Chotzen Syndrome has autosomal dominant inheritance associated with mutations in the *TWIST1* gene located on chromosome 7p21.1. The craniofacial features include variable craniosynostosis, facial asymmetry, midface hypoplasia, hypertelorism and ptosis of eyelids. In de Jong and colleagues' study (2010), OSA was observed in 5% of individuals with Saethre-Chotzen Syndrome.

CONCLUSIONS

Because the etiology of OSA is complex and heterogeneous, it is simplistic to attribute the presence of OSA in every case solely to an anatomical variation, except in the most severe cases (e.g., Pierre Robin sequence). Still, variation in craniofacial morphology that may affect the airway certainly may play a role. Investigating the contribution of genetic variation to craniofacial morphology that may be associated with OSA, as well as the other factors that contribute to OSA, would help to understand its etiology better.

Some may question the validity of reviewing genes associated with selected syndromes that may be associated with OSA if the individual patient does not have the syndrome. There can be a continuum of phenotypic variation from "normal" to syndromic that may be associated with different mutations, or even the same mutation, in a specific gene (Lattanzi et al., 2017). For example, a recent candidate gene study associated *FGFR2* with an increased risk for non-syndromic skeletal Class II and Class III malocclusion, while the *TWIST1* gene also may be involved in non-syndromic craniofacial variation (da Fontoura et al., 2015). A future massive parallel sequencing array for genetic factors in OSA may include genes associated with syndromic and non-syndromic variation in craniofacial growth that may be related to OSA.

REFERENCES

Abass SK, Hartsfield JK Jr. Investigation of genetic factors affecting complex traits using external apical root resorption as a model. Semin Orthod 2008;14(2):115-124.

Al Maaitah E, El Said N, Abu Alhaija ES. First premolar extraction effects on upper airway dimension in bimaxillary proclination patients. Angle Orthod 2012;82(5):853-859.

Almajed A, Viezel-Mathieu A, Gilardino MS, Flores RL, Tholpady SS, Côté A. Outcome following surgical interventions for micrognathia in infants with Pierre Robin Sequence: A systematic review of the literature. Cleft Palate Craniofac J 2017;54(1):32-42.

Amiel J, Laudier B, Attié-Bitach T, Trang H, de Pontual L, Gener B, Trochet D, Etchevers H, Ray P, Simonneau M, Vekemans M, Munnich A, Gaultier C, Lyonnet S. Polyalanine expansion and frameshift mutations of the paired-like homeobox gene PHOX2B in congenital central hypoventilation syndrome. Nat Genet 2003;33(4):459-461.

Balestrazzi P, Corrini L, Villani G, Bolla MP, Casa F, Bernasconi S. The Cohen syndrome: Clinical and endocrinological studies of two new cases. J Med Genet 1980;17(6):430-432.

Balikova I, Lehesjoki AE, de Ravel TJ, Thienpont B, Chandler KE, Clayton-Smith J, Träskelin AL, Fryns JP, Vermeesch JR. Deletions in the VPS13B (COH1) gene as a cause of Cohen syndrome. Hum Mutat 2009;30(9): E845-E854.

Battagel JM, Johal A, Kotecha B. A cephalometric comparison of subjects with snoring and obstructive sleep apnoea. Eur J Orthod 2000;22(4): 353-365.

Berry-Kravis EM, Zhou L, Rand CM, Weese-Mayer DE. Congenital central hypoventilation syndrome: PHOX2B mutations and phenotype. Am J Respir Crit Care Med 2006;174(10):1139-1144.

Bodurtha JN, Mosteller M, Hewitt JK, Nance WE, Eaves LJ, Moskowitz WB, Katz S, Schieken RM. Genetic analysis of anthropometric measures in 11-year-old twins: The Medical College of Virginia Twin Study. Pediatr Res 1990;28(1):1-4.

Bygarski E, Paterson M, Lemire EG. Extreme intra-familial variability of congenital central hypoventilation syndrome: A case series. J Med Case Rep 2013;7:117.

Cade BE, Chen H, Stilp AM, Gleason KJ, Sofer T, Ancoli-Israel S, Arens R, Bell GI, Below JE, Bjonnes AC, Chun S, Conomos MP, Evans DS, Johnson

WC, Frazier-Wood AC, Lane JM, Larkin EK, Loredo JS, Post WS, Ramos AR, Rice K, Rotter JI, Shah NA, Stone KL, Taylor KD, Thornton TA, Tranah GJ, Wang C, Zee PC, Hanis CL, Sunyaev SR, Patel SR, Laurie CC, Zhu X, Saxena R, Lin X, Redline S. Genetic associations with obstructive sleep apnea traits in Hispanic/Latino Americans. Am J Respir Crit Care Med 2016;194(7):886-897.

Cartegni L, Chew SL, Krainer AR. Listening to silence and understanding nonsense: Exonic mutations that affect splicing. Nat Rev Genet 2002; 3(4):285-298.

Chandler KE, Kidd A, Al-Gazali L, Kolehmainen J, Lehesjoki AE, Black GC, Clayton-Smith J. Diagnostic criteria, clinical characteristics, and natural history of Cohen syndrome. J Med Genet 2003;40(4):233-241.

Chen Y, Hong L, Wang CL, Zhang SJ, Cao C, Wei F, Lv T, Zhang F, Liu DX. Effect of large incisor retraction on upper airway morphology in adult bimaxillary protrusion patients. Angle Orthod 2012;82(6):964-970.

Chi L, Comyn FL, Mitra N, Reilly MP, Wan F, Maislin G, Chmiewski L, Thorne-FitzGerald MD, Victor UN, Pack AI, Schwab RJ. Identification of craniofacial risk factors for obstructive sleep apnoea using three-dimensional MRI. Eur Respir J 2011;38(2):348-358.

Chirakalwasan N, Ruxrungtham K. The linkage of allergic rhinitis and obstructive sleep apnea. Asian Pac J Allergy Immunol 2014;32(4):276-286.

Chokroverty S. Overview of sleep & sleep disorders. Indian J Med Res 2010;131:126-140.

Cistulli PA, Sullivan CE. Sleep apnea in Marfan's syndrome: Increased upper airway collapsibility during sleep. Chest 1995;108(3):631-635.

Cohen MM Jr, Hall BD, Smith DW, Graham CB, Lampert KJ. A new syndrome with hypotonia, obesity, mental deficiency, and facial, oral, ocular, and limb anomalies. J Pediatr 1973;83(2):280-284.

Contos JJ, Fukushima N, Weiner JA, Kaushal D, Chun J. Requirement for the lpA1 lysophosphatidic acid receptor gene in normal suckling behavior. Proc Natl Acad Sci U S A 2000;97(24):13384-13389.

Cosentino FI, Bosco P, Drago V, Prestianni G, Lanuzza B, Iero I, Tripodi M, Spada RS, Toscano G, Caraci F, Ferri R. The APOE ε4 allele increases the risk of impaired spatial working memory in obstructive sleep apnea. Sleep Med 2008;9(8):831-839.

da Fontoura CS, Miller SF, Wehby GL, Amendt BA, Holton NE, Southard TE, Allareddy V, Moreno Uribe LM. Candidate gene analyses of skeletal variation in malocclusion. J Dent Res 2015;94(7):913-920.

de Jong T, Bannink N, Bredero-Boelhouwer HH, van Veelen ML, Bartels MC, Hoeve LJ, Hoogeboom AJ, Wolvius EB, Lequin MH, van der Meulen JJ, van Adrichem LN, Vaandrager JM, Ongkosuwito EM, Joosten KF, Mathijssen IM. Long-term functional outcome in 167 patients with syndromic craniosynostosis: Defining a syndrome-specific risk profile. J Plast Reconstr Aesthet Surg 2010;63(10):1635-1641.

de Paula LK, Alvim RO, Pedrosa RP, Horimoto AR, Krieger JE, Oliveira CM, Pereira AC, Lorenzi-Filho G. Heritability of OSA in a rural population. Chest 2016;149(1):92-97.

Deonna T, Arczynska W, Torrado A. Congenital failure of automatic ventilation (Ondine's curse): A case report. TJ Pediatr 1974;84(5):710-714.

Driessen C, Rijken BF, Doerga PN, Dremmen MH, Joosten KF, Mathijssen IM. The effect of early fusion of the spheno-occipital synchondrosis on midface hypoplasia and obstructive sleep apnea in patients with Crouzon syndrome. J Craniomaxillofac Surg 2017;45(7):1069-1073.

Duplomb L, Duvet S, Picot D, Jego G, El-Chehadeh-Djebbar S, Marle N, Gigot N, Aral B, Carmignac V, Thevenon J, Lopez E, Rivière JB, Klein A, Philippe C, Droin N, Blair E, Girodon F, Donadieu J, Bellanné-Chantelot C, Delva L, Michalski JC, Solary E, Faivre L, Foulquier F, Thauvin-Robinet C. Cohen syndrome is associated with major glycosylation defects. Hum Mol Genet 2014;23(9):2391-2399.

Ek J, Urhammer S, Sørensen T, Andersen T, Auwerx J, Pedersen O. Homozygosity of the Pro12Ala variant of the peroxisome proliferation-activated receptor-γ2 (PPAR-γ2): Divergent modulating effects on body mass index in obese and lean Caucasian men. Diabetologia 1999;42(7):892-895.

El H, Palomo JM. Airway volume for different dentofacial skeletal patterns. Am J Orthod Dentofacial Orthop 2011;139(6):e511-e521.

Flores-Mir C, Korayem M, Heo G, Witmans M, Major MP, Major PW. Craniofacial morphological characteristics in children with obstructive sleep apnea syndrome: A systematic review and meta-analysis. J Am Dent Assoc 2013;144(3):269-277.

Franklin KA, Lindberg E. Obstructive sleep apnea is a common disorder in the population: A review on the epidemiology of sleep apnea. J Thorac Dis 2015;7(8):1311-1322.

Friedman E, Sack J. The Cohen syndrome: Report of five new cases and a review of the literature. J Craniofac Genet Dev Biol 1982;2(3):193-200.

Gehrman PR, Keenan BT, Byrne EM, Pack AI. Genetics of sleep disorders. Psychiatr Clin North Am 2015;38(4):667-681.

Germec-Cakan D, Taner T, Akan S. Uvulo-glossopharyngeal dimensions in non-extraction, extraction with minimum anchorage, and extraction with maximum anchorage. Eur J Orthod 2011:33(5):515-520.

Gottlieb DJ, DeStefano AL, Foley DJ, Mignot E, Redline S, Givelber RJ, Young T. APOE ε4 is associated with obstructive sleep apnea/hypopnea: The Sleep Heart Health Study. Neurology 2004;63(4):664-668.

Gungor A. Advanced airway management strategies for severe OSAS and craniofacial anomalies. Am J Otolaryngol 2017;38(1):77-81.

Hui S, Wing YK, Kew J, Chan YL, Abdullah V, Fok TF. Obstructive sleep apnea syndrome in a family with Crouzon's syndrome. Sleep 1998; 21(3):298-303.

Kalra M, Pal P, Kaushal R, Amin RS, Dolan LM, Fitz K, Kumar S, Sheng X, Guha S, Mallik J, Deka R, Chakraborty R. Association of ApoE genetic variants with obstructive sleep apnea in children. Sleep Med 2008; 9(3):260-265.

Khayat A, Bin-Hassan S, Al-Saleh S. Polysomnographic findings in infants with Pierre Robin sequence. Ann Thorac Med 2017;12(1):25-29.

Kolehmainen J, Norio R, Kivitie-Kallio S, Tahvanainen E, de la Chapelle A, Lehesjoki AE. Refined mapping of the Cohen syndrome gene by linkage disequilibrium. Eur J Hum Genet 1997;5(4):206-213.

Kondo I, Nagataki S, Miyagi N. The Cohen syndrome: Does mottled retina separate a Finnish and a Jewish type? Am J Med Genet 1990;37(1): 109-113.

Kryger MH, Roth T, Dement WC. *Principles and Practice of Sleep Medicine.* 5th ed. Philadelphia, PA: Saunders/Elsevier 2011.

Kurihara Y, Kurihara H, Suzuki H, Kodama T, Maemura K, Nagai R, Oda H, Kuwaki T, Cao WH, Kamada N, et al. Elevated blood pressure and craniofacial abnormalities in mice deficient in endothelin-1. Nature 1994;368:703-710.

Kuwaki T, Cao WH, Kurihara Y, Kurihara H, Ling GY, Onodera M, Ju KH, Yazaki Y, Kumada M. Impaired ventilatory responses to hypoxia and hypercapnia in mutant mice deficient in endothelin-1. Am J Physiol 1996;270(6 Pt 2):R1279-R1286.

Larkin EK, Patel SR, Goodloe RJ, Li Y, Zhu X, Gray-McGuire C, Adams MD, Redline S. A candidate gene study of obstructive sleep apnea in European Americans and African Americans. Am J Respir Crit Care Med 2010;182(7):947-953.

Larkin EK, Patel SR, Redline S, Mignot E, Elston RC, Hallmayer J. Apolipoprotein E and obstructive sleep apnea: Evaluating whether a candidate gene explains a linkage peak. Genet Epidemiol 2006;30(2): 101-110.

Larsen AJ, Rindal DB, Hatch JP, Kane S, Asche SE, Carvalho C, Rugh J. Evidence supports no relationship between obstructive sleep apnea and premolar extraction: An electronic health records review. J Clin Sleep Med 2015;11(12):1443-1448.

Lattanzi W, Barba M, Di Pietro L, Boyadjiev SA. Genetic advances in craniosynostosis. Am J Med Genet A 2017;173(5):1406-1429.

Lavezzi AM, Casale V, Oneda R, Gioventù S, Matturri L, Farronato G. Obstructive sleep apnea syndrome (OSAS) in children with Class III malocclusion: Involvement of the PHOX2B gene. Sleep Breath 2013; 17(4):1275-1280.

Lee RW, Vasudavan S, Hui DS, Prvan T, Petocz P, Darendeliler MA, Cistulli PA. Differences in craniofacial structures and obesity in Caucasian and Chinese patients with obstructive sleep apnea. Sleep 2010;33(8): 1075-1080.

Liu Y, Lowe AA, Zeng X, Fu M, Fleetham JA. Cephalometric comparisons between Chinese and Caucasian patients with obstructive sleep apnea. Am J Orthod Dentofacial Orthop 2000;117(4):479-485.

Lombardo RC, Kramer E, Cnota JF, Sawnani H, Hopkin RJ. Variable phenotype in a novel mutation in PHOX2B. Am J Med Genet A 2017; 173(6):1705-1709.

Maspero C, Giannini L, Galbiati G, Rosso G, Farronato G. Obstructive sleep apnea syndrome: A literature review. Minerva Stomatol 2015;64(2): 97-109. [In English and Italian.]

Miyao E, Noda A, Miyao M, Yasuma F, Inafuku S. The role of malocclusion in non-obese patients with obstructive sleep apnea syndrome. Intern Med 2008;47(18):1573-1578.

Nikodemova M, Finn L, Mignot E, Salzieder N, Peppard PE. Association of sleep disordered breathing and cognitive deficit in APOE ε4 carriers. Sleep 2013;36(6):873-880.

Norio R. Finnish disease heritage. I: Characteristics, causes, background. Hum Genet 2003;112(5-6):441-456.

Norio R, Raitta C. Are the Mirhosseini-Holmes-Walton syndrome and the Cohen syndrome identical? Am J Med Genet 1986;25(2):397-398.

North C, Patton MA, Baraitser M, Winter RM. The clinical features of the Cohen syndrome: Further case reports. J Med Genet 1985;22(2):131-134.

Palmer LJ, Buxbaum SG, Larkin EK, Patel SR, Elston RC, Tishler PV, Redline S. Whole genome scan for obstructive sleep apnea and obesity in African-American families. Am J Respir Crit Care Med 2004;169(12):1314-1321.

Patel SR. Shared genetic risk factors for obstructive sleep apnea and obesity. J Appl Physiol (1985) 2005;99(4):1600-1606.

Patel SR, Goodloe R, De G, Kowgier M, Weng J, Buxbaum SG, Cade B, Fulop T, Gharib SA, Gottlieb DJ, Hillman D, Larkin EK, Lauderdale DS, Li L, Mukherjee S, Palmer L, Zee P, Zhu X, Redline S. Association of genetic loci with sleep apnea in European Americans and African-Americans: The Candidate Gene Association Resource (CARe). PLos One 2012; 7(11):e48836.

Polak J, Brock BA, Punjabi NM. Obstructive sleep apnea and glucose metabolism. *Sleep Apnea: Pathogenesis, Diagnosis and Treatment*. 2nd ed. CRC Press 2016:300.

Prasad B, Nyenhuis SM, Weaver TE. Obstructive sleep apnea and asthma: Associations and treatment implications. Sleep Med Rev 2014;18(2):165-171.

Rand CM, Carroll MS, Weese-Mayer DE. Congenital central hypo-ventilation syndrome: A neurocristopathy with disordered respiratory control and autonomic regulation. Clin Chest Med 2014;35(3):535-545.

Redline S, Tosteson T, Tishler PV, Carskadon MA, Milliman RP. Studies in the genetics of obstructive sleep apnea: Familial aggregation of symptoms associated with sleep-related breathing disturbances. Am Rev Respir Dis 1992;145(2 Pt 1):440-444.

Riha RL, Brander P, Vennelle M, Douglas NJ. A cephalometric comparison of patients with the sleep apnea/hypopnea syndrome and their siblings. Sleep 2005;28(3):315-320.

Roedig JJ, Phillips BA, Morford LA, Van Sickels JE, Falcao-Alencar G, Fardo DW, Hartsfield JK Jr, Ding X, Kluemper GT. Comparison of BMI, AHI, and apolipoprotein E ε4 (APOE-ε4) alleles among sleep apnea patients with different skeletal classifications. J Clin Sleep Med 2014;10(4): 397-402.

Saarelainen S, Lehtimäki T, Kallonen E, Laasonen K, Poussa T, Nieminen MM. No relation between apolipoprotein E alleles and obstructive sleep apnea. Clin Genet 1998;53(2):147-148.

Schwartz AR, Patil SP, Laffan AM, Polotsky V, Schneider H, Smith PL. Obesity and obstructive sleep apnea: Pathogenic mechanisms and therapeutic approaches. Proc Am Thorac Soc 2008;5(2):185-192.

Seifert W, Holder-Espinasse M, Spranger S, Hoeltzenbein M, Rossier E, Dollfus H, Lacombe D, Verloes A, Chrzanowska KH, Maegawa GH, Chitayat D, Kotzot D, Huhle D, Meinecke P, Albrecht B, Mathijssen I, Leheup B, Raile K, Hennies HC, Horn D. Mutational spectrum of COH1 and clinical heterogeneity in Cohen syndrome. J Med Genet 2006; 43(5):e22.

Senaratna CV, Perret JL, Lodge CJ, Lowe AJ, Campbell BE, Matheson MC, Hamilton GS, Dharmage SC. Prevalence of obstructive sleep apnea in the general population: A systematic review. Sleep Med Rev 2017;34:70-81.

Skotko BG, Macklin EA, Muselli M, Voelz L, McDonough ME, Davidson E, Allareddy V, Jayaratne YS, Bruun R, Ching N, Weintraub G, Gozal D, Rosen D. A predictive model for obstructive sleep apnea and Down syndrome. Am J Med Genet A 2017;173(4):889-896.

Stefanovic N, El H, Chenin DL, Glisic B, Palomo JM. Three-dimensional pharyngeal airway changes in orthodontic patients treated with and without extractions. Orthod Craniofac Res 2013;16(2):87-96.

Stunkard AJ, Harris JR, Pedersen NL, McClearn GE. The body-mass index of twins who have been reared apart. N Engl J Med 1990;322(21): 1483-1487.

Susarla SM, Abramson ZR, Dodson TB, Kaban LB. Cephalometric measurement of upper airway length correlates with the presence and severity of obstructive sleep apnea. J Oral Maxillofac Surg 2010; 68(11):2846-2855.

Tan HL, Kheirandish-Gozal L, Abel F, Gozal D. Craniofacial syndromes and sleep-related breathing disorders. Sleep Med Rev 2016;27:74-88.

Tauman R, Gozal D. Obesity and obstructive sleep apnea in children. Paediatr Respir Rev 2006;7(4):247-259.

Teodorescu M, Barnet JH, Hagen EW, Palta M, Young TB, Peppard PE. Association between asthma and risk of developing obstructive sleep apnea. JAMA 2015;313(2):156-164.

Tisko R, Sopkova Z, Habalova V, Dorkova Z, Slaba E, Javorsky M, Tkac I, Riha RL, Tkacova R. Effects of apolipoprotein E genotype on serum lipids in obstructive sleep apnoea. Eur Respir J 2014;43(4):1097-1105.

Todd ES, Weinberg SM, Berry-Kravis EM, Silvestri JM, Kenny AS, Rand CM, Zhou L, Maher BS, Marazita ML, Weese-Mayer DE. Facial phenotype in children and young adults with PHOX2B-determined congenital central hypoventilation syndrome: Quantitative pattern of dysmorphology. Pediatr Res 2006;59(1):39-45.

Valiathan M, El H, Hans MG, Palomo MJ. Effects of extraction *versus* nonextraction treatment on oropharyngeal airway volume. Angle Orthod 2010;80(6):1068-1074.

Wachs Pauley LL, Elmahdy M, Morford LA, Roedig J, Phillips B, Wardrop P, et al. Association of APOE-e isoforms and reduced BMI values in obstructive sleep apnea patients with a convex dentofacial profile. Prepared for submission.

Wang Q, Jia P, Anderson NK, Wang L, Lin J. Changes of pharyngeal airway size and hyoid bone position following orthodontic treatment of Class I bimaxillary protrusion. Angle Orthod 2012;82(1):115-121.

Weese-Mayer D, Shannon D, Keens T, Silvestri J. Idiopathic congenital central hypoventilation syndrome: Diagnosis and management. Am J Respir Crit Care Med 1999;160(1):368-373.

Weese-Mayer DE, Berry-Kravis EM, Zhou L, Maher BS, Silvestri JM, Curran ME, Marazita ML. Idiopathic congenital central hypoventilation syndrome: Analysis of genes pertinent to early autonomic nervous system embryologic development and identification of mutations in PHOX2b. Am J Med Genet A 2003;123A(3):267-278.

Westling L, Mohlin B, Bresin A. Craniofacial manifestations in the Marfan syndrome: Palatal dimensions and a comparative cephalometric analysis. J Craniofac Genet Dev Biol 1998;18(4):211-218.

Xu H, Qian Y, Guan J, Yi H, Yin S. No association between the ApoE ε2 and ε4 alleles and the risk of obstructive sleep apnea: A systematic review and meta-analysis. Biomed Rep 2015;3(3):313-318.

Yin T, Li NF, Heizhati M, Zhang J, Zhang J, Zhou L, Chang G. Association of glucose transporter 4 genetic polymorphisms with obstructive sleep apnea syndrome in Han Chinese general population: A cross-section study. Lipids Health Dis 2014;13:12.

Young T, Peppard PE, Gottlieb DJ. Epidemiology of obstructive sleep apnea: A population health perspective. Am J Respir Crit Care Med 2002;165(9):1217-1239.

Yun Z, Maecker HL, Johnson RS, Giaccia AJ. Inhibition of PPAR gamma 2 gene expression by the HIF-1-regulated gene DEC1/Stra13: A mechanism for regulation of adipogenesis by hypoxia. Dev Cell 2002; 2(3):331-341.

Zhuoqiang L, Wu X, Jin X, Peng F, Lin J. Apolipoprotein E e2/e3/e4 variant in association with obstructive sleep apnoea and lipid profile: A meta-analysis. J Int Med Res 2015;44(1):3-14.

CBCT: ITS ROLE IN DETECTING OBSTRUCTIVE AIRWAY PROBLEMS IN ADULTS AND CHILDREN

Sean K. Carlson

ABSTRACT

Obstructive sleep apnea (OSA) is a severe problem for many. Reducing the number of people affected by this disease is a noble goal for all healthcare providers, including orthodontists. This chapter explores how early orthodontic treatment may have a positive effect on the anatomical structures that surround the airway and potentially reduces the risk of developing OSA in the future. There still is a lot to learn; however, with the advent of CBCT technology, we are learning more quickly than ever before. Much of this is due to the fact that the creation of a three-dimensional (3D) virtual patient using CBCT imaging now is a reality.

This chapter addresses airway-centered orthodontics and how clinicians might integrate cone-beam computed tomography (CBCT) into overall patient care. It discusses the importance of airway health in diagnosis and treatment planning. Scientific evidence is promising—showing that jaw development in the transverse dimension has a positive effect on sleep-disordered breathing. The advantages of a 3D approach in diagnosis, airway analysis and treatment planning are essential if clinicians are to discharge their full obligation to their patients.

KEY WORDS: CBCT, OSA, airway, transverse, orthodontics

INTRODUCTION

The 2017 Moyers Symposium was a popular and well-attended event. The subject of sleep apnea and its relationship to orthodontics could not have been timelier. Although I will cite multiple research publications and scientific findings here, I would like to approach this chapter primarily from the perspective of a practicing orthodontist, which I am.

It was a great honor to present at the Symposium. I do not take lightly the impact that this meeting and this book might have in the future, as it represents a groundbreaking publication in orthodontics—

given that it brings airway awareness to center stage in orthodontics for the first time.

When I began my professional career approximately 20 years ago, I did not imagine that I would be focusing on airway and sleep. In fact, I was not even aware that this topic was a "thing" when I started my orthodontic practice. As a resident, I had little exposure to anything "sleep" or "airway" related during my orthodontic education. However, due to the advent of three-dimensional (3D) cone-beam computed technology (CBCT) imaging, my practice changed dramatically after I installed my first CBCT machine in 2008.

The transition of my practice's focus from tooth and jaw into whole health was something that I did not expect. It has been a wonderful experience. Approaching my patients with a broader, 3D perspective on diagnosis not only has allowed me to help many of my patients breathe better, but it also has enlightened me on how malocclusions develop. I have gained insight on the craniofacial region that I never would have experienced in my previous two-dimensional (2D) world.

THE IMPORTANCE OF UNDERSTANDING MULTIPLE CONDITIONS

My practice in the San Francisco Bay area is separated from the newly-famous computer tech world of Silicon Valley by the historically-famous Golden Gate Bridge, construction of which was completed in 1937 (Fig. 1). This bridge symbolizes many things to me as an orthodontist. First, like orthodontics, the Golden Gate Bridge is a wonderful engineering solution to a difficult problem. Second, a fact that many do not know, the construction of this bridge was finished ahead of schedule *and* under budget. This structure is symbolic to me as an orthodontist because most of us would love all our cases to finish ahead of schedule and under budget—but they don't—and some are far from remaining stable during retention.

In contrast to the Golden Gate Bridge, another bridge called the Tacoma Narrows Bridge was constructed further up the west coast at approximately the same time. Unlike the Golden Gate Bridge, the Tacoma Narrows Bridge was built for far less cost and corners were cut during construction. Most importantly, however, there was not enough attention paid to all the conditions surrounding the Tacoma Narrows channel over which the bridge was constructed.

Figure 1. The San Francisco Golden Gate Bridge. Construction completed in 1937.

Due to high winds in the area, the Tacoma Narrows Bridge began to rock and sway during a high windstorm and subsequently fell into the river below in dramatic fashion (Fig. 2). To me, this failure represents a situation in which all the potential problems of the Tacoma Narrows channel were not understood fully. It exemplifies an opportunity where a more complete understanding of the conditions that might lead to failure could have been addressed more thoroughly.

The Tacoma Narrows Bridge story is interesting to me. It relates to orthodontics because many of our cases fail after we have completed them. Why do some of our cases relapse? How might we better understand the "windstorm" that causes relapse? Of equal importance, why are some malocclusions worse than others?

During orthodontic residency, I spent much time learning how to fix malocclusions, but less time asking, "why are they there?" I seem to have learned a lot about how to repair things, but less about how to prevent them. I often use the analogy of orthodontic practices being great auto-body repair shops after a car crash (Fig. 3). We are good at fixing crooked teeth and poor bites *after* they already have developed. As my career progresses, I find myself becoming more interested in learning how to prevent those problems in the first place. I want to learn how to become a "car crash" prevention specialist, not simply a "car crash" repairman; I would like to learn how to prevent malocclusions and relapse.

Figure 2. The collapse of the Tacoma Narrows Bridge. High windstorms were not accounted for carefully, thus resulting in the failure of the project.

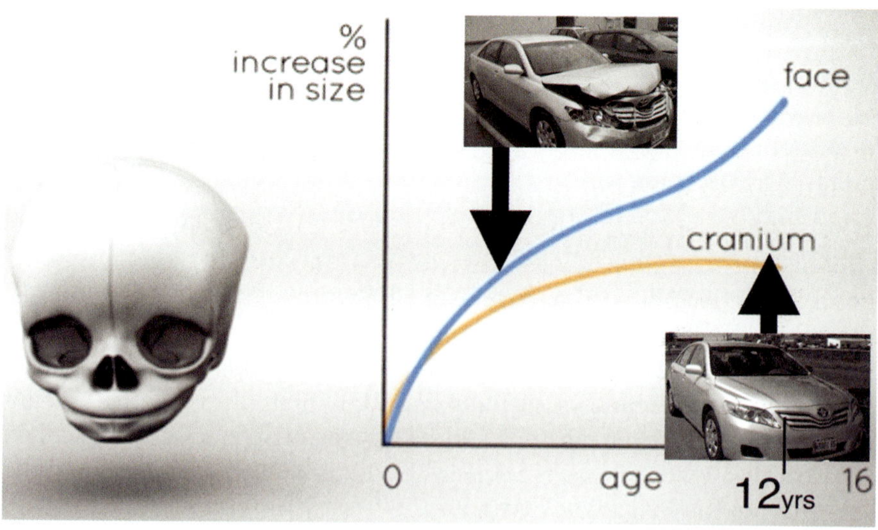

Figure 3. Malocclusions often start forming early. Orthodontics traditionally has focused on "fixing the car crash," in contrast to focusing on "car crash" prevention.

To do this, I obviously need to study in more detail the conditions that cause these things. It makes one think: why are the malocclusions of some patients more difficult than others to treat? Why do the treatment outcomes of some malocclusions fail, where others do not? Why do some teeth become impacted, while others do not? I believe airway and breathing is an important "condition" about which we need to know more.

SMALL AIRWAYS AND NARROW MAXILLAS

The more I became involved in studying 3D CBCT images and evaluating the airway, the more I began to look at my patients as a whole, rather than just as a set of teeth and jaws. The more I study my own cases, the more I realize that the components of entire craniofacial region—including the fragile pharyngeal space—are connected intimately. The area in which we work as orthodontists is connected so closely to the pharyngeal space that it becomes impossible to think that we do not (or cannot) affect it.

What I started to realize the more that I looked at this relationship is that many of my cases—those that had impaction and Class II problems—also had significantly narrow maxillas. Coincidentally, many of these patients also had smaller airways when compared to others, though perhaps this was NOT a coincidence. Could these narrow maxillas and narrow airways somehow be related? My clinical experience (now that I actually am looking from that perspective) tells me that these structures are related. I would like to explore the *potential* of that relationship in this chapter.

Most orthodontists are focused on creating ideal inclusions: well-aligned marginal ridges, absence of rotations, good overbite and good overjet. While an ideal occlusion is a wonderful goal, we also must think about the entire craniofacial region attached to those teeth. As orthodontists began to relate airway to orthodontics, some of us began to use the term "whole-health orthodontics." This phrase simply meant that we were focused not only on the dentition and jaws, but also on the soft tissue structures that surrounded it. We now are paying close attention to the fragile pharyngeal space, nasal airway and sinuses. This focus includes evaluating how our patients talk, eat and breathe. But we realize that these conditions are difficult things to study and that causal relationships likely will be difficult to prove. But it definitely is worth exploring what the research has helped us understand so far.

CBCT REVEALS THE AIRWAY

The arrival of CBCT in the United States in 2001 gave most orthodontists their first 3D look at the structures inside the head and around the teeth. As soon as those early CBCT images were acquired, many of us began to investigate the anatomy of the airway space. Early studies revealed that the anatomical variations in the airway space were something that we could image, isolate and study using CBCT (Schendel and Hatcher, 2010; Alsufyani et al., 2012, 2016; Hatcher, 2012).

Once we started looking, clinicians discovered early on that many of our patients with malocclusions also had airway constrictions in specific anatomical regions. The most common regions included the tonsillar and adenoid areas (Fig. 4). Of course, ear nose and throat (ENT) physicians have been treating these areas for decades, but they rarely were associated with orthodontic problems. As orthodontists began to learn more about airways, we also began to question our patients about their breathing habits and sleep patterns. We realized quickly that some of these patients who had breathing and sleep issues also had narrow jaws and airways, and some of them had sleep apnea. This observation was a major eye opener!

In my practice, we do a fair amount of early Phase I treatment, most often consisting of maxillary expansion and mandibular molar transverse uprighting. I find that once treated, these patients who originally were narrow transversely rarely have impacted teeth following expansion and spontaneous improvements in Class II sagittal problems are seen.

Something that has changed significantly since I started focusing on airway is that of the many variables that typically indicate orthodontic treatment success, creating a patent airway now must be included. Based on previous studies that correlate small airways with OSA in adults (Lowe et al., 1986; Avrahami and Englender, 1995; Ogawa et al., 2007), I often target the largest airway space that I can achieve. According to these authors, positive correlations between OSA and minimal cross-sectional area (MCA) of an airway is observed when the MCA is less than 150 mm^2.

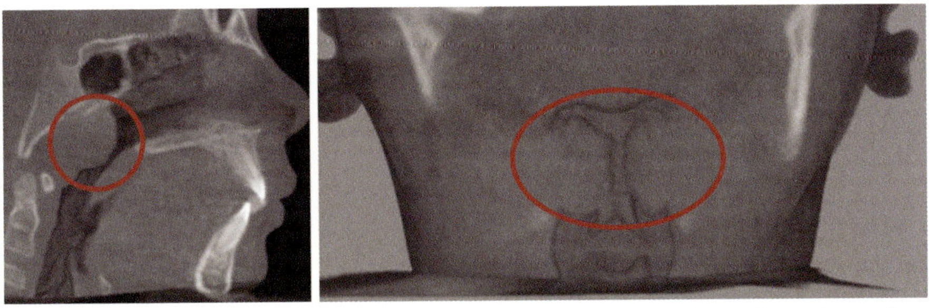

Figure 4. Common areas of airway obstruction, enlarged adenoids and tonsils.

Therefore, I always target achieving MCAs well above 150 mm^2. My efforts to create large airways simply are attempts to help reduce potential structural breathing problems in the future, although we yet have to prove that this strategy is effective. From what I have experienced so far, this approach simply makes practical sense.

Of course, as clinicians began to measure airway with more intention, we needed to understand what normal airways were supposed to look like. We gained a better understanding of this from Schendel and colleague's work (2012). These investigators acquired CBCT images of 1,300 individuals who presented to orthodontic offices. Although this sample might be considered biased in that these subjects likely all had malocclusions that led them to see an orthodontist in the first place, the CBCT sample was large, with subjects ranging in age from 6 to 60 years.

Schendel and coworkers (2012) were interested primarily in reporting the measured MCA for each age group. The MCA represents the region of the smallest circumference of the airway and likely would indicate the area of greatest airflow resistance in the pharynx. Schendel and coworkers measured MCA on all 1,300 individuals. They found that the MCA increased from age 6 to age 20 (Fig. 5). After age 20, however, the MCA tended to decrease into old age (Fig. 6).

What these findings indicate is that airways enlarge naturally during childhood and adolescence, but then worsen as we go through adulthood. The gradual narrowing of the airway in adulthood likely is one of the reasons that many patients do not experience sleep apnea until later in life. Many adults likely begin their 20s with slightly larger airways than what they end up with by middle and old age.

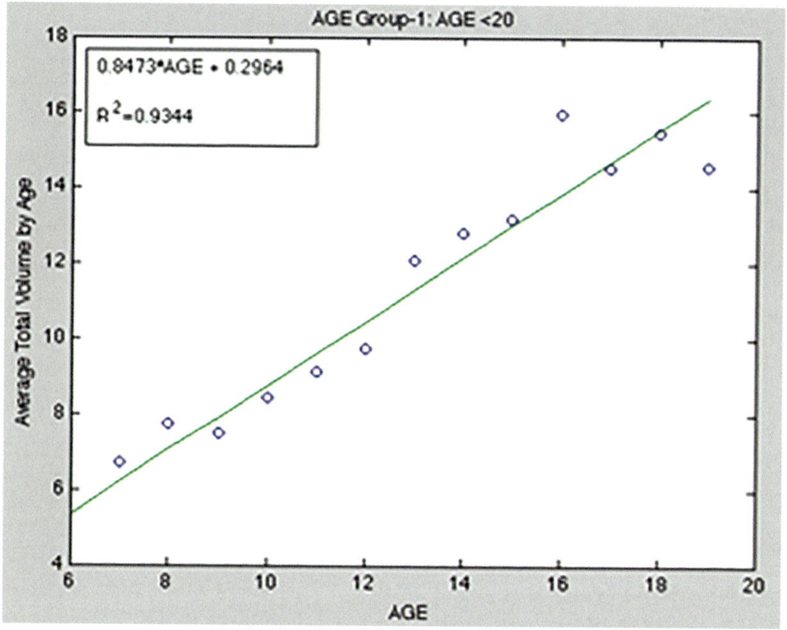

Figure 5. Schendel and associates' CBCT study (2012) of MCA, which increases from age 6 to 20 by a factor of roughly 10 mm^2 per year. Reprinted with permission from Elsevier.

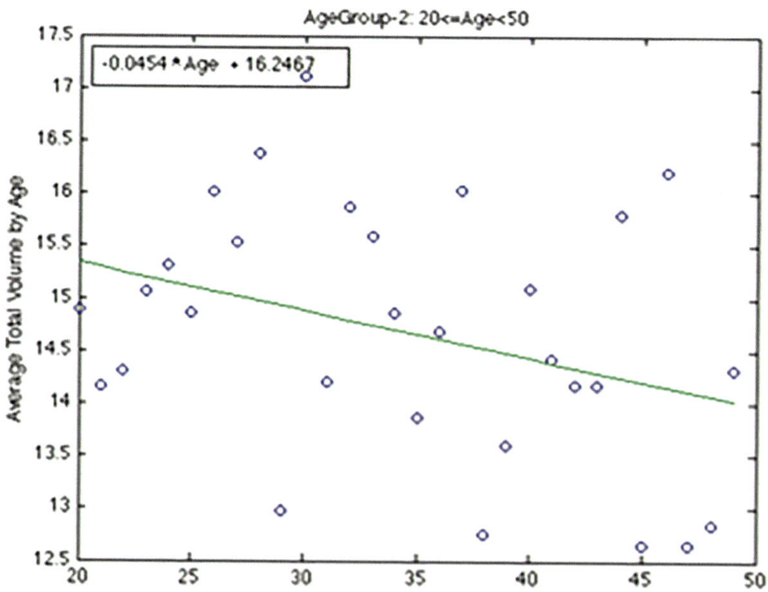

HOW SMALL IS TOO SMALL?

Correlation studies have been performed surrounding the measurement of MCA (Lowe et al., 1986; Avrahami and Englender, 1995; Ogawa et al., 2007), some of which were based on medical CT images and others on CBCT images. The data from these studies show that patients with MCA measurements less than 50 mm^2 also have a high risk of having OSA (Fig. 7). Those patients with MCA measurements of 50-100 mm^2 have a moderate risk of having OSA. Patients with an MCA measurement of 100-150 mm^2 have a low risk of having OSA. Patients with MCA measurements above 150 mm^2 show no correlation with risk of having OSA. Although these thresholds should be tempered by the clinical presentation of the individual patient, these values do give us targets from which to begin.

AIRWAY IMAGE SHAPE

Ever since we learned that the airways were something we could image and measure, clinicians have become curious about what these measurements mean. Measuring the airway with imaging is not without controversy (Alsufyani et al., 2012, 2016), yet there have been interesting concepts that have emerged from various studies.

Chen and associates (2002) evaluated the airway using volumetric medical CT images in patients with obstructive sleep apnea (OSA). The research team was interested in measuring the MCA and its position for all subjects. Using polysomnography (PSG), they divided their sample into two groups: those with OSA and those who were snorers without OSA. The pharyngeal space was divided into two areas (Fig. 8), with the upper pharyngeal space (or *retropalatal area*) defined as that area above the

50/100/150

High risk OSA = 0-50 mm2
Mod risk OSA = 50-100 mm2
Low risk OSA = 100-150 mm2

Figure 7. Correlation of MCA with OSA

← Figure 6. Schendel and coworkers' CBCT study (2012) of MCA, which decreases during adulthood. Reprinted with permission from Elsevier.

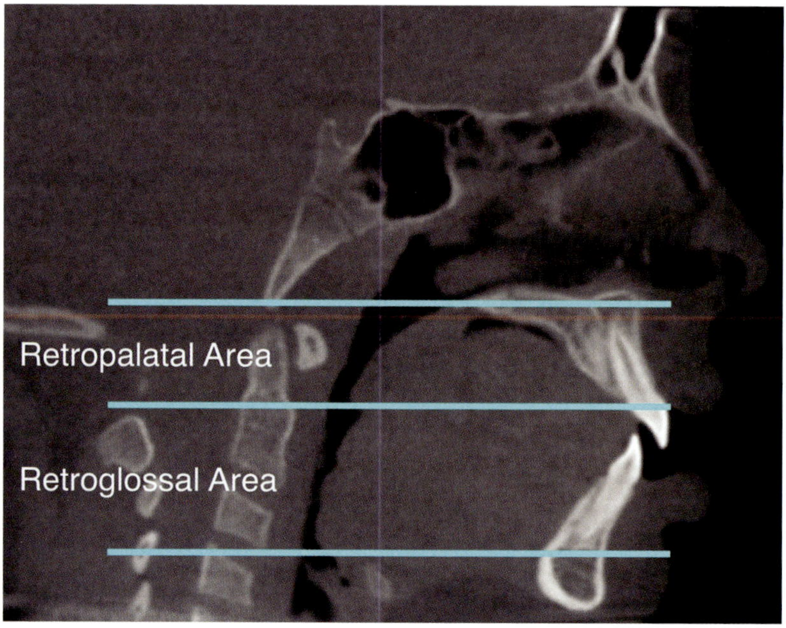

Figure 8. The divisions of retropalatal area and retroglossal area.

uvula up to the hard palate and the lower pharyngeal space (or *retro-glossal area*) as inferior to the uvula down to the epiglottis. What they discovered was that there was a stronger correlation with OSA when the MCA was in the upper part of the pharynx (retropalatal area).

What is interesting about Chen and associates' study (2002) is that common sense would have us think that the retroglossal area (lower pharynx) might be more correlated with sleep apnea due to the thought that the base of the tongue might fall back against the posterior pharynx during sleep, making it collapse and block the airway. In that the significant correlation instead was associated with the retropalatal area (upper pharynx), however, this finding tells us that narrow measurements in the retropalatal area (upper pharynx) may be more important when it comes to assessing risk for OSA.

Encisco and coworkers (2010) had similar findings. For all OSA subjects in their study, the MCA was located in the retropalatal area (upper pharynx), which again was different relative to what one might expect. This study also used CBCT airway measurements to understand better the parameters of the airway volume and their correlations with answers from sleep questionnaires. Like the study by Chen and coworkers (2002), this study divided their sample into two separate groups using PSG. Therefore, one group consisted of patients who had OSA, while the other consisted of snorers without OSA. In agreement with many of the studies mentioned above, this study also found patients with OSA had significantly smaller MCAs than those who snored but did not have OSA.

Encisco and coworkers (2010) also found that smaller lateral (right-left) dimensions of the MCAs were correlated more significantly with OSA than were the horizontal (anteroposterior; A-P) dimensions (Fig. 9); this finding goes against common sense. Many would assume that OSA most likely occurs when the tongue falls back against the posterior of the lower pharyngeal airway, therefore making small A-P measurements appear more at risk.

That small lateral measurements of the airway correlated more significantly with OSA, however, is somewhat puzzling. It leads us to think that OSA is considerably more complex than simply an A-P collapse of the airspace. My theory, which might explain this finding, is that wider airways (right to left) may result in more lateral airflow (along the sides) when the tongue falls back against the posterior pharynx. But narrower transverse airways may not allow for this lateral flow; thus, those airways become obstructed completely and the result is OSA.

Solutions for adult OSA are not simple (Fig. 10). The less invasive solutions include weight loss, positional therapy, nasal decongestion, continuous positive airway pressure (CPAP) and oral appliances. Unfortunately, even oral appliances have significant side effects (Lowe, 1999, 2012). More invasive solutions to adult OSA include uvulopalatopharyngoplasty (UPPP), maxilla-mandibular advancement (MMA) and tracheotomy. CPAP tends to have low compliance.

Many oral surgeons who are involved in treating adult OSA report that the long-term effects of soft tissue modification have little effect on OSA (Aronovich and Helman, 2018). However, MMAs seem to be one of the best treatment approaches for adult OSA, though the procedure is invasive. If we can avoid having to do these procedures, it would save our patients a lot of morbidity. It would be ideal if we could help our young patients avoid developing OSA later in life. What if we could prevent the "car crash"? Wouldn't it be great if we could develop the airway?

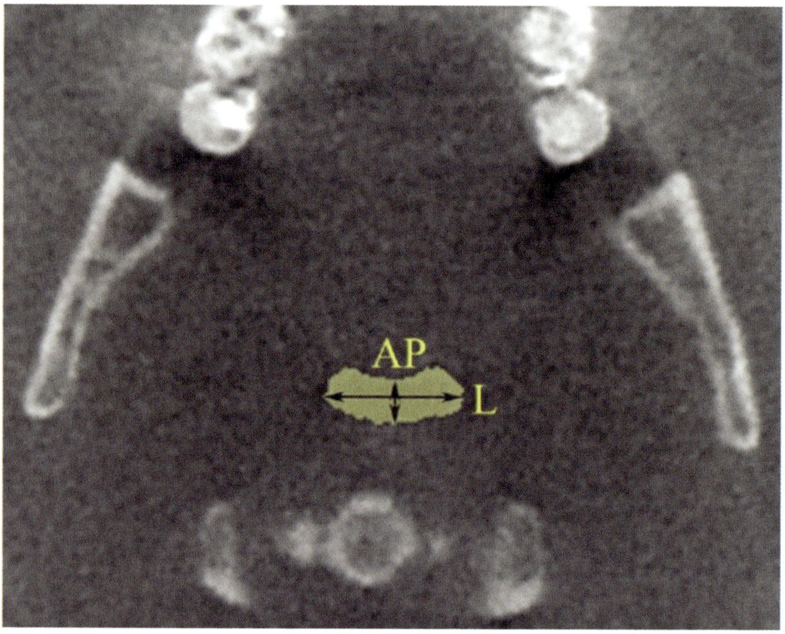

Figure 9. Chen and colleagues (2002) found transverse measurements (L) to be more correlated with OSA than AP measurements.

• Weight Loss	• Continuous positive airway pressure (CPAP)
• Positional therapy	• Uvulopalatopharyngoplasty (UPPP)
• Nasal decongestion	• Maxillomandibular advancement (MMA)
• Oral appliances	• Tracheotomy

Figure 10. Common solutions for adult OSA.

DEVELOPING AIRWAYS

It is important to note that airway size is not indicative of OSA. In fact, even what might appear to be structural narrowing of the airway is not always indicative of OSA or breathing disorders. However, it makes sense to most clinicians that a larger airway likely would result in increased ease of breathing, if all other circumstances were kept constant. Even without acquiring sleep studies on every patient, however, it makes sense to make airways as large as possible during orthodontic intervention.

What if orthodontists could play a role in preventing OSA by developing the airway? Numerous research publications have reported findings that suggest this approach is possible. For example, Smith and associates (2012) published a 3D CBCT analysis of airway volume changes after rapid maxillary expansion. They divided the airway into different sections (Fig. 11). They reported that following rapid maxillary expansion, the nasal cavity volume and the nasopharynx volume increased for all subjects (Fig. 12). This increase was significant statistically. According to their measurements, they saw a 15% increase with expansion alone.

What I find in my practice is that airway obstruction often is related to increased adenoid and tonsillar tissue (Fig. 4). Obstruction of the airway space in these areas is far more common than I previously thought. So, what happens when we can clear up these obstructive areas through surgery and expand at the same time?

A group from Stanford looked at this in detail (Guilleminault et al., 2011). They studied patients who were scheduled to undergo adenotonsillectomy as well as rapid maxillary expansion. The study divided the sample into two groups: the first group would receive the adenotonsillectomy, followed by rapid maxillary expansion; and the second would receive rapid maxillary expansion, followed by the adenotonsillectomy. The researchers performed PSG at each time point, before and after each procedure. They measured the AHI and Respiratory Disturbance Index (RDI; Figs. 13-14). RDI is calculated as the number of apnea events/hour plus the number of hypopnea events/hour plus the number of respiratory-effort related arousals (RERAs) per hour of sleep.

The findings of Guilleminault and associates (2011) were interesting. Following expansion, there was a 60% decrease in RDI score. Once

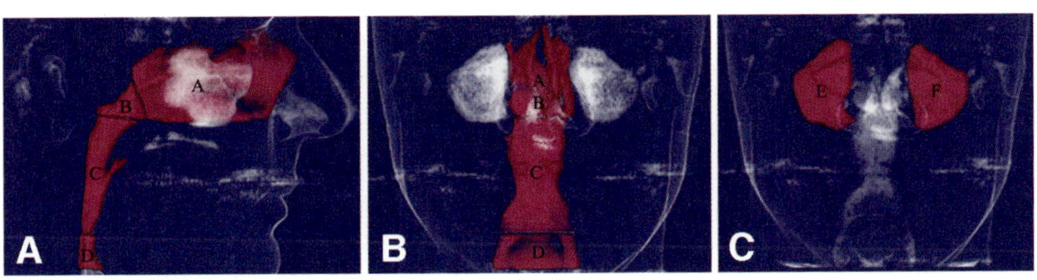

Figure 11. Smith and colleagues' study (2012) showed airway volume changes after palatal expansion. *A*: Sagittal view of multiple airway regions with regions A and B representing the nasal cavity volume and the nasopharynx volume, respectively. *B*: Coronal view of these same regions. *C*: Positions of the right and left maxillary sinuses. Reprinted with permission from Elsevier.

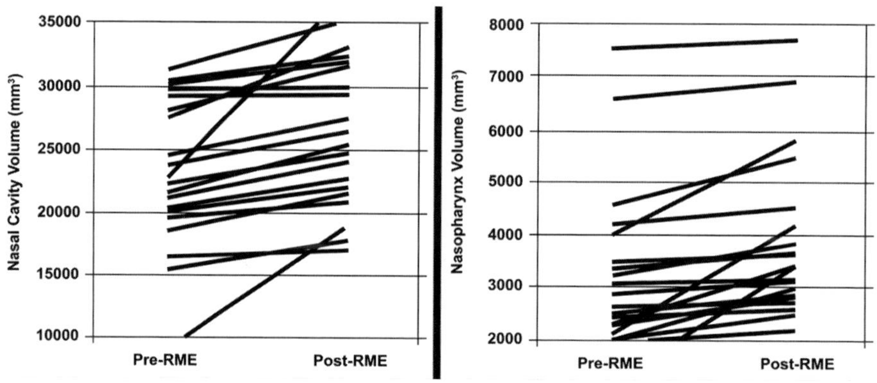

Figure 12. Smith and coworkers (2012) found airway volume increases for all subjects. Reprinted with permission from Elsevier.

APNEA-HYPOPNEA INDEX (AHI)
• Calculated by dividing the number of events (apneas with hypopneas) by the number of hours of sleep • Events = 10 second pauses in breathing with a decrease in blood oxygenation

Figure 13. Calculating AHI score. 5-15/hour = mild; 15-30/hour = moderate; > 30/hour = severe.

RESPIRATORY DISTURBANCE INDEX (RDI)
• Includes respiratory-effort related arousals (RERAs) • A RERA is characterized by increasing respiratory effort for 10 seconds or more leading to an arousal from sleep without hypopnea or apnea • Calculated by dividing total number of events (RERAs + apneas + hypopneas) by the number of hours of sleep

Figure 14. Calculating RDI score. 5-15/hour = mild; 15-30/hour = moderate; > 30/hour = severe.

they added the adenotonsillectomy, there was a 90% decrease in RDI. This is a significant change in the quality of sleep for this sample of patients. What is most interesting about this for the clinical orthodontist is that even if you are not looking for airway obstructions, your expansion efforts are improving sleep for many of your patients, without you even knowing. I believe that if orthodontists began looking more closely at airways and identifying areas of airway obstruction, then we would be able to improve the sleep and breathing quality for countless patients that otherwise would be overlooked.

In my practice, I notice that airway size increases more than expected following early arch development. Our early arch development procedures primarily include rapid maxillary expansion and mandibular molar uprighting in the transverse dimension (Fig. 15).

To understand this treatment effect better in my practice, I recently completed an in-house study of 26 consecutively-treated patients (Figs. 16-18). All subjects were treated with maxillary expansion and lower molar uprighting. If we use Schendel and associates' work (2012) describing the MCA at various ages, we might expect our patients to fall somewhat close to Schendel's numbers regarding MCA.

In our sample of 26 consecutively-treated patients, the MCA at T1 was 77.9 mm^2, compared with Schendel's sample where the MCA at the same average age of 9.5 was approximately 95 mm^2. Following Phase I arch development, the MCA for our sample increased to 150.3 mm^2, compared to Schendel's "expected" MCA of 121 mm^2. Following completion of Phase II treatment, the MCA for our sample increased to 219.77 mm^2, compared to Schendel's "expected" MCA of 136.1 mm^2. Our sample of consecutively-treated patients showed an increase in MCA of > 2.5x that of Schendel's group. Could it be that our group is growing significantly more than Schendel's group? Or has our transverse development improved the airway growth significantly? I think this type of change is worth further investigation.

MY PERSONAL STORY

For most of us, clinical treatment comes down to our own experience. Once you have experienced a single airway improvement patient

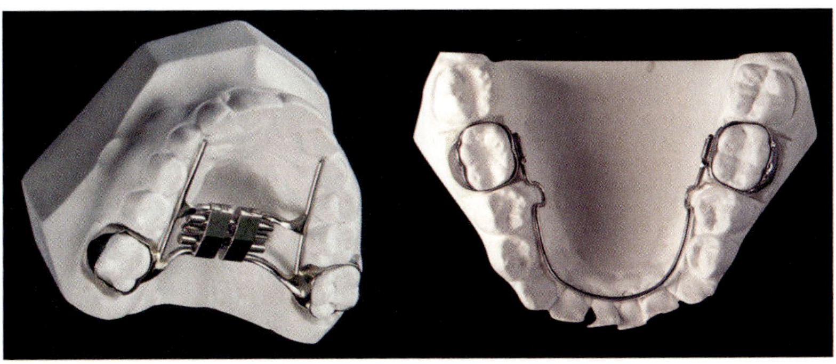

Figure 15. Arch development appliances for maxillary expansion and mandibular molar transverse uprighting.

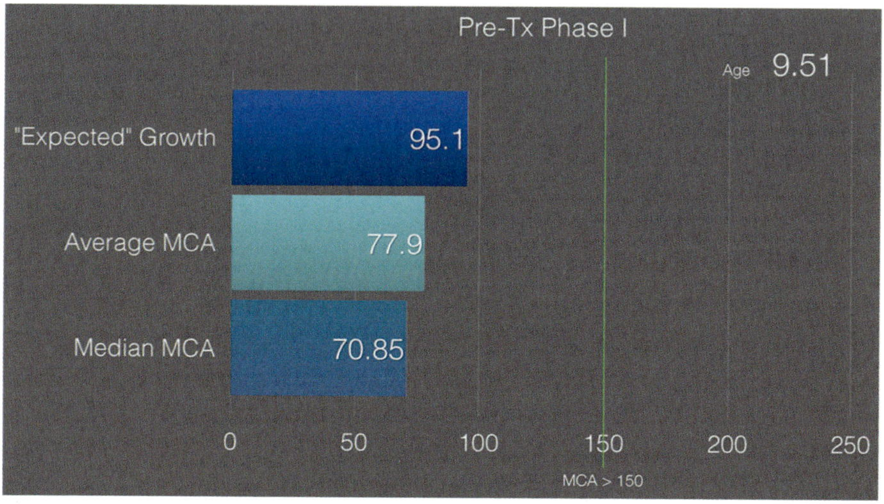

Figure 16. Our in-house practice study. The average MCA for our patients at T1 was 77.9 mm^2.

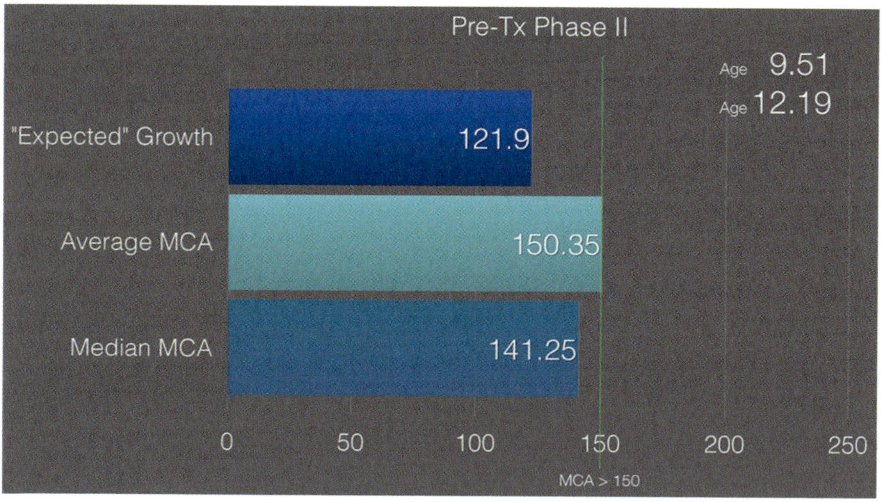

Figure 17. Our in-house practice study. The average MCA for our patients at T2 was 150.4 mm^2.

in your practice, then you realize the significant life-changing value you can bring to your patients. For me, this experience came in the form of a family member.

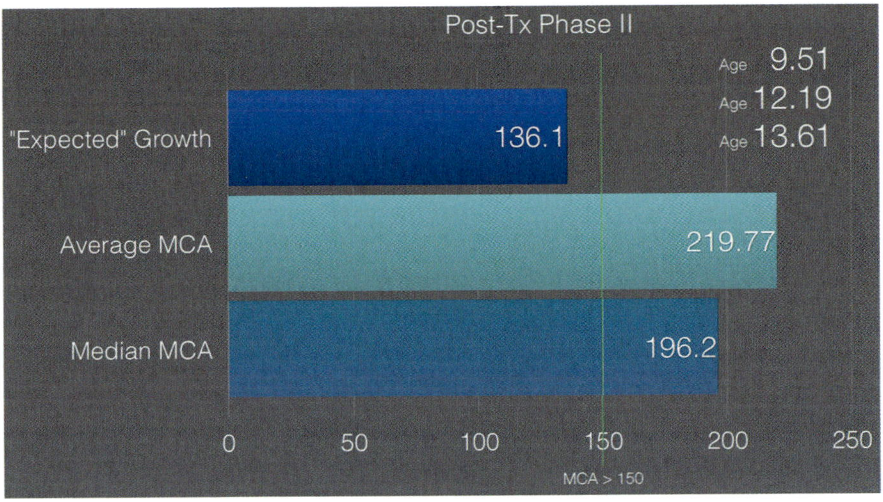

Figure 18. Our in-house practice study. The average MCA for our patients at T3 was 219.8 mm^2.

Years ago, over a family holiday, the mother of my four-year-old niece reported that my niece was snoring at night. Given that I was just beginning to study airway and breathing in my own practice, I instantly dug deeper into her signs and symptoms. I immediately asked her mother to send me a video of my niece sleeping. The video was shocking as she was struggling to breathe and obviously was suffering from OSA. I knew that her mother might run into significant roadblocks while pursuing treatment; therefore, I requested that we obtain objective evidence of her problem. I first requested a CBCT scan to assess her airway, the findings of which were shocking. The adenoid tissue and tonsillar tissue were obstructing nearly the entire airway (Fig. 19).

It is worth mentioning that her narrow airway was something that had been overlooked by the pediatrician for the first four years of my niece's life. Therefore, knowing that we would need significant evidence to convince the pediatrician and the ENT physician to investigate further, we continued gathering evidence. We obtained a radiology report that further indicated these compromised airway findings from the perspective of a radiologist. Gathering all information together—sleep video, CBCT scan of airway and radiology report—I sent her mother to

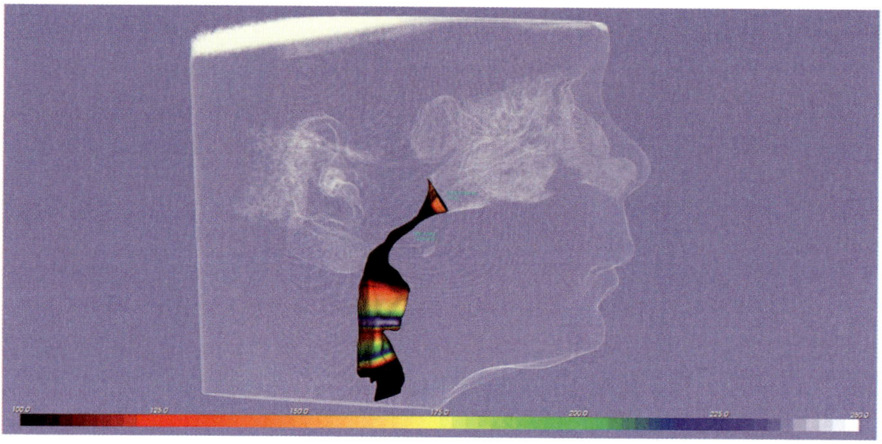

Figure 19. Author's niece's very small airway.

present my niece's case to her ENT physician. To my surprise, my niece was scheduled *immediately* for an adenotonsillectomy—no PSG sleep study or nasal scoping needed. I cannot help think that the power of the CBCT images helped significantly in expediting proper treatment.

During my niece's adenotonsillectomy procedure, the physician noted that the adenoid tissue was obstructing 75% of the nasal airway and that one tonsil was resting on the other when she was lying down. The surgeon noted that these were some of the biggest tonsils she had ever seen. How could this have been overlooked? How many patients like my niece are being overlooked every year? As orthodontists, it is important to realize that we can play a significant role in discovering these problems and truly changing our patient's lives for the better.

AIRWAYS AND TEETH ARE RELATED

Orthodontists should be focusing on teeth, right? How can we be expected to focus on everything? Well, the more I look at, the more I see. I believe that our attention to airway unexpectedly helped me understand more about malocclusion prevention. Malocclusions and airway problems seem to go hand in hand. The transverse development studies on maxillary expansion discussed earlier in this chapter describe how increasing maxillary width can improve airway volume. But increasing maxillary width also seems to work in preventing malocclusions.

As I mentioned in the beginning of the chapter, my goal is to prevent as many "car crashes" from happening, as opposed to simply fixing them after they have happened. It took a few of my own cases for me to discover a better way to prevent the "car crash." These cases involved early canine impaction detection and natural resolution of the ectopic canine eruption. I discovered that for patients whom we identified early as having ectopic canines, resolution of the ectopic canines occurred easily when expansion was performed early.

Although canine impaction prevention was the "aha" moment for me, it actually was the undiagnosed transverse problem that became visible. Miner and colleagues (2012) seemed to describe exactly what I had been missing early in my career. It was something that only CBCT imaging would allow me to see. They reported a transverse CBCT study that looked at skeletal width of the maxilla and mandible (Fig. 20). It is important to note, this study was a *skeletal* transverse study, not a *dental* one. The researchers divided their sample into five groups: normal controls, unilateral crossbites, bilateral crossbites, superior convergent cases and inferior convergent cases. Inferior convergent cases described the situation where the maxillary molars were leaning outward toward the mandibular molars, thus resulting in converging roots of the maxillary molars. Inferior convergent cases did the opposite.

These researchers measured the skeletal width at the molar site in the coronal slices of the CBCT image. What they discovered was that the normal controls showed and approximately 1 mm difference between the maxillary and mandibular skeletal width, with the mandible being slightly wider. What was most interesting about the study, however, was that the superior conversion cases had skeletal discrepancies that were *greater* than that of the unilateral cross-bite cases and almost as large as the bilateral crossbite cases.

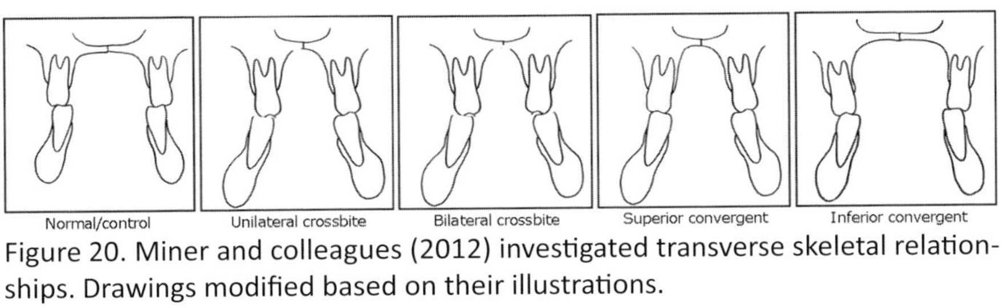

Figure 20. Miner and colleagues (2012) investigated transverse skeletal relationships. Drawings modified based on their illustrations.

The conclusion drawn from this paper was extremely meaningful: "patients without crossbites can have significant discrepancies that might warrant treatment" (Miner et al., 2012). These findings should be an eye opener for all orthodontists and pediatric dentists. It is essential that we to look beyond the dental relationships to discover from where our malocclusions are coming. In my opinion, focusing on the transverse relationship not only will improve airway development, but it also will improve the likelihood of normal eruption.

IS AIRWAY IMAGING THE ANSWER?

It is important to remember that investigations into airway using CBCT imaging still are relatively new. We must be cautious and not jump to conclusions about airway measurements and their significance in the development of OSA. We do not have proven causal relationships yet between airway measurements and OSA. As this chapter describes, however, there are significant correlations between airway measurements and OSA to which we must open our eyes. For me, now having treated patients with an airway-focused thought process for the last seven years, I believe the transverse relationship is important to airway development.

Many clinicians are hoping to find a simple airway measurement that will tell them what to do when it comes to managing OSA. Unfortunately, my feeling is that will not be possible with CBCT airway imaging. Although CBCT airway imaging is a helpful tool, it is not the panacea for diagnosing or treating OSA. It is simply yet another bit of information that must go into our decision-making process.

Therefore, I encourage clinicians not simply to look at airway volume or MCA in the hopes of gathering a diagnostic number, but rather to take these numbers into account when making clinical treatment decisions. In addition to gaining a better understanding of airway, paying closer attention to the transverse relationship also will reveal significant answers to our most puzzling malocclusion questions.

CONCLUSIONS

As clinicians, it is important for us to think about all possible condition that might affect our treatment efficacy and treatment stability. Therefore, like the Tacoma Narrows Bridge, we should use everything in our power to avoid making the same mistakes twice. I encourage clinicians to think of each patient as a single unique construction project. Examine all the variables. Examine all the conditions. Examine the patient as a whole.

CBCT likely will play a significant role in our understanding of OSA in the future, but it will not be the *only* tool. It will not give us *all* the answers. Introducing CBCT has changed my practice significantly and I am sure it has changed the lives of many of my patients who otherwise may have been overlooked.

REFERENCES

Alsufyani NA, Flores-Mir C, Major PW. Three-dimensional segmentation of the upper airway using cone beam CT: A systematic review. Dentomaxillofac Radiol 2012;41(4):276-284.

Alsufyani NA, Noga ML, Witmans M, Major PW. Upper airway imaging in sleep-disordered breathing: Role of cone-beam computed tomography. Oral Radiol 2017. [Epub before print.]

Aronovich S, Helman JI. Surgical management of obstructive sleep apnea: From childhood to adulthood. In: McNamara JA Jr, Shelgikar AV, eds. *Sleep Apnea: What Every Clinician (and Patient) Should Know.* Craniofacial Growth Series, Center for Human Growth and Development, The University of Michigan, Ann Arbor, MI 2018;54:155-174.

Avrahami E, Englender M. Relation between CT axial cross-sectional area of the oropharynx and obstructive sleep apnea syndrome in adults. AJNR Am J Neuroradiol 1995;16(1):135-140.

Chen NH, Li KK, Li SY, Wong CR, Chuang ML, Hwang CC, Wu YK. Airway assessment by volumetric computed tomography in snorers and subjects with obstructive sleep apnea in a Far-East Asian population (Chinese). Laryngoscope 2002;112(4):721-726.

Enciso R, Nguyen M, Shigeta Y, Ogawa T, Clark GT. Comparison of cone-beam CT parameters and sleep questionnaires in sleep apnea patients and control subjects. Oral Surg Oral Med Oral Pathol Oral Radiol Endod 2010;109(2):285-293.

Guilleminault C, Monteyrol PJ, Huynh NT, Pirelli P, Quo S, Li K. Adenotonsillectomy and rapid maxillary distraction in pre-pubertal children: A pilot study. Sleep Breath 2011;15(2):173-177.

Hatcher DC. Cone beam computed tomography: Craniofacial and airway analysis. Dent Clin North Am 2012;56(2):343-357.

Lowe AA, Gionhaku N, Takeuchi K, Fleetham JA. Three-dimensional CT reconstructions of tongue and airway in adult subjects with obstructive sleep apnea. Am J Orthod Dentofacial Orthop 1986; 90(5):364-374.

Miner RM, Al Qabandi S, Rigali PH, Will LA. Cone-beam computed tomography transverse analysis. Part I: Normative data. Am J Orthod Dentofacial Orthop 2012;142(3):300-307.

Ogawa T, Enciso R, Shintaku WH, Clark GT. Evaluation of cross-section airway configuration of obstructive sleep apnea. Oral Surg Oral Med Oral Pathol Oral Radiol Endod 2007;103(1):102-108.

Schendel SA, Hatcher D. Automated 3-dimensional airway analysis from cone-beam computed tomography data. J Oral Maxillofac Surg 2010; 68(3):696-701.

Schendel SA, Jacobson R, Khalessi S. Airway growth and development: A computerized 3-dimensional analysis. J Oral Maxillofac Surg 2012; 70(9):2174-2183.

Smith T, Ghoneima A, Stewart K, Liu S, Eckert G, Halum S, Kula K. Three-dimensional computed tomography analysis of airway volume changes after rapid maxillary expansion. Am J Orthod Dentofacial Orthop 2012;141(5):618-626.

CRITICAL CONCEPTS IN THE DIAGNOSIS OF THE AIRWAY USING 3D IMAGES

Lucia H.S. Cevidanes, Cauby M. Chaves, Jr., Tung Nguyen, Alexandre Moro, Paula P. Spada, Suelen W. Borges, Alexandre T. Motta, Yoon-Ji Kim, Marilia Yatabe, Marcos Ioshida, Antonio C.O. Ruellas

Remember to breathe. It is after all, the secret of life.
-Gregory Maguire

ABSTRACT

The effect of mode of breathing on craniofacial growth has been a widely-debated and controversial issue within orthodontics for decades, indicating a need for improved understanding of this relationship. Both medical and surgical treatments continue to be performed to modify sleep-related respiration. Over the last decade, there have been numerous airway studies using cone-beam computed tomography (CBCT) scans to diagnose airway volume. Three-dimensional (3D) assessments using CBCT, however, fail to assess dynamic aspects of the physiology of respiration that are essential for the diagnosis of obstructive sleep apnea (OSA). The upper airway is a complex, multi-functional, dynamic neuromechanical system. This chapter describes three critical concepts to be considered when using 3D imaging to diagnose airway morphology:

1. The need for dynamic assessments of respiration that are modified by time and function;
2. Improved descriptions of pharyngeal space boundaries and head posture; and
3. How the patency of the pharynx is affected by the dynamic neuromechanical system in breathing.

A strict cooperative work between radiologists and respiratory physiologists is highly desirable to integrate clinical and functional data together with morphologic and morphometric findings. Future implementation of the analysis of exhaled gases, dynamic airflow measures, specific image analysis methods for muscle assessments, post-processing imaging techniques and genetic analysis hopefully will close gaps that currently exist in radiographic airway assessments.

KEY WORDS: cone-beam computed tomography, airway, dynamic, boundaries, head posture

---

INTRODUCTION

Controversies regarding the diagnosis of sleep-related disorders and treatment options to modify sleep-related respiration have been debated for decades (Kluemper et al., 1995; Peltomäki, 2007). Examples of treatment approaches include surgical removal of the tonsils and adenoids, reduction of the nasal turbinates, septal deviation correction, allergy medication, rapid palatal expansion and orthognathic surgery (Dayal and Phillipson, 1985; Stradling et al., 1990; Kluemper et al., 1995; Kim et al., 2004; Pang et al., 2009; Koinis-Mitchell et al., 2012; Pirelli et al., 2015). The need to modify respiration and the choice of therapies are confounded by the dynamics of respiratory physiology, boundaries of what clinicians define as airway and the normal developmental processes in growth, development and aging.

Enlarged adenoids, which can be seen on lateral cephalograms, classically have been thought to obstruct nasal breathing partially, leading to mouth breathing and so-called adenoid face (Subtelny, 1954; Gwynne-Evans, 1956). Landmark studies began with Todd's original work (1936) that described the concept that adenoid overgrowth mechanically impedes breathing. Later, Subtelny (1954) summarized a method for airway assessment in lateral cephalograms and documented the growth cycle of the adenoids.

Linder-Aronson (1970) defined the adenoid face as characterized by an incompetent lip seal, a narrow upper dental arch, retroclined mandibular incisors, increased anterior face height, a steep mandibular plane angle and retrognathic mandible compared with faces of healthy controls. Similar craniofacial morphology also has been described in children with enlarged tonsils (Behlfelt et al., 1990).

The complex association between mechanical obstruction of the airway and facial growth, as observed in 2D cephalometric studies, has been discussed by the intriguing works of McNamara (1981), Kluemper and colleagues (1995), Vig (1998) and Peltomäki (2007). Furthermore, reduction of muscular tone during sleep in children with large adenoids and tonsils, or with other underlying abnormal upper airway anatomy, may lead to sleep-related airway obstruction and eventually to obstructive sleep apnea (OSA). These children have been found to have similar craniofacial characteristics (e.g., adenoid face) and their first treatment is removal of adenoids and tonsils (Guilleminault et al., 2004; Andersen et al., 2016).

As removal of adenoids and tonsils often does not resolve OSA, further diagnostic approaches for better understanding of mechanisms underlying the disease are needed to improve therapeutic strategies and reduce the associated negative consequences of OSA. Polysomnography (PSG) represents the gold standard to confirm the clinical suspicion of OSA, assess its severity and guide therapeutic choices. Behavioral, medical and surgical options are available for OSA treatment. Continuous positive airway pressure (CPAP) represents the treatment of choice in most adult patients. CPAP has been demonstrated to be effective in reducing symptoms, cardiovascular morbidity and mortality and neurocognitive sequelae (Yu et al., 2017), but can be tolerated poorly. Further, the results of clinical studies do not support surgery and pharmacological therapy as first-line treatment, but these approaches might be useful in carefully selected patients (Mannarino et al., 2012).

Over the last decade, there have been numerous airway studies using cone-beam computed tomography (CBCT) scans to diagnose airway volume, minimal cross-sectional area and upper airway shape variability (Guijarro-Martínez and Swennen, 2011; Alsufyani et al., 2012; Hatcher, 2012; van Vlijmen et al., 2012; Zimmerman et al., 2016). Three-dimensional (3D) assessments using CBCT, however, fail to assess the dynamic aspects of the physiology of sleep-related respiration that are essential for the diagnosis of OSA.

It is critical to understand airway development beyond the "mechanistic" way proposed by Solow and associates (1984), as occurring by changes in the muscular balance. Such mechanical obstruction of the airway would lead to mouth breathing, low tongue position in the oral cavity, imbalance between forces from the cheeks, lower mandibular position, extended head posture and the dental and skeletal consequences characteristic of adenoid facies (Linder-Aronson, 1970).

Taking into account recent evidence from children with OSA, it has become clear that the upper airway is a complex, multi-functional, dynamic neuromechanical system. Airway patency during breathing requires moment-to-moment coordination of neural and mechanical behavior and varies with posture. Failure to recruit and coordinate dilator muscles continuously to counterbalance the forces that act to close the airway results in hypopneas or apneas. The presence of at least five hypopneas and/or apneas per hour during sleep meets diagnostic criteria for OSA in adults.

Obesity and anatomical variations (e.g., mandibular retrognathism) increase the likelihood of upper airway collapse by altering the passive behavior of the upper airway. In OSA patients, airway dilation appears less coordinated compared with that in healthy subjects with comparable body mass index. How neural drive to the airway dilators relates to the biomechanical behavior of the upper airway (movement and stiffness) still is understood poorly. Bilston and Gandevia (2014) highlighted that the biomechanical behavior of the upper airway cannot be simply predicted from electromyographic activity of its muscles.

THREE CRITICAL CONCEPTS WHEN USING 3D IMAGE TO DIAGNOSE AIRWAY MORPHOLOGY

Understand Time and Function

For our pediatric patients, sleep-disordered breathing (SDB) is a clinical diagnosis. If a child has signs and symptoms of SDB and results of a polysomnogram (PSG) confirm an obstructive breathing pattern, the correct diagnosis is OSA. Without a PSG, healthcare professionals cannot diagnose OSA, and CBCT is not an alternative imaging tool for such sleep physiology assessment. Positive PSG results are more likely to be of clinical significance in children predisposed to OSA (e.g., those with certain genetic syndromes, morbid obesity, or neuromuscular or craniofacial disease), in those with significant associated significant morbidity (e.g., pulmonary hypertension), or in patients with severe OSA (Ingram and Friedman, 2015).

OSA represents the severe end of the spectrum of SDB, which ranges from primary snoring to upper airway resistance syndrome to OSA. The primary etiologic factor for pediatric OSA is adenotonsillar hypertrophy, especially between the ages of 2 and 7, when the tonsils undergo a growth phase. The success rates for OSA resolution with adenotonsillectomy was on average 58% in a meta-analysis of more than 1,000 children (Friedman et al., 2009).

It is equally important to understand that the etiology, manifestation and treatment modalities of OSA in growing children might be vastly different from adults. A recent study shows 91% of children diagnosed with OSA between the ages of 8 and 11 underwent spontaneous remission at 16 to 19 years of age (Spilsbury et al., 2015). It is possible that the continued development and growth of the airway into late adolescence contributes to remission of OSA in growing patients (Schendel et al., 2012).

For some adult OSA patients who do not tolerate CPAP treatment and for whom other more conservative approaches to treat OSA have failed, maxillomandibular advancement surgery (MMA) may be pursued. MMA has been shown to increase linear, cross-sectional plane and volumetric measurements of pharyngeal airways as detected in CBCT imaging, while reducing the Apnea-Hypopnea Index (AHI) and the respiratory disturbance index (Tan et al., 2017). However, while the skeletal displacements with jaw surgery can be maintained with rigid fixation, long-term, conclusive results are lacking in regard to post-MMA upper airway muscular adaptation (Fig. 1).

Nasopharynx and Oropharynx: Space Boundaries and Posture

From a respiratory physiologic perspective, there are three major components of the respiratory system: the airway, the lungs and the muscles of respiration. The airway—which includes the nose, mouth, pharynx, larynx, trachea, bronchi and bronchioles—carries air between the lungs and the body's exterior. The lungs act as the functional units of the respiratory system by passing oxygen into the body and carbon dioxide out of the body. The muscles of respiration, including the diaphragm and intercostal muscles, work together to act as a pump, pushing air into and out of the lungs during breathing.

The limited volumetric assessments in CBCT images require arbitrarily "standardized" boundaries, consistently defined within the field of view in which the CBCT scans are acquired (Fig. 2). The upper boundary often attempts to limit only the nasopharynx area, but the anterior boundary includes communication with the oral cavity, where tongue posture is an unpredictable boundary. An inferior boundary for CBCT

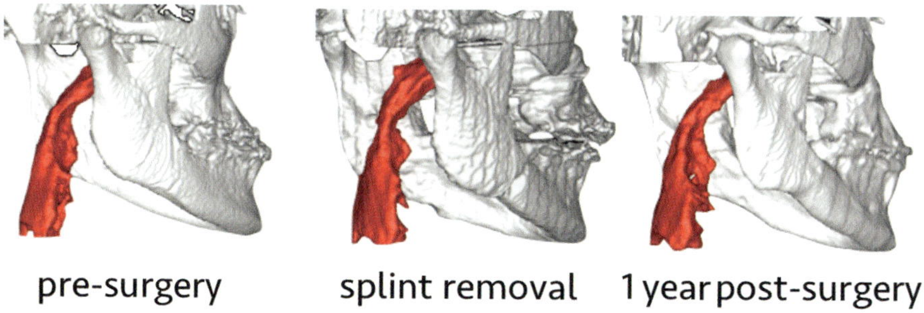

Figure 1. Skeletal and airway changes in one-year follow-up of two-jaw surgery. Note that the maxillary advancement led to an increase in the upper portion of the pharynx in the short term. While the maxillary advancement is stable one-year post-surgery, however, the nasopharynx changes does not seem to remain stable. This patient also had mandibular setback surgery and the lower portion of the pharynx is not constricted one year post-surgery.

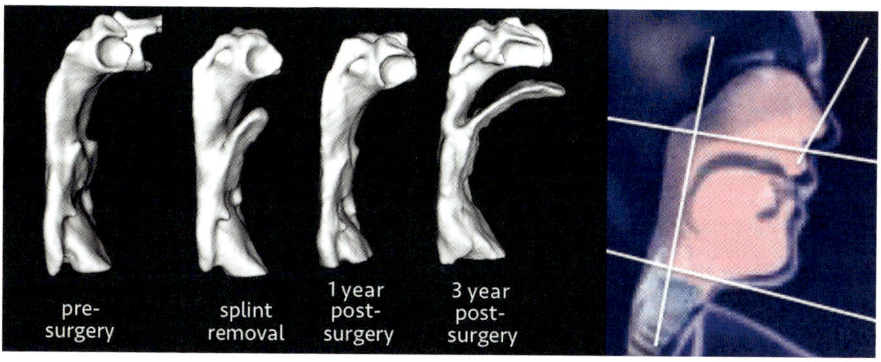

Figure 2. Pharynx boundaries are arbitrary and tongue posture naturally may vary in longitudinal series of CBCT for a patient who underwent mandibular advancement surgery.

assessments can be based on either the vertebras as moving references or the epiglottis, the position of which also will vary during the physiology of respiration. The muscular walls of the airway also fluctuate in an analogous way as an air-filled balloon, depending on whether the patient is inspiring or expiring as part of the dynamic process of respiration during image acquisition (Fig. 3).

Furthermore, head posture markedly alters the muscular boundaries of the airway, not only when we compare supine or upright acquisitions (Van Holsbeke et al., 2014), but also if clinicians attempt to correct or standardize head position after acquisition of CBCT (Fig. 4).

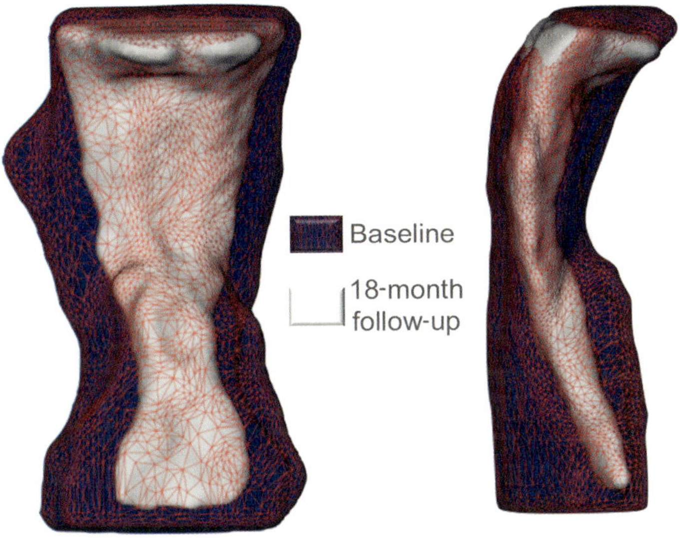

Figure 3. Airway changes during an 18-month follow-up period for an untreated child during the pubertal growth spurt. One would expect the airway space to be larger in the 18-month follow-up, but the stage of respiration (i.e., inspiration or expiration) and head posture may have affected the size and shape of the airway.

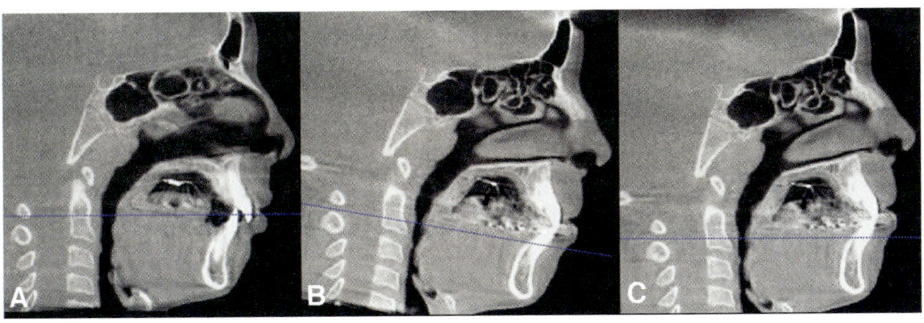

Figure 4. *A:* The pharynx appears narrower after treatment with an appliance to advance the mandible (Herbst). *B-C:* The patient's head was tilted down after treatment. While image analysis software allows clinicians to standardize head orientation (*C*), head position during image acquisition (*B*) cannot be corrected.

Time Versus Space

The assessment of the factors involved in patency of the upper airway (nasopharynx) using CBCT imaging needs to be interpreted in light of the complexity of the dynamic neuromechanical system in breathing. While some factors have been studied, many others have not and the interactions between biomechanics of the tissues, neural control of the muscles and airflow yet have to be understood fully.

In young, normal-weight, healthy humans, the upper airway is "dynamically stable" and responds to external stimuli in order to maintain patency throughout the respiratory cycle. Obesity increases demands on the upper airway as a dynamic system, but in healthy subjects, patency is maintained. However, in OSA patients, small changes in conditions (e.g., posture, pressure, sleep stage) cause instability and upper airway narrowing or collapse (Bilston and Gandevia, 2014).

In OSA, it seems that there is a dissociation between increases in neural drive to the upper airway muscles and evoked contractile force. Certainly, the neural drive in OSA is insufficient to compensate for unfavorable passive upper airway mechanics arising from a narrow airway. Evolutionary changes in the upper airway are adapted poorly to the vicissitudes of obesity and aging (Occhipinti et al., 2017).

While outcomes of skeletal counterparts of the pharyngeal airway can be improved by jaw surgery changes (Fig. 5; Motta et al., 2011),

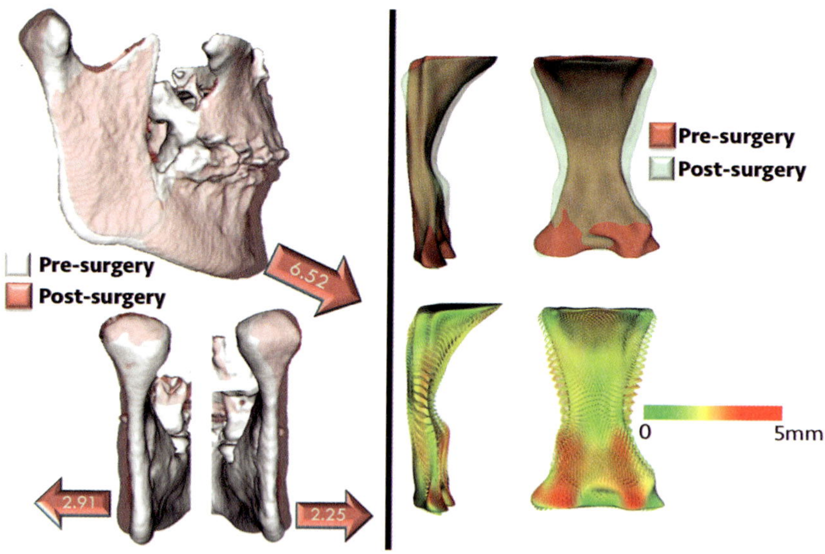

Figure 5. Pharynx changes one year following mandibular advancement surgery. Skeletal and soft tissue changes with mandibular advancement. Notably with mandibular advancement surgery, lateral rami displacements often were observed. Interestingly, this intervention also led to an increased transverse dimensions of the nasopharynx.

little is known about the factors that affect pharyngeal muscles adaptation and stability of airway patency over time (Fig. 6). Submental ultrasonographic (Bilici et al., 2017) parameters among OSA patients are a promising imaging modality to measure tongue, geniohyoid muscle and lateral pharyngeal wall dimensions precisely by integrating such information (Nguyen et al., 2016).

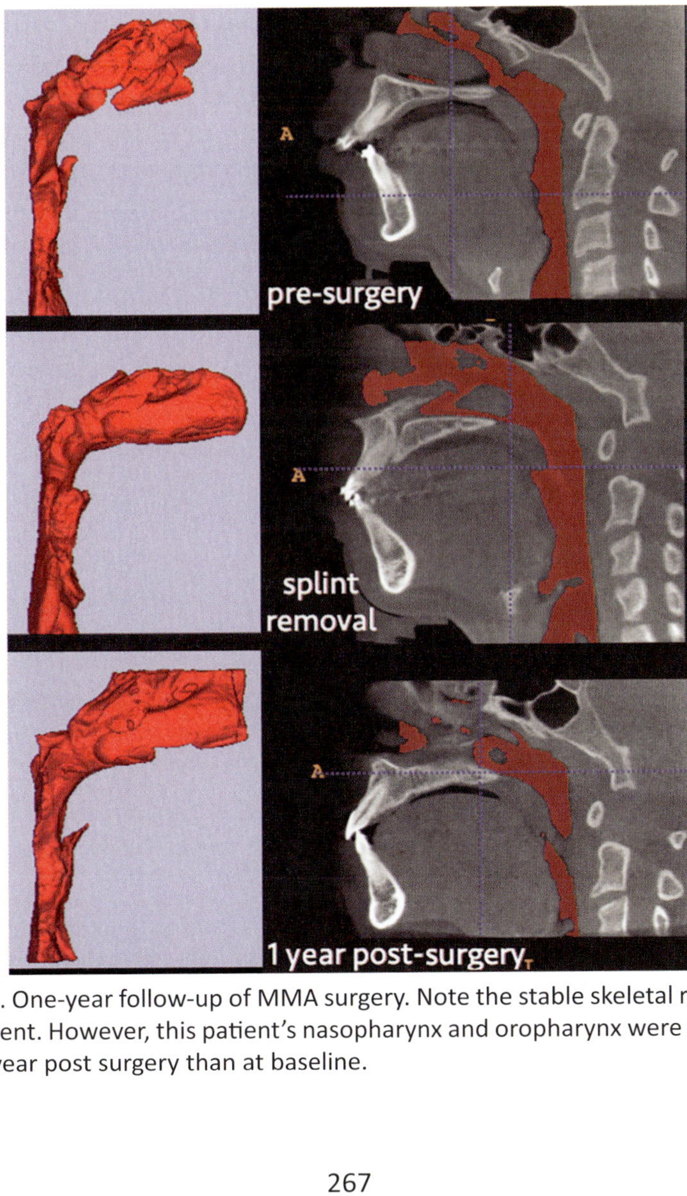

Figure 6. One-year follow-up of MMA surgery. Note the stable skeletal results for this patient. However, this patient's nasopharynx and oropharynx were narrower at one-year post surgery than at baseline.

Other advances also have been brought by magnetic resonance elastography that relies on the principle that the propagation of a vibration wave through tissue depends on the viscoelastic properties of the tissue for tongue and soft palate assessments (Bilston and Gandevia, 2014). As we consider obesity and aging, such multi-dimensional assessments and interpretation of the aging airways need to be considered in our daily clinical practice. Thus, strict cooperative work between radiologists and respiratory physiologists is highly desirable to integrate clinical and functional data together with morphologic and morphometric findings. Future implementation of the analysis of exhaled gases, dynamic airflow measures, specific image analysis methods for muscle assessments, post-processing imaging techniques and genetic analysis hopefully will close the gaps for airway assessments in the care of OSA.

CONCLUSIONS

Three critical concepts need to be considered for proper diagnosis of sleep apnea using 3D images:

1. The dynamics assessments of respiration is modified by time and function;
2. Pharyngeal space boundaries and head posture require standardization; and
3. The patency of the pharynx is affected by the neuromechanical system in breathing.

REFERENCES

Alsufyani NA, Flores-Mir C, Major PW. Three-dimensional segmentation of the upper airway using cone beam CT: A systematic review. Dentomaxillofac Radiol 2012;41(4):276-284.

Andersen IG, Holm JC, Homøe P. Obstructive sleep apnea in obese children and adolescents, treatment methods and outcome of treatment: A systematic review. Int J Pediatr Otorhinolaryngol 2016;87:190-197.

Behlfelt K, Linder-Aronson S, McWilliam J, Neander P, Laage-Hellman J. Cranio-facial morphology in children with and without enlarged tonsils. Eur J Orthod 1990;12(3):233-243.

Bilici S, Engin A, Ozgur Y, Ozlem Onerci C, Ahmet Gorkem Y, Aytul Hande Y. Submental ultrasonographic parameters among patients with obstructive sleep apnea. Otolaryngol Head Neck Surg 2017;156(3):559-566.

Bilston LE, Gandevia SC. Biomechanical properties of the human upper airway and their effect on its behavior during breathing and in obstructive sleep apnea. J Appl Physiol (1985) 2014;116(3):314-324.

Dayal VS, Phillipson EA. Nasal surgery in the management of sleep apnea. Ann Otol Rhinol Laryngol 1985;94(6 Pt 1):550-554.

Friedman M, Wilson M, Lin HC, Chang HW. Updated systematic review of tonsillectomy and adenoidectomy for treatment of pediatric obstructive sleep apnea/hypopnea syndrome. Otolaryngol Head Neck Surg 2009;140(6):800-808.

Guijarro-Martínez R, Swennen GR. Cone-beam computerized tomography imaging and analysis of the upper airway: A systematic review of the literature. Int J Oral Maxillofac Surg 2011;40(11):1227-1237.

Guilleminault C, Li KK, Khramtsov A, Pelayo R, Martinez S. Sleep disordered breathing: Surgical outcomes in prepubertal children. Laryngoscope 2004;114(1):132-137.

Gwynne-Evans E. The so-called adenoid facies. Postgrad Med 1956; 19(3):253-255.

Hatcher DC. Cone beam computed tomography: Craniofacial and airway analysis. Dent Clin North Am 2012;56(2):343-357.

Ingram DG, Friedman NR. Toward adenotonsillectomy in children: A review for the general pediatrician. JAMA Pediatr 2015;169(12):1155-1161.

Kim ST, Choi JH, Jeon HG, Cha HE, Kim DY, Chung YS. Polysomnographic effects of nasal surgery for snoring and obstructive sleep apnea. Acta Otolaryngol 2004;124(3):297-300.

Kluemper GT, Vig PS, Vig KW. Nasorespiratory characteristics and craniofacial morphology. Eur J Orthod 1995;17(6):491-495.

Koinis-Mitchell D, Craig T, Esteban CA, Klein RB. Sleep and allergic disease: A summary of the literature and future directions for research. J Allergy Clin Immunol 2012;130(6):1275-1281.

Linder-Aronson S. Adenoids: Their effect on mode of breathing and nasal airflow and their relationship to characteristics of the facial skeleton and the dentition: A biometric, rhino-manometric and cephalometro-radiographic study on children with and without adenoids. Acta Oto-laryngol Suppl 1970;265:1-132.

Mannarino MR, Di Filippo F, Pirro M. Obstructive sleep apnea syndrome. Eur J Intern Med 2012;23(7):586-593.

McNamara JA Jr. Influence of respiratory pattern on craniofacial growth. Angle Orthod 1981;51(4):269-300.

Motta AT, Cevidanes LH, Carvalho FA, Almeida MA, Phillips C. Three-dimensional regional displacements after mandibular advancement surgery: One year of follow-up. J Oral Maxillofac Surg 2011;69(5): 1447-1157.

Nguyen T, Phillips C, Paniagua B. The use of SPHARM-PDM and mean latitude axis to evaluate airway changes. Angle Orthod 2016;86(6): 943-948.

Occhipinti M, Larici AR, Bonomo L, Incalzi RA. Aging airways: Between normal and disease: A multidimensional diagnostic approach by combining clinical, functional, and imaging data. Aging Dis 2017; 8(4):471-485.

Pang KP, Tan R, Puraviappan P, Terris DJ. Anterior palatoplasty for the treatment of OSA: Three-year results. Otolaryngol Head Neck Surg 2009;141(2):253-256.

Peltomäki T. The effect of mode of breathing on craniofacial growth: Re-visited. Eur J Orthod 2007;29(5):426-429.

Pirelli P, Saponara M, Guilleminault C. Rapid maxillary expansion (RME) for pediatric obstructive sleep apnea: A 12-year follow-up. Sleep Med 2015;16(8):933-935.

Schendel SA, Jacobson R, Khalessi S. Airway growth and development: A computerized 3-dimensional analysis. J Oral Maxillofac Surg 2012; 70(9):2174-2183.

Solow B, Siersbæk-Nielsen S, Greve E. Airway adequacy, head posture, and craniofacial morphology, Am J Orthod 1984;86(3);214-223.

Spilsbury JC, Storfer-Isser A, Rosen CL, Redline S. Remission and incidence of obstructive sleep apnea from middle childhood to late adolescence. Sleep 2015;38(1):23-29.

Stradling JR, Thomas G, Warley AR, Williams P, Freeland A. Effect of adenotonsillectomy on nocturnal hypoxaemia, sleep disturbance, and symptoms in snoring children. Lancet 1990;335(8684):249-253.

Subtelny JD. The significance of adenoid tissue in orthodontia. Angle Orthod 1954;24(2):59-69.

Tan SK, Leung WK, Tang ATH, Zwahlen RA. How does mandibular advancement with or without maxillary procedures affect pharyngeal airways? An overview of systematic reviews. PLoS One 2017;12(7):e0181146.

Todd TW. Integral growth of the face: Nasal area. Int J Orthod Oral Surg 1936;22(4);321-334.

Van Holsbeke CS, Verhulst SL, Vos WG, De Backer JW, Vinchurkar SC, Verdonck PR, van Doorn JW, Nadjmi N, De Backer WA. Change in upper airway geometry between upright and supine position during tidal nasal breathing. J Aerosol Med Pulm Drug Deliv 2014;27(1):51-57.

van Vlijmen OJ, Kuijpers MA, Bergé SJ, Schols JG, Maal TJ, Breuning H, Kuijpers-Jagtman AM. Evidence supporting the use of cone-beam computed tomography in orthodontics. J Am Dent Assoc 2012; 143(3):241-252.

Vig KWL. Nasal obstruction and facial growth: The strength of evidence for clinical assumptions. Am J Orthod Dentofacial Orthop 1998; 113(6):603-611.

Yu J, Zhou Z, McEvoy RD, Anderson CS, Rodgers A, Perkovic V, Neal B. Association of positive airway pressure with cardiovascular events and death in adults with sleep apnea: A systematic review and meta-analysis. JAMA 2017;318(2):156-166.

Zimmerman JN, Lee J, Pliska BT. Reliability of upper pharyngeal airway assessment using dental CBCT: A systematic review. Eur J Orthod 2017;39(5):489-496.

CHARACTERIZATION OF THE UPPER AIRWAY STARTING FROM CBCT DATA

Paolo M. Cattaneo, Gabriele Di Carlo, Sirwan Fernandez Gurani

ABSTRACT

The introduction of cone-beam computed tomography (CBCT) and the increased awareness on the detrimental effects generated by sleep-disordered breathing (SDB) has revived interest in studying the upper airway (UA). Nevertheless, despite the high number of publications studying UA based on CBCT scans, there is no standard procedure shared by all researchers in assessing UA morphology. In this chapter, an overview of the challenges and limitations related to the study of the upper airway will be presented.

KEY WORDS: upper airway, CBCT, imaging, three-dimensional

INTRODUCTION

Upper Airway Morphology

The upper airway (UA) is a complex three-dimensional (3D) structure that involves skeletal, cartilaginous and soft tissues that are adapted to functions involving respiration, deglutition and phonation. It is part of the airway, which stretches from the nose or mouth to the alveoli of the lungs, a conduit that includes the nose, mouth, trachea, pharynx, bronchi, bronchioles and alveoli. The UA is divided anatomically into four regions: the area from the nasal turbinates to the hard palate (the nasopharyngeal region); the retropalatal (velopharyngeal) region; the retroglottal (oropharyngeal) region; and finally the hypopharyngeal region (located between the base of the tongue and the larynx; Schwab, 1998; McCrillis et al., 2009).

The UA is the part of the total airway that is analyzed most frequently in orthodontics and maxillofacial surgery, as it often is imaged by common radiographic investigations (e.g., lateral cephalograms and, more recently, cone-beam computed tomography [CBCT]) in its entirety as part of routine diagnosis and treatment planning.

Sleep-disordered Breathing (SDB)

According to the American Thoracic Society, sleep-disordered breathing (SDB) is an umbrella term for several chronic conditions in which partial or complete cessation of breathing occurs many times while the patient is asleep. The European Respiratory Society Task Force has defined obstructive SDB as "a syndrome of UA dysfunction during sleep, characterized by snoring and/or increased respiratory effort secondary to increased UA resistance and pharyngeal collapsibility" (Kaditis et al., 2016). The impact on health caused by SDB, particularly in its most severe form—obstructive sleep apnea (OSA)—increasingly has been recognized in both adults and children (Marcus, 2001; Casale et al., 2009). OSA is characterized by an anatomically small pharyngeal airway, hypoxemia (abnormally low concentration of oxygen in the blood), hypercapnia (abnormally elevated carbon dioxide [CO_2] levels) and obstructive hypoventilation (White, 2001). OSA also can lead to cortical arousals and sleep fragmentation.

SDB is believed to have a large impact on the general population, as the prevalence of SDB in adults in the United States has been estimated to be 4% for men and 2% for women (Young et al., 1993). Yet, these estimates can vary widely, as they are dependent of the methodology used in assessing the condition. In the general pediatric population, the prevalence of SDB has been reported to be as low as 1% and up to 25% for primary snoring; thus, prevalence is dependent on the geographical area of the assessment and the methodology used. Some authors reported that more than 11% of children are habitual snorers (Archbold et al., 2002; Schlaud et al., 2004; Bixler et al., 2009). Surveys on prevalence of childhood snoring in several countries at different age intervals have shown the following for SDB:

- Hong Kong: a range 0.7% to 13% has been reported (Ng et al., 2002);
- Thailand: 0.7% (age range = 6 to 13 years; Anuntaseree et al., 2001);
- Greece: 4.3% (age range = 1 to 18 years; Kaditis et al., 2004);

- United States: 5.7% (age range = 5 to 7 years; O'Brien et al., 2003); and
- Italy: 13% (age range = 3 to 6 years; Castronovo et al., 2003).

Consequently, the need for standardization of methodology and age range across studies is obvious (Lumeng and Chervin, 2008).

Characterization of UA Morphology

The UA from the posterior end of nasal septum to the epiglottis lacks a fixed rigid structural support; thus, the shape and size of the airway are dependent strongly on the position of the soft tissue structures (e.g., soft palate, tongue and walls of the oropharynx). Consequently, UA morphology can be influenced by gravity; when patients are in a supine position, the tongue and soft palate tend to move posteriorly, thus reducing the oropharyngeal area (Ayappa and Rapoport, 2003; Gurani et al., 2016).

The patency of the pharynx also depends on the activity of pharyngeal dilator muscles that help control the UA lumen dimension. A decreased activity of these muscle will lead to a reduction in the size of the UA lumen. In individuals with an anatomically large pharynx, a decrease in the activity of the dilator muscle most likely will lead to small increment in pharyngeal airway resistance; however, in individuals characterized by a pharynx with small anteroposterior and transverse dimensions, the same decrease in the activity of these muscles may lead to a large increment in pharyngeal airway resistance. Therefore, indirectly the anatomy of the pharynx seems to play a role in UA patency (White, 2005).

In the 1970s, the study of UA morphology was central among orthodontists. With a multi-disciplinary approach, the relationship between growth and UA was studied from various points of view (e.g., neuromuscular adaptations, nasopharyngeal obstruction, growth, breathing and speech; McNamara, 1981). Among the many researches who interested themselves in this topic, Linder-Aronson played an important role in the study of the UA. Based on his studies that demonstrated positive changes in growth after adenoidectomy (Linder-Aronson, 1970; Linder-Aronson et al., 1986), the recommendation of tonsillectomy and adenoidectomy to improve facial growth and mode of breathing was followed broadly. Although of limited clinical relevance, this research had a high impact on the orthodontic community. However, Fields and colleagues (1991) recognized that the relationship between oronasal breathing and growth in the long face pattern patients was not elucidated fully.

More recently, the possibility that craniofacial disharmony might be an important predisposing factor in the development and progression of SDB in children was suggested in a study performed in non-syndromic children (Katyal et al., 2013). On the other hand, in patients without signs of SDB, no correlations between UA and craniofacial morphology were found (Di Carlo et al., 2015).

There are several reasons behind these incongruences; however, the methodology adopted in measuring UA seems to play an important role. Warren and Spalding (1990) stated that the relationship between nasorespiratory function and dentofacial development was controversial, mostly because this relationship was based on two-dimensional (2D) measurements, typically performed only on lateral and frontal cephalograms. The 2D approach might overlook most of the anatomical information necessary to make a proper evaluation of the UA.

Indeed, the three-dimensional (3D) complex anatomical structure of the UA cannot be depicted by a 2D approach: 2D headfilms images—besides being influenced by artifacts (e.g., distortion, differences in magnification and superimposition of bilateral craniofacial structures)—do not render information on cross-sectional area and volume (Baumrind and Frantz, 1971*a,b;* Ahlqvist et al., 1986). The validity of examining the anatomical features of UA and nasopharyngeal obstruction by using lateral cephalograms also was questioned by Vig and Hall (1980), while Guilleminault and colleagues (1984) did not find a direct correlation between the posterior airway space measured on lateral cephalograms and the severity of OSA.

Computed tomography (CT) and magnetic resonance imaging (MRI) allowed clinicians and researchers to overcome the limitations of 2D investigations, as they are able to depict the actual 3D morphology of the UA, though, their use is limited by high irradiation (CT), movement artifacts (MRI), techniques both expensive and with restricted accessibility (Schwab, 1998).

CBCT and UA Measurements

CBCT was introduced into the dental community in 1998 (Mozzo et al., 1998). Following its introduction and thanks to adequate availability due to a more affordable cost compared to CT and MRI, easier acquisition protocols—together with the relative low ionizing radiation to the patients (Cattaneo and Melsen, 2008; Noar and Pabari, 2013; Kiljunen et al., 2015; Ludlow et al., 2015)—interest in studies of UA has surged again.

The increased interest in UA was evident particularly within the orthodontic literature, due to the fact that for orthodontic treatment planning the area imaged often encompasses the UA. The reason that interest in studying UA has surged is related to the fact that using CBCT images allows for a quantitative evaluation of both cross-sectional areas and volumes of UA. Aboudara and coworkers (2009) stated that "CBCT is a simple and effective method to evaluate UA."

The above statement has been corroborated by two systematic reviews. In the first, van Vlijmen and coworkers (2012) concluded that the use of CBCT represents a more valuable diagnostic tool to study UA than conventional plane radiography. In the second, Kapila and Nervina (2015) suggested that volume or the cross-sectional area might be a better method of identifying UA constriction, thus indicating that CBCT would represent a better examination than conventional images.

In the last decade, CBCT has gained considerable acclaim in connection with orthognathic surgery as a viable 3D imaging modality, as a replacement of medical CT. This interest increased even further in combination with the fact that 3D virtual orthognathic surgery planning incorporating CAD/CAM technology has gained acceptance. Indeed, CBCT has become the first choice to gather the necessary information for the planning of the complex movement associated with orthognathic surgery.

The study of the effects of orthognathic surgery on UA patency has gained interest, as corrective jaw surgery possibly can alter the reciprocal position of bone structures and soft tissues (Scarfe and Farman, 2008; de Souza Carvalho et al., 2012; Noar and Pabari, 2013; Gokce et al., 2014; Christovam et al., 2016). Even though there is no clear relationship between the possible increase of SDB following anatomical modification of UA morphology subsequent to orthognathic surgical interventions, it has been suggested that SDB can arise in patients with predisposing risk factors (Susarla et al., 2010).

CHALLENGES IN MEASURING THE UPPER AIRWAY

Since the introduction of CBCT, many papers studying the upper airway have been published (a simple search on Web of Knowledge performed on June 1, 2017 using "CBCT" and "Airway*" limited to the fields of "Dentistry Oral Surgery Medicine," "Otorhinolaryngology," "Respiratory Problem," resulted in 133 hits). The quality in terms of methodology, standardization and risk of bias of the published literature, however, seems to be far from optimal (Guijarro-Martinez and Swennen, 2011; Alsufyani et al., 2012; Zimmerman et al., 2016). Therefore, different challenges that can be encountered when measuring UA on CBCT scans will be underlined and analyzed throughout this chapter.

Hounsfield Unit (HU) and Grey Scale in CBCT Scans

In medical CTs, the radiographic density (i.e., the x-ray attenuation), which is represented by each voxel in a dataset, is expressed by the CT number. The density then is expressed using a predefined scale that converts the CT number, which is specific to each CT scanner type. For medical CTs, the Hounsfield Unit (HU) scale is used, a value that is calculated from the linear attenuation coefficient value of each tissue and calibrated using the reference of the linear attenuation coefficient of water and air (Hounsfield, 1980). The HU scale—though not fully accurate as the linear attenuation coefficient of a material depends on the x-ray energy (Table I; Hubbel and Seltzer, 1996)—was introduced in order to standardize the reading of density among the different medical CT units.

For CBCT-generated datasets, calculations are slightly different. CBCTs are generated images (datasets) that at first glance seem similar to those produced with medical CTs; however, contrary to the images generated from medical CT, the CBCT-generated images are not density calibrated (Lagravère et al., 2008*b*; Molteni, 2013). The reason behind the discrepancy of the images produced by medical CT and CBCT scanners is related to the large difference in both data acquisition technology, as well as to the different x-ray generators and detectors used (Molteni, 2013);

Table 1. Table of x-ray attenuation coefficients for various tissue. Adapted from the values reported on the NIST homepage.

MATERIAL	PHOTON ENERGY LEVEL			
	40 KeV	60 KeV	80 KeV	100 KeV
Fat	0.228	0.188	0.171	0.160
Water	0.268	0.206	0.184	0.171
Cortical bone	1.280	0.604	0.428	0.356

medical CT scanners use high output and high current x-ray generator and a stack of linear detectors, while in CBCT scanners, the x-ray generators are similar to the ones used in conventional panoramic x-ray systems and the detector typically is a flat panel.

The fact that CBCT images are not calibrated for density constitutes one of the big challenges when dealing with CBCT scans. The reliability and consistency of determining the absolute HU and, therefore, the actual density of each voxel in CBCT-generated datasets are equivocal; many studies assessing the values of bone density in relation to HUs (or better grey values or pseudo-HU) did not provide evidence of a reliable relationship between them.

Therefore, the grey values provided in CBCT scans must be taken with caution and cannot be used as absolute (Hua et al., 2009; Hohlweg-Majert et al., 2011; Nackaerts et al., 2011). Moreover, the reading of the grey value for each specific voxel retrieved from CBCT-generated datasets should not be used quantitatively (Molteni, 2013; Pauwels et al., 2013). Furthermore, it has been shown that even between two consecutive scans taken with the same CBCT unit, large variations in the voxel grey values can occur (Spin-Neto et al., 2015).

As an overall consequence, even though some authors have proposed methods for converting the grey values obtained with CBCT scanners into HU values (Lagravère et al., 2008*a*; Reeves et al., 2012), the proposed conversions are far from precise and are based on unrealistic assumptions, which make them practically unusable in clinical settings (Molteni, 2013).

Thresholding and Segmentation of the Upper Airway

Because CBCT datasets are not density calibrated, it is not possible to use a fixed pseudo-HU level to segment the UA. This issue was described in a study by Yamashina and colleagues (2008), who scanned both a phantom (simulating the UA) and a real patient to evaluate the reliability of CBCT values for measuring the oropharynx in comparison with the results from a medical CT scanner. In this study, the investigators concluded that the grey values produced by the CBCT scanner to represent the UA were different from the HU units obtained from the medical CT. Moreover, confirming what has been reported for the full range of grey values (Spin-Neto et al., 2015), the standard variation of the CBCT-produced values representing the air range was much larger when compared to the CT data.

Moreover, as explained above, the grey-level scales generated by the same CBCT unit are not consistent from one scan to the succeeding one (Spin-Neto et al., 2015). Thus, even in this setting, it is not possible to determine a unique threshold value that can be used to segment the UA in a series of CBCT scans taken from one CBCT scanner.

Unfortunately, there are several articles in the literature in which either a unique gray value for a series of CBCT scans or the use of a value generated automatically by the software were used to segment the UA and, based on the so-produced segmentation, the volume measurements were performed. Some authors have suggested correctly that instead of using a unique thresholding value or a value automatically generated by software as part of manual segmentation, a more accurate segmentation of the UA and the segmentation is under control of the operator (El and Palomo, 2010, 2014; Lenza et al., 2010; Di Carlo et al., 2015).

Despite the improvement in the available software, manual segmentation still is time consuming and errors can be generated if manual segmentation is not applied correctly. Moreover, the possibility of altering the threshold value manually typically is available only in research-oriented software, while most clinical software does not have this capability. These two factors might explain why the automatic segmentation approach represents the choice made by most of the researchers studying UA.

The main drawbacks of choosing automatic segmentation (i.e., letting the software deciding the threshold value) is poor accuracy, yet the process sometimes is characterized by high correlation (i.e., the software is consistent with itself), thus introducing systematic errors in the measurements of UA (El and Palomo, 2010). On the other hand, investigators have shown that by using manual segmentation, a high reliability is achieved (El and Palomo, 2010, 2014; Lenza et al., 2010; Di Carlo et al., 2015).

To achieve high reliability in the manual segmentation approach, it is of utmost importance to follow the right technique and protocol to find the appropriate threshold value. Moreover, as the grey value scale in various CBCT scans is not constant, it is important to determine the correct threshold value to segment the UA in each specific CBCT scan. The method for choosing the right threshold value is critical; if the methodology is not described properly, a single operator may make a subjective decision, thus affecting measurement accuracy and reliability (El and Palomo, 2010; Weissheimer et al., 2012).

One of the possible approaches to determine the appropriate threshold consists of drawing an intensity profile line passing through the UA. The profile line is defined at the set of grey or HU values taken from regularly spaced points along a line segment. A plot of the grey or HU values along the drawn line segment, thus, is generated.

In Figure 1, a profile line has been traced along the midsagittal image through the UA starting from basion to the posterior nasal spine. On the generated histogram, it is possible to follow the variation of the grey values along the line (Fig. 2). The user can draw a series of profile lines throughout the length of the UA, so that the correct threshold value can be selected and the segmentation of the UA can be accomplished. This approach has been followed in a series of research projects with a good intra- and inter-operator reliability (Aboudara et al., 2009; El and Palomo, 2010; Lenza et al., 2010; Di Carlo et al., 2015).

Looking at the profile line drawn along the UA (Figs. 3-4), it can be seen that the grey values along the line are not constant, but that some variation with a trend to decrease from the nasopharynx to the oropharynx exists. Based on the profile line(s), a mask can be created (Fig. 5) and the corresponding 3D segmented volume can be generated and depicted (Fig. 6). To take into account this variation, a dynamic threshold approach can be used; with this approach, the threshold value is not unique, but follows the variations seen on the CBCT images.

In other words, with a dynamic threshold approach, the segmentation of an object is based on the connectivity of grey values in a certain

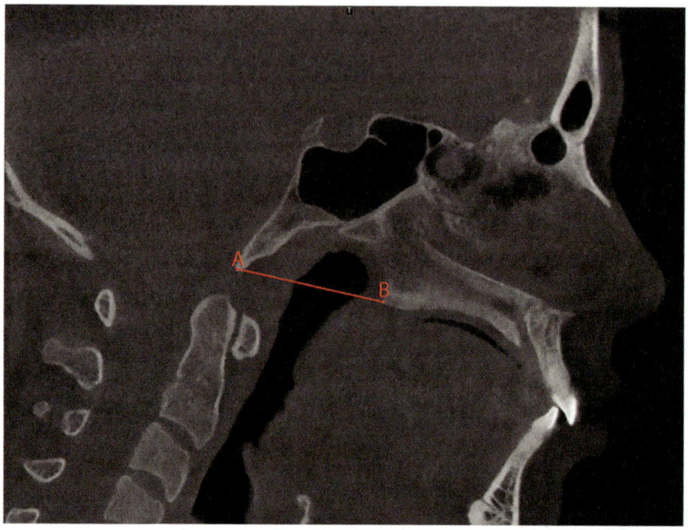

Figure 1. Profile line drawn from A (basion) to B (posterior nasal spine).

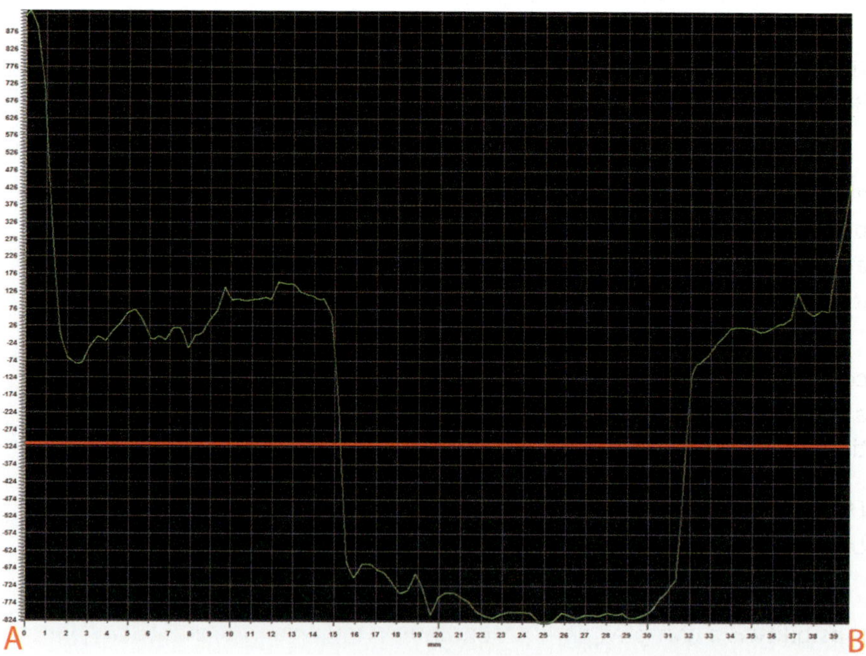

Figure 2. Histogram showing the grey values (green curve) along the profile line drawn in Figure 1 (from A to B). On the x-axis, the distance between A and B is represented and the grey values are reported. The red line represents the threshold chosen level to segment the UA.

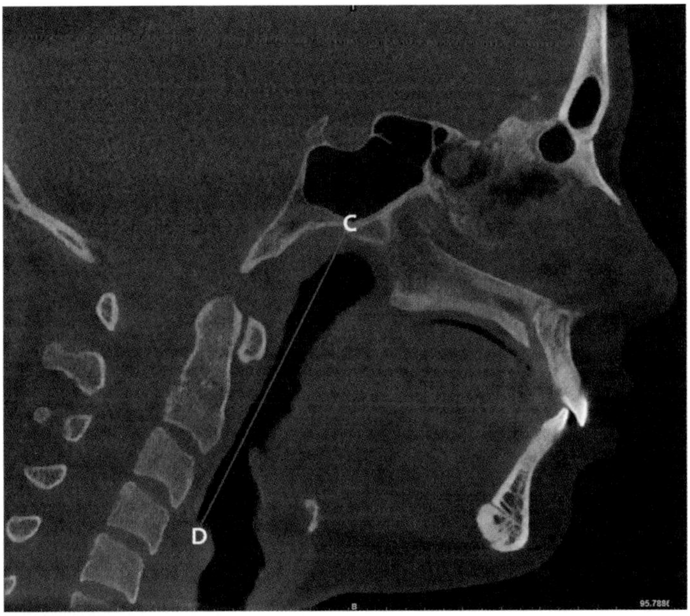

Figure 3. Profile line drawn from along the UA from C (nasopharynx level) to D (oropharynx level).

grey value range: an algorithm is comparing the grey values in the neighboring voxels, following the maximal deviation set by looking at the variation seen on the profile line. This approach rarely is chosen, however, as it requires an advanced software program and is more time consuming.

Note: All the images used here to illustrate the thresholding procedure are obtained with a large field of view scan (16 x 18 cm) taken with a NewTom 5G scanner, with 110 KeV, mA of 2.57-5.20 and an exposure time of 3.6 sec).

UA Patency: The Influence of Morphology (Lumen Dimension and Shape, UA Length), Patient Position and Respiration Phase

Morphology. The patency of the UA is multi-faceted and its collapsibility is a complex phenomenon, in which many predisposing factors can play a role (Abramson et al., 2010). The collapse of the UA happens only when the UA is not supported by rigid structures. The physical behavior of UA has been compared to a Starling resistor, where the collapsible segment is represented by the pharyngeal tract (Gold and Schwartz,

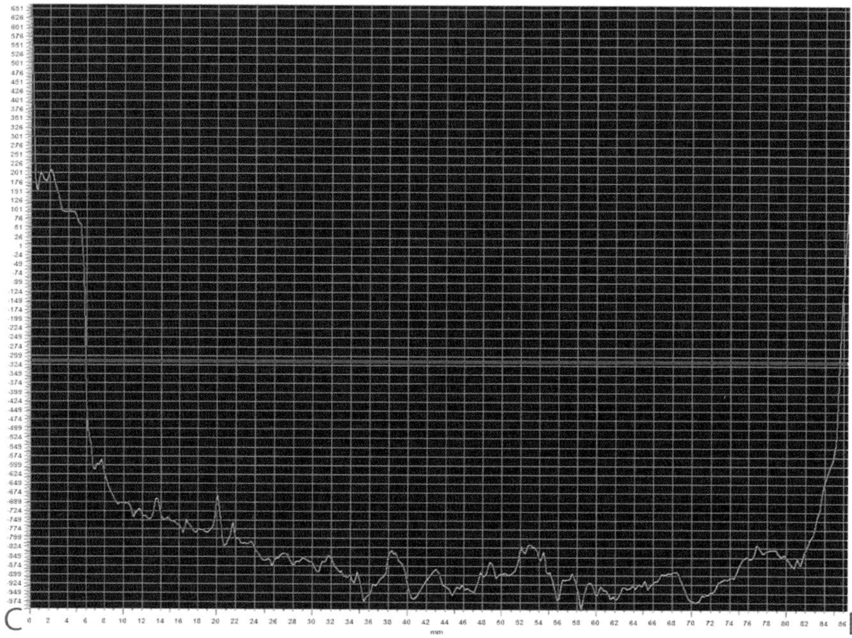

Figure 4. Histogram showing the grey values (green curve) along the profile line drawn in Figure 3 (from C to D). On the X-axis, the distance between A and B is represented; on the Y-axis, the grey values are reported. It is worth noting that the grey values representing air in the middle of the curve (so, in theory, displaying the same grey level) show a trend to decrease from C to D. The blue line represents the chosen threshold level to segment the UA.

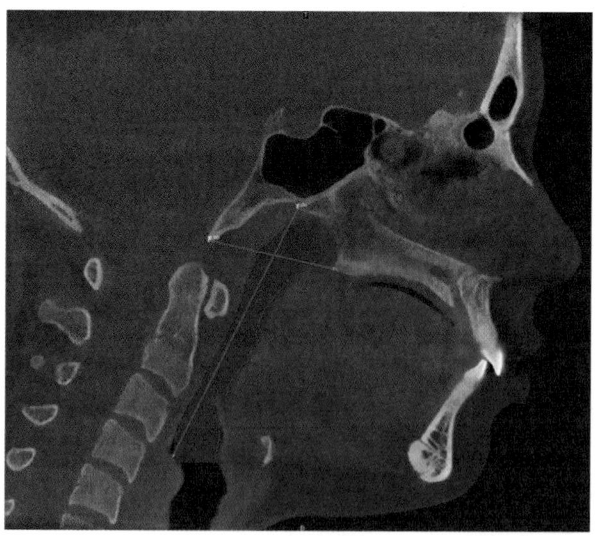

Figure 5. Profile lines as in Figures 1 and 3 and the segmented UA (in red).

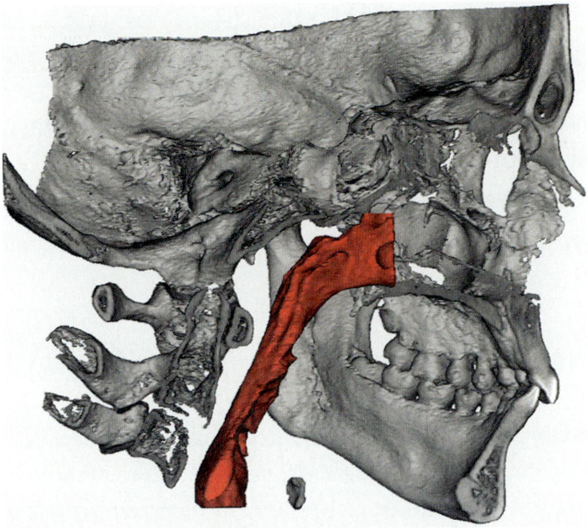
Figure 6. Rendering of the UA.

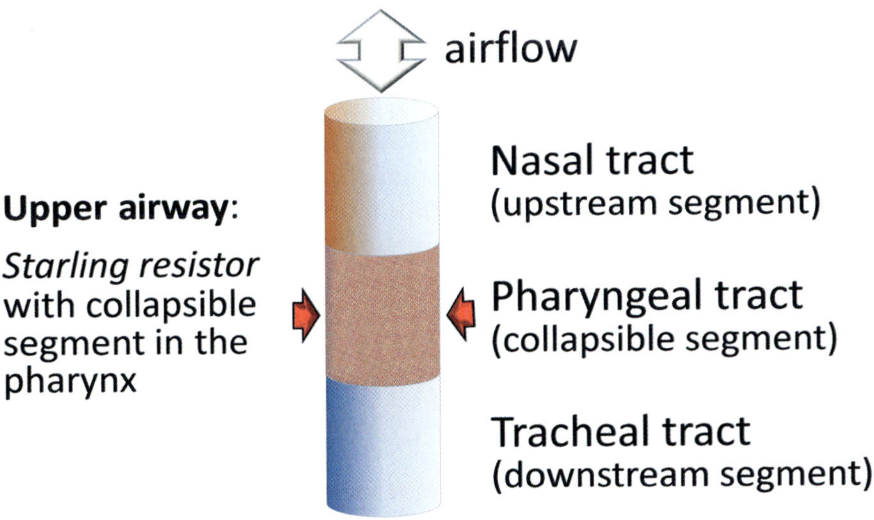

Figure 7. Representation of the UA as a Starling resistor.

1996; Fig. 7). The pharyngeal tract can collapse when the balance between the internal and external pressure become negative (e.g., the transmural pressure falls below zero); when the intra-airway pressure decreases to less than that of the external pressure, the pharyngeal tract is prone to collapse (Schwartz and Smith, 2013).

The factors contributing to the collapse of the UA can be divided into: factors that decrease the intraluminal pressures; factors that increase the external pressure (e.g., obesity and sleeping position); and factors that decrease the resistance to collapse of the pharyngeal wall (e.g., decrease in dilator muscle activity; Isono et al., 1993).

The resistance to flow (R) for a fluid can be determined using Ohm's law:

$$R = \frac{\Delta P}{V}$$

where "ΔP" is the difference in pressure (also known as driving pressure) and "V" is the flow.

According to Poiseuille's law (this law is valid for an incompressible fluid; however, as small ΔP are present in the normal breathing cycle, this approximation also can be accepted for airflow) for a round duct, "R" is calculated following the equation:

$$R = \frac{8nl}{\pi r^4}$$

where "n" is the viscosity of the fluid, "l" is the length of the duct and "r" is radius of the duct.

It is worth noting that the radius appears in the denominator at the fourth power, so that even a small variation of radius has a big influence on the resistance. Thus, it is not surprising that the dimension of the UA lumen is important for predicting OSA and that the successful treatment of OSA depends on the amelioration of the lumen dimension (Ogawa et al., 2007; Haskell et al., 2009). The size of the lumen is not the only critical factor, however; the shape of the lumen also plays an important role (Conley et al., 2014). Indeed, the R is influenced greatly by the shape; in most of the pharyngeal tract, the shape of the lumen does not resemble a circle, but can assume various forms (Fig. 8).

To calculate the flow in a non-circular duct, the hydraulic diameter often is used, especially in case of turbulent flow. Figure 9 illustrates the importance of shape: the round and the rectangular shapes both are characterized by the same hydraulic diameter (i.e., they allow the same flow, given the same boundary conditions); however, note that the cross section of the rectangular shape is more than 2x the section of the round shape.

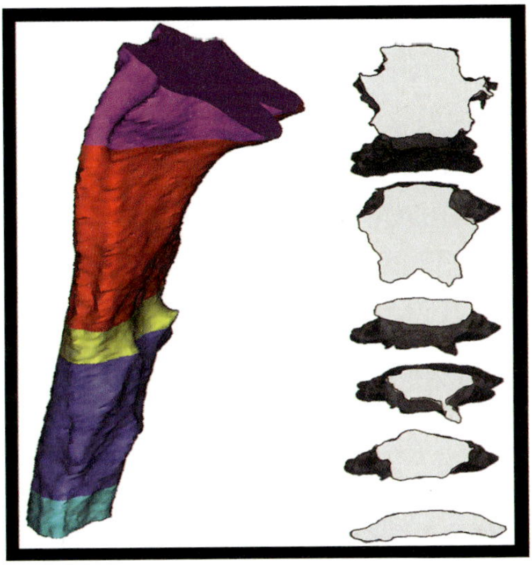

Figure 8. UA lumen dimension and shape at different levels.

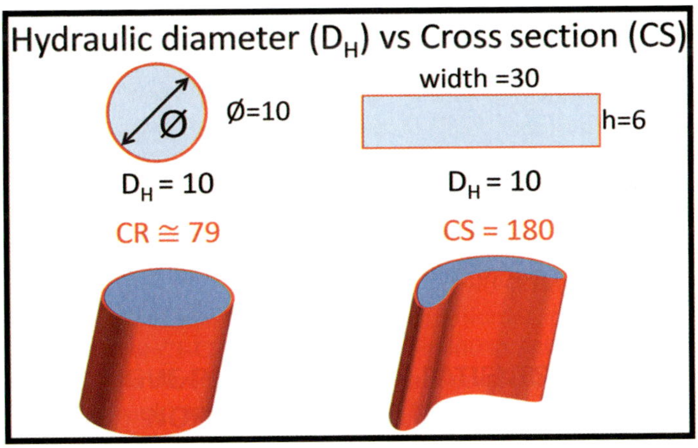

Figure 9. The hydraulic diameters for the round and the rectangular shapes are the same; however, the corresponding cross sections areas are different (more than 2x for the rectangular shape). This fact plays a critical role in the airflow rate, especially as UA configuration can change from a tube to an elliptical shape.

When looking at the Poiseuille's law, it is important to note that the length of the duct plays a role, though minor in comparison with the radius. Thus, it is not surprising that the airway length was associated positively and significantly with an increased respiratory disturbance index (Abramson et al., 2010) and that a significant correlation was reported between the length of UA and the severity of the OSA (Segal et al., 2008).

Patient Position, Head Posture and Tongue Position During Imaging. As mentioned earlier, the pharyngeal part of the UA from the posterior end of nasal septum to the epiglottis lacks a fixed rigid structural support. Thus, the shape and size of the pharynx are dependent on the position of the soft tissue structures including the soft palate, tongue and walls of the oropharynx (Ayappa and Rapoport, 2003; Gurani et al., 2016). Hence, the morphology of UA can be influenced by gravity (Pae et al., 1994; Miyamoto et al., 1997; Battagel et al., 2002; Ingman et al., 2004; Sutthiprapaporn et al., 2008) with the tongue and soft palate moving posteriorly when the patient is in a supine position, thus reducing the oropharyngeal area.

In an older study from Safar and associates (1959), it was reported that closure or altered size of the UA has been observed in anesthetized but spontaneously breathing patients, following neck flexion in both the supine and prone positions. This finding was confirmed by Hellsing's study (1989), during which the patients were imaged twice: the first radiograph was taken with a natural posture; the second with a voluntary head extension of 20°, both with the patient standing and not anesthetized.

The influence of the patient's position on UA patency and thus on respiration is reflected by the fact that on average 56% of patients diagnosed with OSA display a position-dependent OSA, which is defined as a difference of 50% or more in the apnea index between the supine and non-supine positions (Ravesloot et al., 2013). This finding has influenced the present therapy of OSA, where the position of the patient plays a role (Deacon et al., 2016).

Head posture and patient's position during 3D imaging of the UA (including CT, MRI and CBCT scans) seem to play a role in the dimension of the UA (Ono et al., 2000; Zhang et al., 2011; Pirila-Parkkinen et al., 2012). However, Gurani and associates' systematic review (2016) pointed out that few studies are taking these parameters into consideration, while an earlier one by Guijarro-Martinez and Swennen (2011) concluded that the influence of tongue position constitutes an obstacle to reliably measure the UA. Though there are studies underlining the importance of standardizing head posture during vertical 3D image acquisition (Lenza et al., 2010; Guijarro-Martinez and Swennen, 2011; Di Carlo et al., 2015), overall quality of the studies looking at head posture was limited and with a low level of evidence.

Zimmermann and coworkers (2016) noted that the majority of studies on UA based on CBCT were performed with the patients imaged in a vertical or seated position, thus introducing an extra bias to the reported measurements.

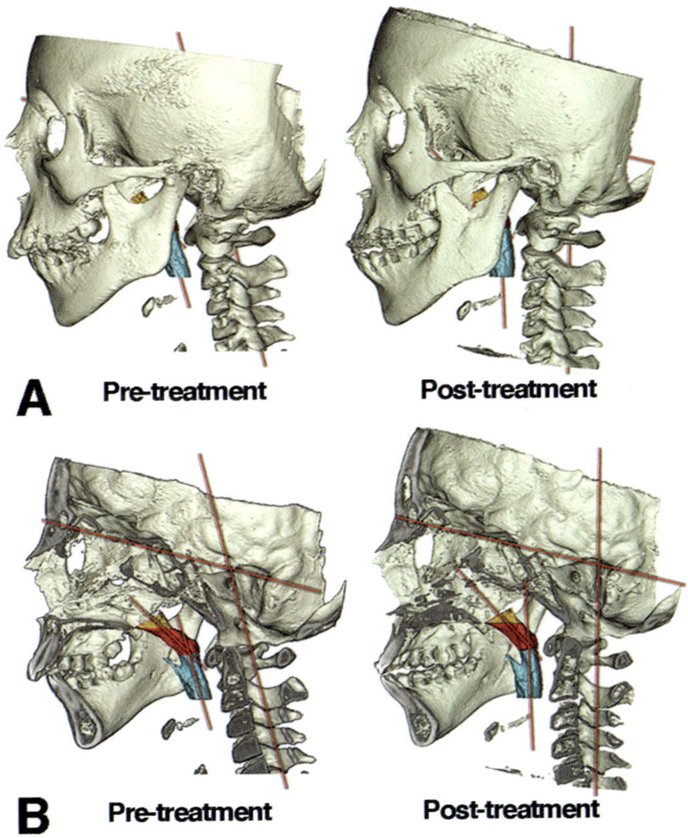

Figure 10. Functional appliance therapy. Airway morphology and position are known to be affected by head posture, as previously demonstrated in 2D cephalometric studies. Pre-treatment (*A*) and post-treatment (*B*) 3D segmented models of the skull and upper airway from CBCT scans after functional appliance therapy are depicted. It is striking to see how the morphology of the UA is related closely to head posture (*B*).

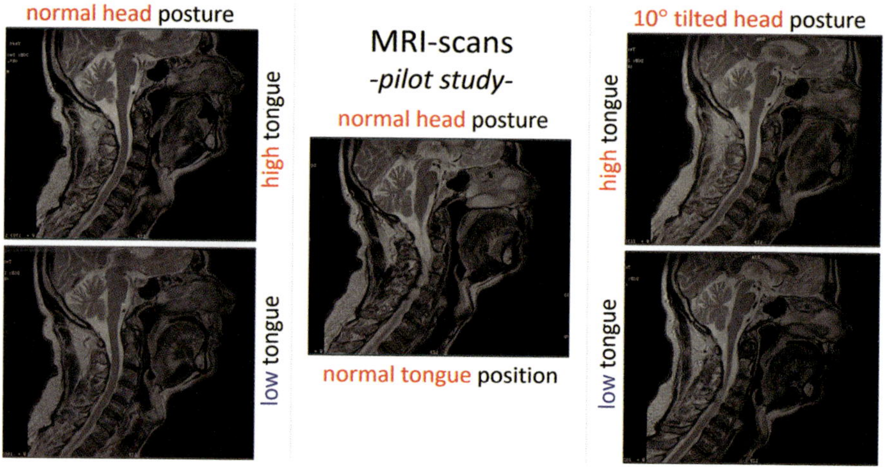

Figure 11. MRI scan of a healthy volunteer from a pilot study. In the central image, the subject was positioned with the head in a neutral posture and the tongue in the normal position. In the images on the left, the head was kept in the same neutral posture, but the subject was asked to position the tip of the tongue on the palate (*top*) and at the bottom of the mouth (*bottom*). In the images on the right, the head of the patient was tilted 10° forward and the tongue was in the same two positions as described above.

The influence of head posture and tongue position can be seen in Figures 10 and 11. In Figure 10, the images were taken with a CBCT scanner (NewTom 5G, QR srl, Italy; scanning protocol: 110 KeV, mA 2.57-5.20, exposure time = 3.6 seconds). In Figure 11, the images were taken with a MRI scanner (General Electric Discovery MR750w 3.0T). These images are part of a pilot study (unpublished data) in which healthy subjects were positioned with the head in two different positions (normal position and tilted 10° forward) and were asked to move the tongue from the normal position to either the bottom of the mouth or with the tip touching the palate.

Respiration Phase. Radiographic measurements have demonstrated that when patients are awake, the patency of the UA is maintained well even when the patient assumes different postures. During sleep, however, the airway lumen tends to decrease (Safar et al., 1959; Trudo et al., 1998). Therefore, even in studies where the patients were 3D imaged in a supine position, the UA measurements might be biased as they suffer from the differences in UA morphology existing between the sleep and awake status (Ayappa and Rapoport, 2003; Abramson et al., 2010).

Another issue, not well documented, is related to the phase of respiration during image acquisition (Haponik et al., 1983; Lowe et al., 1986; Bhattacharyya et al., 2000; Abbott et al., 2004). A typical large field of view CBCT scan takes about 18 seconds (i-Cat, Imaging Sciences International, PA; NewTom 5G, QR srl, Verona, Italy). Therefore, even when the patients are instructed to avoid head movement and swallowing, as well as to breathe quietly through the nose, some movement still is present during scanning (Spin-Neto and Wenzel, 2016). Furthermore, the position of the epiglottis and tongue might not be consistent from one acquisition to the next, also in relation to the head position of the patient; these differences in position may have an impact when assessing the UA volumes and dimensions in conjunction with a treatment outcome analysis of a specific intervention (Gurani et al., 2016). From the same systematic review, it was noted that valid information about the effect of an altered head posture from natural head position during 3D imaging on UA dimensions and morphology is lacking in the current published literature.

Structure of Reference, Division of the UA

Orthodontic Patients. Some experts have suggested the division of the UA in three distinct portions: nasopharynx, oropharynx and hypopharynx (Hudgel, 1992; Schwab et al., 1995; Schwab, 1998). Still, no consensus exists about the determination of the boundaries of the total UA in the literature and a clear and standardized definition of these boundaries is missing (Guijarro-Martinez and Swennen, 2011; Zimmerman et al., 2016).

The same applies to the definition of the sub-regions in UA: different definitions for the UA sub-region's boundaries have been proposed, thus rendering the comparison among the results of these studies difficult (El and Palomo, 2010; Lenza et al., 2010; Hong et al., 2011; Ghoneima and Kula, 2013; Guijarro-Martinez and Swennen, 2013). In particular, Guijarro-Martinez and Swennen (2011) depicted this issue well; Table 8 in their article documents how the boundary definitions of the airway from the nasal and oral cavities to the larynx are extremely variable.

To select and divide the total UA in sub-regions in 3D studies, landmarks on lateral cephalograms are located on the cranial base and cervical spine, following what first was proposed by Linder-Aronson (1970), who considered them as the most reliable points. Concerns were raised about the choice of landmarks on the cervical spine, however, as these are affected significantly by head posture (Muto et al., 2002; Goncalves et al., 2006; Mattos et al., 2011). Other investigators, instead of using these internal landmarks, used planes parallel to the horizontal plane of the CBCT datasets, thus relying on the preliminary orientation of the images (Zimmerman et al., 2016).

Orthognathic Surgery Patients. As mentioned previously, a wide discrepancy between the general definition of the superior and inferior border of UA exists. When a skeletal modification occurs following orthognathic surgery, this aspect is complicated further, as the typical structures (e.g., the posterior nasal spine and the occlusal plane) are displaced during surgery (Gonçales et al., 2014; Fernandez-Ferrer et al., 2015; Christovam et al., 2016). Thus, the selection of anatomical structures that are not affected by surgery is essential.

In a recent technical report of Di Carlo and colleagues (2017), the authors have proposed a new 3D analysis to study UA in patients undergoing orthognathic surgery. All the landmarks and planes are based on stable, reproducible anatomical structures not involved in the modifications occurring during orthognathic surgery (Fig. 12). To delimit the upper border of the UA, a plane passing through the midpoint of the sella-basion line and perpendicular to both the coronal plane through sella and basion and to the sagittal plane was defined. To delimit the boundary between the retropalatal and the oropharyngeal region, a plane passing through basion and parallel to the upper border plane was defined. To delimit the inferior border of the UA, a plane passing through the tip of the epiglottis and parallel to the upper border plane was identified. Based on these references planes, the retropalatal and the oropharyngeal volumes could be calculated, as well as the area of the four cross-sectional areas limiting these volumes.

By following this procedure, a good intra- and inter-observer reliability was achieved and no systematic errors were seen. The procedure could be applied successfully to analyze the UA in patients both pre- and post-orthognathic surgery. The adoption of this novel method thus will

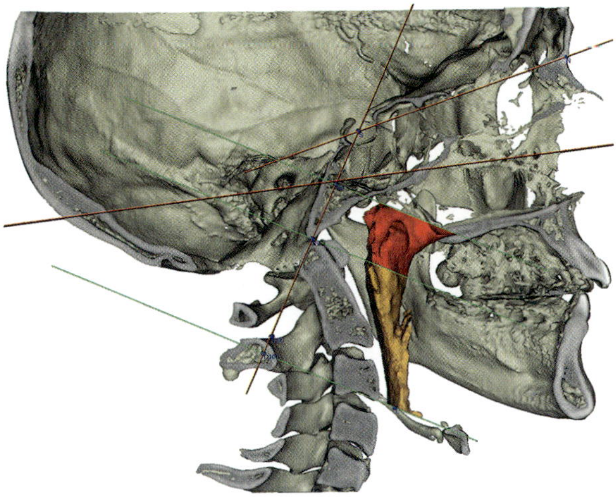

Figure 12. The reference planes placed on the untouched structures in the cranial base are reported in red; the planes used to delimit UA in green; the retropalatal partial volume is depicted in red; and the oropharynx volume in orange.

contribute to overcome the limitations of the previously proposed analyses and may increase the consistency in measuring the UA.

CONCLUSIONS

There is no question that the development of CBCT has provided the opportunity to depict the UA three dimensionally with an accessible, non-invasive and relative low-radiation imaging modality. The study of the UA is not trivial, as it involves many structures and physiological aspects that transcend the competences typical of orthodontics. Moreover, the morphology of the UA plays only one role in the multi-faceted chronic conditions known as SDB.

Nevertheless, it is of the utmost importance that efforts should be made to characterize the morphology of the UA properly, recognizing the importance of respiration phase, the influence of the soft tissue, the correct definition of the anatomical boundaries of the UA and the proper segmentation technique. Moreover, it is important to remember that the patient's clinical information (e.g., BMI, neck circumference and head posture) are as significant as the methodology applied.

CBCT scans have become an important source of 3D data and the number of scans that have and will be collected might prove to be essential in studying SDB in the future.

ACKNOWLEDGEMENTS

The authors would like express their gratitude to Professor Else Marie Pinholt for her sparring during the last years, Dr. Jens Jørgen Thorn and Dr. Janne Ingerslev from the University of Southern Denmark for performing the orthognathic surgeries. Appreciation also is extended to the post-graduate students from the Section of Orthodontics at the Department of Dentistry, Aarhus University, Denmark; without their help and dedication, this study would not have been possible to complete. Finally, gratitude is given to Marie Cornelis for the many positive discussions, constant encouraging and support, and for sharing the interest in research in orthodontics.

REFERENCES

Abbott MB, Donnelly LF, Dardzinski BJ, Poe SA, Chini BA, Amin RS. Obstructive sleep apnea: MR imaging volume segmentation analysis. Radiology 2004;232(3):889-895.

Aboudara C, Nielsen I, Huang JC, Maki K, Miller AJ, Hatcher D. Comparison of airway space with conventional lateral headfilms and 3-dimensional reconstruction from cone-beam computed tomography. Am J Orthod Dentofacial Orthop 2009;135(4):468-479.

Abramson Z, Susarla S, August M, Troulis M, Kaban L. Three-dimensional computed tomographic analysis of airway anatomy in patients with obstructive sleep apnea. J Oral Maxillofac Surg 2010;68(2):354-362.

Ahlqvist J, Eliasson S, Welander U. The effect of projection errors on cephalometric length measurements. Eur J Orthod 1986;8(3):141-148.

Alsufyani NA, Flores-Mir C, Major PW. Three-dimensional segmentation of the upper airway using cone beam CT: A systematic review. Dentomaxillofac Radiol 2012;41(4):276-284.

Anuntaseree W, Rookkapan K, Kuasirikul S, Thongsuksai P. Snoring and obstructive sleep apnea in Thai school-age children: Prevalence and predisposing factors. Pediatr Pulmonol 2001;32(3):222-227.

Archbold KH, Pituch KJ, Panahi P, Chervin RD. Symptoms of sleep disturbances among children at two general pediatric clinics. J Pediatr 2002;140(1):97-102.

Ayappa I, Rapoport DM. The upper airway in sleep: Physiology of the pharynx. Sleep Med Rev 2003;7(1):9-33.

Battagel JM, Johal A, Smith AM, Kotecha B. Postural variation in oropharyngeal dimensions in subjects with sleep disordered breathing: A cephalometric study. Eur J Orthod 2002;24(3):263-276.

Baumrind S, Frantz RC. The reliability of head film measurements: 1. Landmark identification. Am J Orthod 1971*a*;60(2):111-127.

Baumrind S, Frantz RC. The reliability of head film measurements: 2. Conventional angular and linear measures. Am J Orthod 1971*b*;60(5): 505-517.

Bhattacharyya N, Blake SP, Fried MP. Assessment of the airway in obstructive sleep apnea syndrome with 3-dimensional airway computed tomography. Otolaryngol Head Neck Surg 2000;123(4):444-449.

Bixler EO, Vgontzas AN, Lin HM, Liao D, Calhoun S, Vela-Bueno A, Fedok F, Vlasic V, Graff G. Sleep disordered breathing in children in a general population sample: Prevalence and risk factors. Sleep 2009;32(6):731-736.

Casale M, Pappacena M, Rinaldi V, Bressi F, Baptista P, Salvinelli F. Obstructive sleep apnea syndrome: From phenotype to genetic basis. Curr Genomics 2009;10(2):119-126.

Castronovo V, Zucconi M, Nosetti L, Marazzini C, Hensley M, Veglia F, Nespoli L, Ferini-Strambi L. Prevalence of habitual snoring and sleep-disordered breathing in preschool-aged children in an Italian community. J Pediatr 2003;142(4):377-382.

Cattaneo PM, Melsen B. The use of cone-beam computed tomography in an orthodontic department in between research and daily clinic. World J Orthod 2008;9(3):269-282.

Christovam IO, Lisboa CO, Ferreira DM, Cury-Saramago AA, Mattos CT. Upper airway dimensions in patients undergoing orthognathic surgery: A systematic review and meta-analysis. Int J Oral Maxillofac Surg 2016;45(4):460-471.

Conley RS, Cattaneo PM, Haskell BS. Characterization of the upper airway morphology and Its changes in the apneic patient using cone-beam computed tomography. In: Kapila SD, ed. *Cone-Beam Computed*

Tomography in Orthodontics: Indications, Insights, and Innovations. Wiley Blackwell 2014;273-291.

de Souza Carvalho AC, Magro Filho O, Garcia IR Jr, Araujo PM, Nogueira RL. Cephalometric and three-dimensional assessment of superior posterior airway space after maxillomandibular advancement. Int J Oral Maxillofac Surg 2012;41(9):1102-1111.

Deacon NL, Jen R, Li Y, Malhotra A. Treatment of obstructive sleep apnea: Prospects for personalized combined modality therapy. Ann Am Thorac Soc 2016;13(1):101-108.

Di Carlo G, Fernandez Gurani S, Pinholt EM, Cattaneo PM. A new simple three-dimensional method to characterize upper airway in orthognathic surgery patient. Dentomaxillofac Radiol 2017;46(8):20170042.

Di Carlo G, Polimeni A, Melsen B, Cattaneo PM. The relationship between upper airways and craniofacial morphology studied in 3D: A CBCT study. Orthod Craniofac Res 2015;18(1):1-11.

El H, Palomo JM. Measuring the airway in 3 dimensions: A reliability and accuracy study. Am J Orthod Dentofacial Orthop 2010;137(4 Suppl): S50.e1-e9.

El H, Palomo JM. Three-dimensional evaluation of upper airway following rapid maxillary expansion: A CBCT study. Angle Orthod 2014;84(2): 265-273.

Fernández-Ferrer L, Montiel-Company JM, Pinho T, Almerich-Silla JM, Bellot-Arcis C. Effects of mandibular setback surgery on upper airway dimensions and their influence on obstructive sleep apnoea: A systematic review. J Craniomaxillofac Surg 2015;43(2):248-253.

Fields HW, Warren DW, Black K, Phillips CL. Relationship between vertical dentofacial morphology and respiration in adolescents. Am J Orthod Dentofacial Orthop 1991;99(2):147-154.

Ghoneima A, Kula K. Accuracy and reliability of cone-beam computed tomography for airway volume analysis. Eur J Orthod 2013;35(2):256-261.

Gokce SM, Gorgulu S, Gokce HS, Bengi AO, Karacayli U, Ors F. Evaluation of pharyngeal airway space changes after bimaxillary orthognathic surgery with a 3-dimensional simulation and modeling program. Am J Orthod Dentofacial Orthop 2014;146(4):477-492.

Gold AR, Schwartz AR. The pharyngeal critical pressure: The whys and hows of using nasal continuous positive airway pressure diagnostically. Chest 1996;110(4):1077-1088.

Gonçales ES, Duarte MA, Palmieri C Jr, Zakhary GM, Ghali GE. Retrospective analysis of the effects of orthognathic surgery on the pharyngeal airway space. J Oral Maxillofac Surg 2014;72(11):2227-2240.

Goncalves JR, Buschang PH, Goncalves DG, Wolford LM. Postsurgical stability of oropharyngeal airway changes following counter-clockwise maxillo-mandibular advancement surgery. J Oral Maxillofac Surg 2006; 64(5):755-762.

Guijarro-Martinez R, Swennen GR. Cone-beam computerized tomography imaging and analysis of the upper airway: A systematic review of the literature. Int J Oral Maxillofac Surg 2011;40(11):1227-1237.

Guijarro-Martinez R, Swennen GR. Three-dimensional cone beam computed tomography definition of the anatomical subregions of the upper airway: A validation study. Int J Oral Maxillofac Surg 2013;42(9): 1140-1149.

Guilleminault C, Riley R, Powell N. Obstructive sleep apnea and abnormal cephalometric measurements: Implications for treatment. Chest 1984;86(5):793-794.

Gurani SF, Di Carlo G, Cattaneo PM, Thorn JJ, Pinholt EM. Effect of head and tongue posture on the pharyngeal airway dimensions and morphology in three-dimensional imaging: A systematic review. J Oral Maxillofac Res 2016;7(1):e1.

Haponik EF, Smith PL, Bohlman ME, Allen RP, Goldman SM, Bleecker ER. Computerized tomography in obstructive sleep apnea: Correlation of airway size with physiology during sleep and wakefulness. Am Rev Respir Dis 1983;127(2):221-226.

Haskell JA, McCrillis J, Haskell BS, Scheetz JP, Scarfe WC, Farman AG. Effects of Mandibular Advancement Device (MAD) on airway dimensions assessed with cone-beam computed tomography. Semin Orthod 2009;15(2):132-158.

Hellsing E. Changes in the pharyngeal airway in relation to extension of the head. Eur J Orthod 1989;11(4):359-365.

Hohlweg-Majert B, Pautke C, Deppe H, Metzger MC, Wagner K, Schulze D. Qualitative and quantitative evaluation of bony structures based on DICOM dataset. J Oral Maxillofac Surg 2011;69(11):2763-2770.

Hong JS, Oh KM, Kim BR, Kim YJ, Park YH. Three-dimensional analysis of pharyngeal airway volume in adults with anterior position of the mandible. Am J Orthod Dentofacial Orthop 2011;140(4):e161-169.

Hounsfield GN. Computed medical imaging. Nobel lecture, December 8, 1979. J Comput Assist Tomogr 1980;4(5):665-674.

Hua Y, Nackaerts O, Duyck J, Maes F, Jacobs R. Bone quality assessment based on cone beam computed tomography imaging. Clin Oral Implants Res 2009;20(8):767-771.

Hubbel JH, Seltzer SM. X-ray mass attenuation coefficients. In: *Commerce NIoSaT-USDo.* In: NIST Standard Reference Database 126, 1996. https://www.nist.gov/pml/x-ray-mass-attenuation-coefficients.

Hudgel DW. The role of upper airway anatomy and physiology in obstructive sleep apnea. Clin Chest Med 1992;13(3):383-398.

Ingman T, Nieminen T, Hurmerinta K. Cephalometric comparison of pharyngeal changes in subjects with upper airway resistance syndrome or obstructive sleep apnoea in upright and supine positions. Eur J Orthod 2004;26(3):321-326.

Isono S, Morrison DL, Launois SH, Feroah TR, Whitelaw WA, Remmers JE. Static mechanics of the velopharynx of patients with obstructive sleep apnea. J Appl Physiol 1993;75(1):148-154.

Kaditis AG, Alonso Alvarez ML, Boudewyns A, Alexopoulos EI, Ersu R, Joosten K, Larramona H, Miano S, Narang I, Trang H, Tsaoussoglou M, Vandenbussche N, Villa MP, Van Waardenburg D, Weber S, Verhulst S. Obstructive sleep disordered breathing in 2- to 18-year-old children: Diagnosis and management. Eur Respir J 2016;47(1):69-94.

Kaditis AG, Finder J, Alexopoulos EI, Starantzis K, Tanou K, Gampeta S, Agorogiannis E, Christodoulou S, Pantazidou A, Gourgoulianis K, Molyvdas PA. Sleep-disordered breathing in 3,680 Greek children. Pediatr Pulmonol 2004;37(6):499-509.

Kapila SD, Nervina JM. CBCT in orthodontics: Assessment of treatment outcomes and indications for its use. Dentomaxillofac Radiol 2015; 44(1):20140282.

Katyal V, Kennedy D, Martin J, Dreyer C, Sampson W. Paediatric sleep-disordered breathing due to upper airway obstruction in the orthodontic setting: A review. Aust Orthod J 2013;29(2):184-192.

Kiljunen T, Kaasalainen T, Suomalainen A, Kortesniemi M. Dental cone beam CT: A review. Phys Med 2015;31(8):844-860.

Lagravère MO, Carey J, Ben-Zvi M, Packota GV, Major PW. Effect of object location on the density measurement and Hounsfield conversion in a NewTom 3G cone beam computed tomography unit. Dentomaxillofac Radiol 2008*a*;37(6):305-308.

Lagravère MO, Carey J, Toogood RW, Major PW. Three-dimensional accuracy of measurements made with software on cone-beam computed tomography images. Am J Orthod Dentofacial Orthop 2008*b*; 134(1):112-116.

Lenza MG, Lenza MM, Dalstra M, Melsen B, Cattaneo PM. An analysis of different approaches to the assessment of upper airway morphology: A CBCT study. Orthod Craniofac Res 2010;13(2):96-105.

Linder-Aronson S. Adenoids: Their effect on mode of breathing and nasal airflow and their relationship to characteristics of the facial skeleton and the denition. A biometric, rhino-manometric and cephalometro-radiographic study on children with and without adenoids. Acta Otolaryngol Suppl 1970;265:1-132.

Linder-Aronson S, Woodside DG, Lundström A. Mandibular growth direction following adenoidectomy. Am J Orthod 1986;89(4):273-284.

Lowe AA, Gionhaku N, Takeuchi K, Fleetham JA. Three-dimensional CT reconstructions of tongue and airway in adult subjects with obstructive sleep apnea. Am J Orthod Dentofacial Orthop 1986;90(5):364-374.

Ludlow JB, Timothy R, Walker C, Hunter R, Benavides E, Samuelson DB, Scheske MH. Effective dose of dental CBCT: A meta analysis of published data and additional data for nine CBCT units. Dentomaxillofac Radiol 2015;44(1):20140197.

Lumeng JC, Chervin RD. Epidemiology of pediatric obstructive sleep apnea. Proc Am Thorac Soc 2008;5(2):242-252.

Marcus CL. Sleep-disordered breathing in children. Am J Respir Crit Care Med 2001;164(1):16-30.

Mattos CT, Vilani GN, Sant'Anna EF, Ruellas AC, Maia LC. Effects of orthognathic surgery on oropharyngeal airway: A meta-analysis. Int J Oral Maxillofac Surg 2011;40(12):1347-1356.

McCrillis JM, Haskell J, Haskell BS, Brammer M, Chenin D, Scarfe WC, Farman AG. Obstructive sleep apnea and the use of cone beam computed tomography in airway imaging: A review. Semin Orthod 2009;15(1):63-69.

McNamara JA Jr. Influence of respiratory pattern on craniofacial growth. Angle Orthod 1981;51:269-300.

McNamara JA Jr, ed. *Naso-respiratory Function and Craniofacial Growth.* Craniofacial Growth Series, Center for Human Growth and Development, The University of Michigan, Ann Arbor, MI 1979;9:1-332.

Miyamoto K, Ozbek MM, Lowe AA, Fleetham JA. Effect of body position on tongue posture in awake patients with obstructive sleep apnoea. Thorax 1997;52(3):255-259.

Molteni R. Prospects and challenges of rendering tissue density in Hounsfield units for cone beam computed tomography. Oral Surg Oral Med Oral Pathol Oral Radiol 2013;116(1):105-119.

Mozzo P, Procacci C, Tacconi A, Martini PT, Andreis IA. A new volumetric CT machine for dental imaging based on the cone-beam technique: Preliminary results. Eur Radiol 1998;8(9):1558-1564.

Muto T, Takeda S, Kanazawa M, Yamazaki A, Fujiwara Y, Mizoguchi I. The effect of head posture on the pharyngeal airway space (PAS). Int J Oral Maxillofac Surg 2002;31(6):579-583.

Nackaerts O, Maes F, Yan H, Couto Souza P, Pauwels R, Jacobs R. Analysis of intensity variability in multislice and cone beam computed tomography. Clin Oral Implants Res 2011;22(8):873-879.

Ng DK, Kwok KL, Poon G, Chau KW. Habitual snoring and sleep bruxism in a paediatric outpatient population in Hong Kong. Singapore Med J 2002;43(11):554-556.

Noar JH, Pabari S. Cone beam computed tomography: Current understanding and evidence for its orthodontic applications? J Orthod 2013;40(1):5-13.

O'Brien LM, Holbrook CR, Mervis CB, Klaus CJ, Bruner JL, Raffield TJ, Rutherford J, Mehl RC, Wang M, Tuell A, Hume BC, Gozal D. Sleep and neurobehavioral characteristics of 5- to 7-year-old children with parentally reported symptoms of attention-deficit/hyperactivity disorder. Pediatrics 2003;111(3):554-563.

Ogawa T, Enciso R, Shintaku WH, Clark GT. Evaluation of cross-section airway configuration of obstructive sleep apnea. Oral Surg Oral Med Oral Pathol Oral Radiol Endod 2007;103(1):102-108.

Ono T, Otsuka R, Kuroda T, Honda E, Sasaki T. Effects of head and body position on two- and three-dimensional configurations of the upper airway. J Dent Res 2000;79(11):1879-1884.

Pae EK, Lowe AA, Sasaki K, Price C, Tsuchiya M, Fleetham JA. A cephalometric and electromyographic study of upper airway structures in the upright and supine positions. Am J Orthod Dentofacial Orthop 1994;106(1):52-59.

Pauwels R, Nackaerts O, Bellaiche N, Stamatakis H, Tsiklakis K, Walker A, Bosmans H, Bogaerts R, Jacobs R, Horner K; SEDENTEXCT Project Consortium. Variability of dental cone beam CT grey values for density estimations. Br J Radiol 2013;86(1021):20120135.

Pirilä-Parkkinen K, Pirttiniemi P, Pääkkö E, Tolonen U, Nieminen P, Löppönen H. Pharyngeal airway in children with sleep-disordered breathing in relation to head posture. Sleep Breath 2012;16(3):737-746.

Ravesloot MJ, van Maanen JP, Dun L, de Vries N. The undervalued potential of positional therapy in position-dependent snoring and obstructive sleep apnea: A review of the literature. Sleep Breath 2013;17(1):39-49.

Reeves TE, Mah P, McDavid WD. Deriving Hounsfield units using grey levels in cone beam CT: A clinical application. Dentomaxillofac Radiol 2012;41(6):500-508.

Safar P, Escarraga LA, Chang F. Upper airway obstruction in the unconscious patient. J Appl Physiol 1959;14:760-764.

Scarfe WC, Farman AG. What is cone-beam CT and how does it work? Dent Clin North Am 2008;52(4):707-730.

Schlaud M, Urschitz MS, Urschitz-Duprat PM, Poets CF. The German study on sleep-disordered breathing in primary school children: Epidemiological approach, representativeness of study sample, and preliminary screening results. Paediatr Perinat Epidemiol 2004;18(6):431-440.

Schwab RJ. Upper airway imaging. Clin Chest Med 1998;19(1):33-54.

Schwab RJ, Gupta KB, Gefter WB, Metzger LJ, Hoffman EA, Pack AI. Upper airway and soft tissue anatomy in normal subjects and patients with sleep-disordered breathing: Significance of the lateral pharyngeal walls. Am J Respir Crit Care Med 1995;152(5 Pt 1):1673-1689.

Schwartz AR, Smith PL. CrossTalk proposal: The human upper aiway does behave like a Starling resistor during sleep. J Physiol 2013;591(9): 2229-2232.

Segal Y, Malhotra A, Pillar G. Upper airway length may be associated with the severity of obstructive sleep apnea syndrome. Sleep Breath 2008;12(4):311-316.

Spin-Neto R, Gotfredsen E, Wenzel A. Standardized method to quantify the variation in voxel value distribution in patient-simulated CBCT data sets. Dentomaxillofac Radiol 2015;44(2):20140283.

Spin-Neto R, Wenzel A. Patient movement and motion artefacts in cone beam computed tomography of the dentomaxillofacial region: A systematic literature review. Oral Surg Oral Med Oral Pathol Oral Radiol 2016;121(4):425-433.

Susarla SM, Thomas RJ, Abramson ZR, Kaban LB. Biomechanics of the upper airway: Changing concepts in the pathogenesis of obstructive sleep apnea. Int J Oral Maxillofac Surg 2010;39(12):1149-1159.

Sutthiprapaporn P, Tanimoto K, Ohtsuka M, Nagasaki T, Iida Y, Katsumata A. Positional changes of oropharyngeal structures due to gravity in the upright and supine positions. Dentomaxillofac Radiol 2008;37(3):130-135.

Trudo FJ, Gefter WB, Welch KC, Gupta KB, Maislin G, Schwab RJ. State-related changes in upper airway caliber and surrounding soft-tissue structures in normal subjects. Am J Respir Crit Care Med 1998;158(4): 1259-1270.

van Vlijmen OJ, Kuijpers MA, Bergé SJ, Schols JG, Maal TJ, Breuning H, Kuijpers-Jagtman AM. Evidence supporting the use of cone-beam computed tomography in orthodontics. J Am Dent Assoc 2012;143(3): 241-252.

Vig PS, Hall DJ. The inadequacy of cephalometric radiographs for airway assessment. Am J Orthod 1980;77(2):230-233.

Warren DW, Spalding PM. Dentofacial morphology and breathing: A century of controversy. In: Melsen B, ed. *Controversies in Orthodontics.* Berlin: Quintessence-Verlag 1990.

Weissheimer A, Menezes LM, Sameshima GT, Enciso R, Pham J, Grauer D. Imaging software accuracy for 3-dimensional analysis of the upper airway. Am J Orthod Dentofacial Orthop 2012;142(6):801-813.

White DP. Pathogenesis of obstructive and central sleep apnea. Am J Respir Crit Care Med 2005;172(11):1363-1370.

White DP. Sleep-related breathing disorder: 2. Pathophysiology of obstructive sleep apnoea. Thorax 1995;50(7):797-804.

Yamashina A, Tanimoto K, Sutthiprapaporn P, Hayakawa Y. The reliability of computed tomography (CT) values and dimensional measurements of the oropharyngeal region using cone beam CT: Comparison with multidetector CT. Dentomaxillofac Radiol 2008;37(5):245-251.

Young T, Palta M, Dempsey J, Skatrud J, Weber S, Badr S. The occurrence of sleep-disordered breathing among middle-aged adults. N Engl J Med 1993;328(17):1230-1235.

Zhang W, Song X, Masumi SI, Tanaka T, Zhu Q. Effects of head and body positions on 2- and 3-dimensional configuration of the oropharynx with jaw protruded: A magnetic resonance imaging study. Oral Surg Oral Med Oral Pathol Oral Radiol Endod 2011;111(6):778-784.

Zimmerman JN, Lee J, Pliska BT. Reliability of upper pharyngeal airway assessment using dental CBCT: A systematic review. Eur J Orthod 2017;39(5):489-496.

UPPER AIRWAY CHARACTERISTICS OF PATIENTS WITH CLEFT LIP AND PALATE

Marilia Yatabe, Ivy Kiemle Trindade-Suedam,
Inge Elly Kiemle Trindade

ABSTRACT

Patients with cleft lip and palate often show significant anatomical alterations of the face and, as a result, of the upper airway. The rehabilitation surgeries needed to correct the anatomical defects may have a direct influence over the upper airway. This chapter aims to describe the techniques used to assess the upper airway characteristics—acoustic rhinometry, nasometry and rhinomanometry—and correlate these diagnostic tools to clinical interventions commonly performed in patients with clefts—rapid maxillary expansion, orthognathic surgery, rhinoseptoplasty and pharyngeal flap.

KEY WORDS: cleft palate, cleft lip, acoustic rhinometry, rhinomanometry, nasopharynx

---

INTRODUCTION

The Hospital for Rehabilitation of Craniofacial Anomalies (HRCA) of the University of São Paulo (USP) is known for the treatment of patients with orofacial clefts and other craniofacial anomalies. Since its inauguration in 1967, the HRCA has provided assistance to over 100,000 patients, helped educate students from all over Latin America and performed joint research projects with institutions all over the world.

Research was a consequence of the need to understand the anatomical changes and the influence of these changes better over the physiological functions, growth and treatment outcomes. Since the beginning, speech and respiratory disorders were diagnosed in patients with craniofacial anomalies, leading to the development of the Laboratory of Physiology (LOP). Its aims were to assess the impact of different surgical and dental procedures of the upper airway, the stomatognathic system and the sleep quality of individuals with cleft lip and palate.

Different methods of assessment are used in the LOP. They may be grouped in two categories: direct and indirect. Direct methods allow the visualization of the velopharyngeal anatomy and its movement during speech (e.g., nasoendoscopy [Fig. 1] and videofluoroscopy).

Indirect methods provide acoustic repercussion of the velopharyngeal functioning (e.g., acoustic rhinometry, rhinomanometry and nasometry). Acoustic rhinometry is a diagnostic tool to evaluate nasal patency, providing information on nasal cross-sectional areas and volumes by analyzing sound pulse reflections. Results are given in numerical values of the area, distance from the nostril and volume (Fig. 2).

Rhinomanometry, also known as flow-pressure technique, provides quantitative information on velopharyngeal function (Figs. 3 and 4). The technique is based on the principle that the cross-sectional area of a constriction (or orifice) may be estimated by the simultaneous measurement of the differential pressure between the two sides of the constriction and the rate of airflow through it.

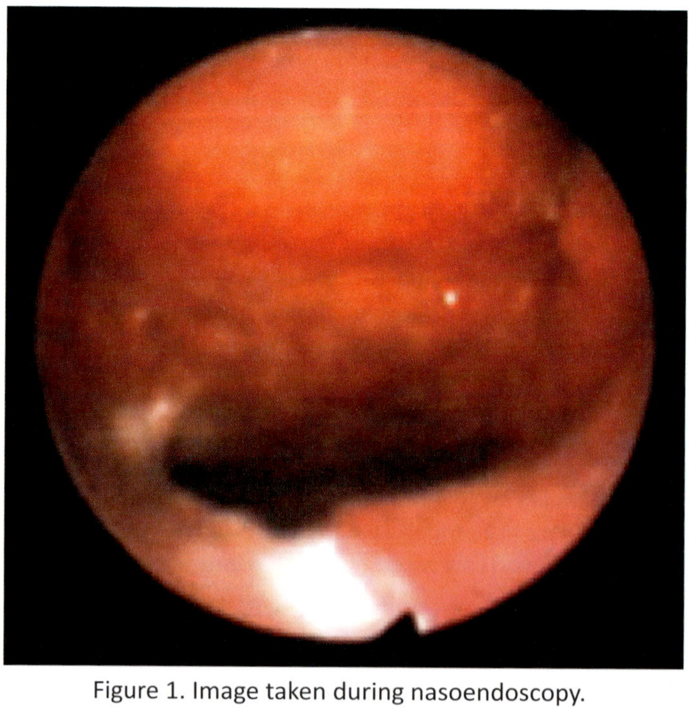

Figure 1. Image taken during nasoendoscopy.

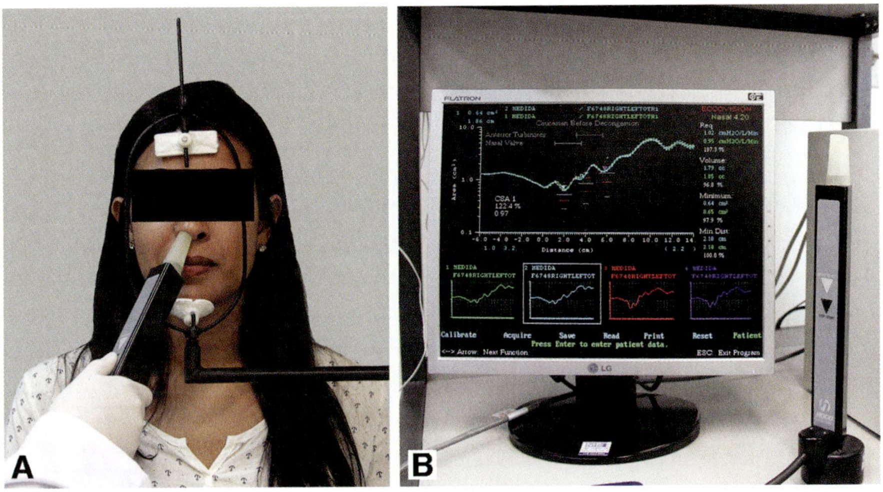

Figure 2. Acoustic rhinometry. *A:* The rhinometer positioned for the assessment of the right nasal cavity. *B:* The resulting graph.

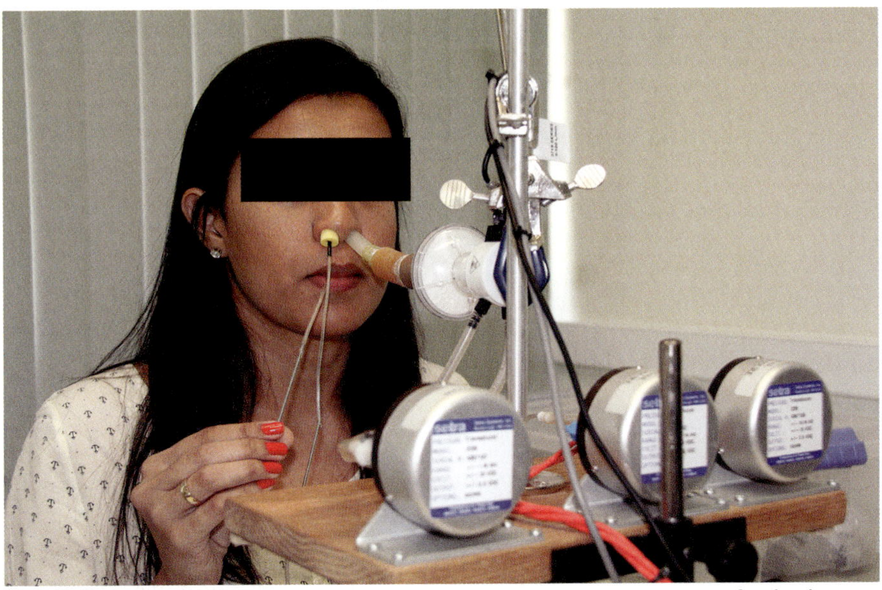

Figure 3. Rhinomanometry: tubing in position for the assessment of velopharyngeal orifice area during rest and speech.

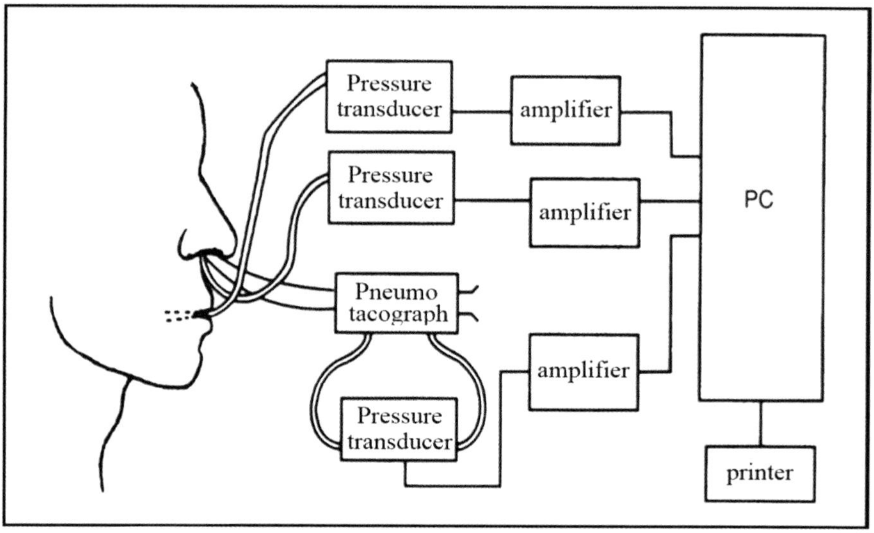

Figure 4. Schematic instrumentation for estimating the velopharyngeal cross-sectional area (PERCI-SARS System, Microtronics Corp., Chapel Hill, NC).

The velopharyngeal orifice area is determined during rest, breathing and during the production of a plosive sound by positioning a catheter within the oral cavity and one of the nostrils, which is held in position by an obturator. Both catheters measure static air pressures that are transmitted to pressure transducers. Nasal airflow is measured by means of a plastic tube adapted to the other nostril, which is connected to a heated pneumotachograph that also is connected to a pressure transducer. The signals of the three transducers (nasal pressure, nasal oral pressure and flow) are sent to the PERCI-SARS system (Microtonics Corp., Chapel Hill, NC) for analysis. The pressure flow technique also may be used for the assessment of minimum nasal cross-sectional areas unilateral or bilaterally.

Nasometry measures nasalance, the acoustic correlation of nasality (Figs. 5 and 6). The technique estimates the speech resonance by measuring the relative amount of acoustic energy that emerges from the nasal cavity during production of nasal and oral syllables (Dalston, 1992).

The cleft itself may affect the lip, the palate or both. The first step of rehabilitation is lip surgical repair, which ideally is performed at three months of age, followed by the palate repair around twelve months of age. These surgeries reconstruct skin, muscles and oral mucosa and, as a

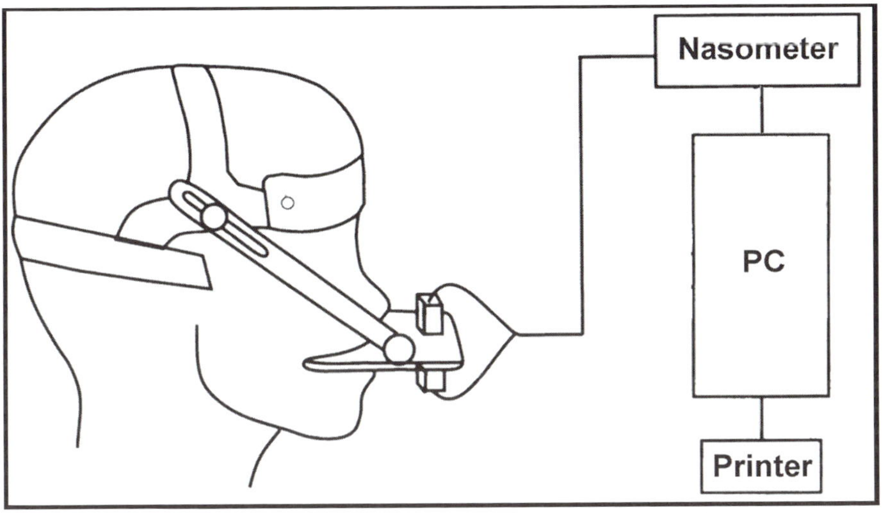

Figure 5. Schematic instrumentation for measuring nasalance, the acoustic correlate of nasality.

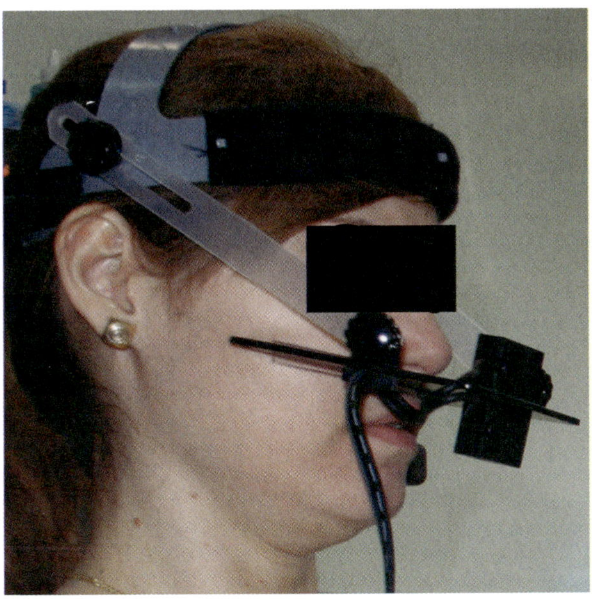

Figure 6. Nasometry: nasometer in position for the nasalance.

consequence, create a fibrous tissue with the potential to restrict maxillary development and anterior displacement (Mars and Houston, 1990; Yoshida et al., 1992; Capelozza Filho et al., 1996).

As a result of the restricted maxillary development, anterior displacement and the scars over the soft tissue, the typical characteristics of patients with clefts are:

1. Atresic maxilla: the restricted forces of the scars from the palatoplasty along with the buccinator forces, which lead to a more atresic maxilla.
2. Retropositioned maxilla: the restricted forces of the scars from the palatoplasty and cheiloplasty (lip reduction) act against the natural anterior and inferior displacement of the maxilla. Associated with a normal development/displacement of the mandible, it creates a Class III malocclusion by maxillary deficiency.
3. Deviated nasal septum: mainly in patients with unilateral cleft lip and palate, the nasal septum deviates to the cleft side, resulting in a smaller airway volume in the cleft side compared to the non-cleft side (Fukushiro and Trindade, 2005).
4. Velopharyngeal Insufficiency: palatoplasty can result in a shorter soft palate and, as a consequence, whenever the patient speaks, the soft palate is not able to close the gap between the oral and nasopharynx, resulting in a hypernasal speech.

To correct these alterations, patients with cleft undergo life-time rehabilitation, from infancy until adulthood, as seen in the timeline in Figure 7. The following topics will address the most common procedures that may affect the airways of patients with clefts.

RAPID MAXILLARY EXPANSION

Rapid maxillary expansion is an orthodontic treatment step performed during early mixed dentition to correct the position of the segments of the maxilla. As a result, the occlusion relationship not only will be improved, but the maxillary segments also will be aligned by the time of the alveolar bone graft surgery (Fig. 8).

A previous study from the LOP involved an acoustic rhinometry assessment before and after rapid maxillary expansion of patients with clefts. As exemplified in Table 1, it showed an overall enlargement of the

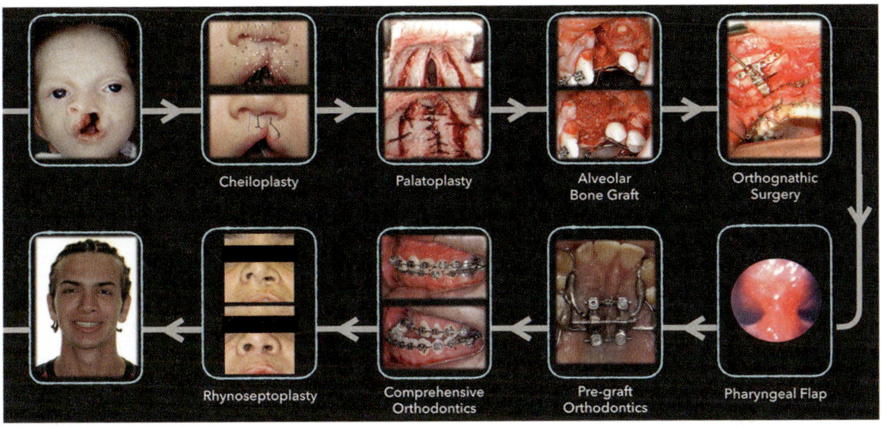

Figure 7. Timeline of procedures commonly performed in patients with cleft lip and palate at the HRCA.

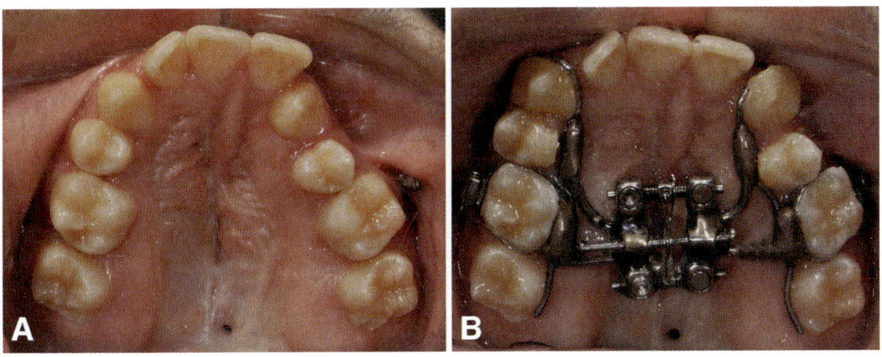

Figure 8. Example of a constricted maxilla before (*A*) and after (*B*) expansion.

Table 1. Volume and cross-sectional area of the nasal valve before and after rapid maxillary expansion (RME). CSA = cross-sectional area.

	Pre-RME	Post-RME
CSA	1.19 mm^2	1.41 mm^2
Volume	3.68 mm^3	4.79 mm^3

nasal area, being significant only in the cleft side of patients with unilateral cleft lip and palate (Trindade et al., 2010; Trindade-Suedam et al., 2016).

ORTHOGNATHIC SURGERY (OS)

Orthognathic surgery (OS) is indicated in cases with moderate-to-severe maxilla-mandibular discrepancy, when comprehensive orthodontic treatment alone would not be able to correct the malocclusion. This surgery not only improves the occlusion and facial esthetics, but also shows a marked influence in the upper airway dimensions and speech.

A study from the LOP (Trinidade et al., 2003) showed that maxillary advancement promotes an increase in the nasopharyngeal space, which also improves breathing during sleep. In cases that the discrepancy is too severe, however, the mandible can be set back. This surgical procedure could compromise the oropharyngeal airway of these patients (Trindade et al., 2003).

On the other hand, depending on the soft palate morphology before surgery, maxillary advancement may develop a velopharyngeal insufficiency (Fig. 9), resulting in hypernasal sound during speech (Table 2).

RHINOSEPTOPLASTY

In patients with unilateral clefts, it typically is common to observe an asymmetry of the nose due to a deviation of the nasal pyramid to the non-cleft side and a flattening of the nostril on the cleft side (Fig. 10; Freitas et al., 2013). The functional and esthetical correction of these nasal deformities involves rhinoseptoplasty and columella lengthening surgeries. Trindade and colleagues' study (2009) showed that after rhinoseptoplasty, there was an increase of the internal nasal dimensions and, consequently, improvement of nasal patency, with no changes in speech resonance (Tables 3 and 4).

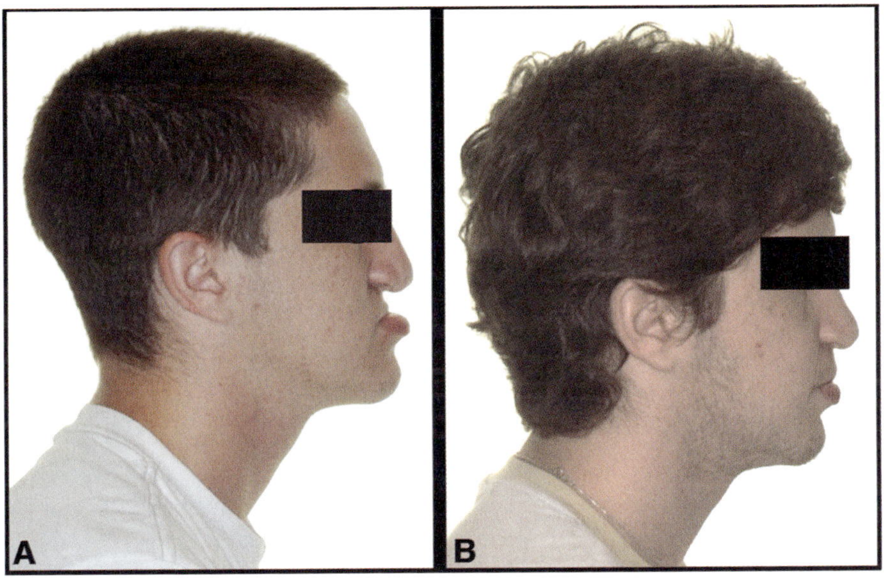

Figure 9. Right soft-tissue profile of a patient with Class III malocclusion, before (*A*) and after (*B*) OS.

Table 2. Nasalance, which is the acoustic correlation of nasality changes, before and after OS.

	Normality	Pre-OS	Post-OS
Oral syllables	Normal ≤ 27%	27.4 %	39.8 %
Nasal syllables	Normal ≥ 43%	44.2 %	51.4 %

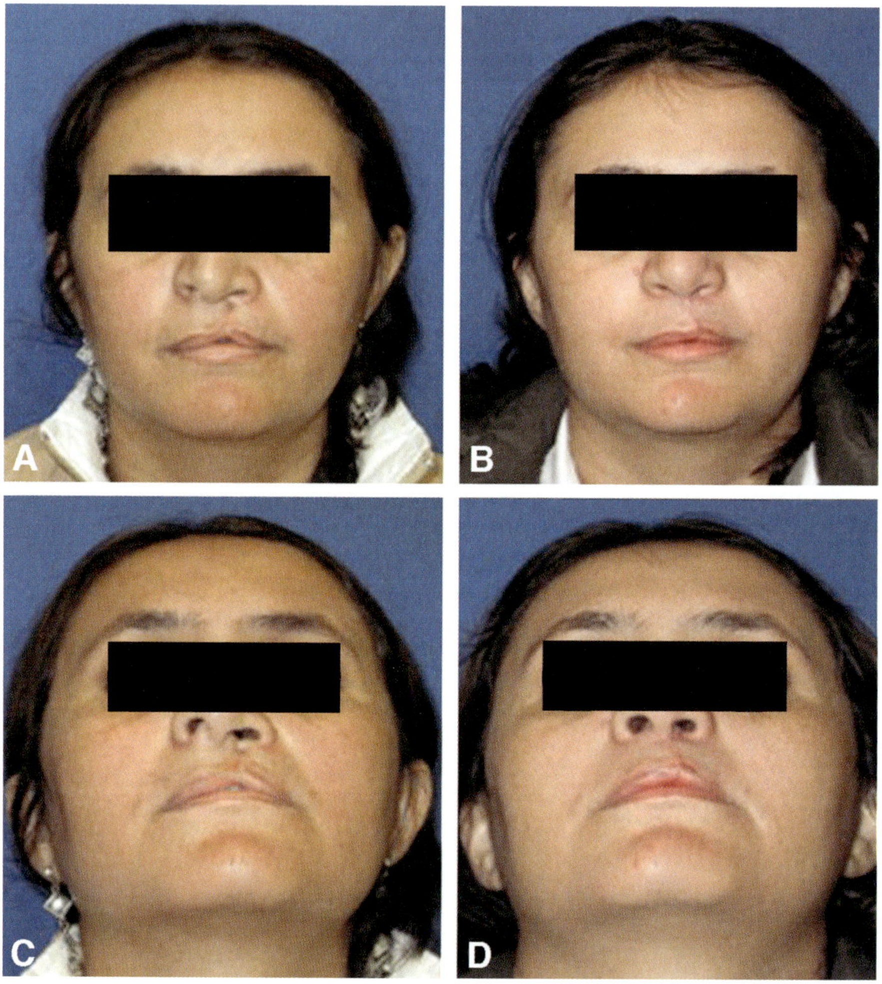

Figure 10. *A-B:* Pre- and post-rhinoseptoplasty in an anterior view of the alar base of the nose. *C-D:* Pre- and post-rhinoseptoplasty in an inferior view of the alar base of the nose.

PHARYNGEAL FLAP

In patients with a short palate and/or pharyngeal wall impairment, pharyngeal flap surgery is indicated because the soft tissues are unable to close the gap between the oro- and nasopharynx during speech.

Table 3. Nasal area and volume changes before and after rhinoseptoplasty.

		Pre-rhinoseptoplasty		Post-rhinoseptoplasty	
		Area	Volume	Area	Volume
Nasal valve	**Cleft side**	0.28 cm^2	0.95 cm^3	0.76 cm^2	2.02 cm^3
	Non-cleft side	0.51 cm^2	1.48 cm^3	0.52 cm^2	1.58 cm^3
Anterior region of the nose	**Cleft side**	0.81 cm^2	3.21 cm^3	1.15 cm^2	5.09 cm^3
	Non-cleft side	0.85 cm^2	3.50 cm^3	1.00 cm^2	4.20 cm^3
Posterior region of the nose	**Cleft side**	1.03 cm^2	12.23 cm^3	1.78 cm^2	19.29 cm^3
	Non-cleft side	0.97 cm^2	18.00 cm^3	1.16 cm^2	19.94 cm^3

Table 4. Nasal area and volume changes before and after rhinoseptoplasty.

	Normality	Pre-rhinoseptoplasty	Post-rhinoseptoplasty
Oral syllables	**Normal ≤ 27%**	10.95 %	12.81 %
Nasal syllables	**Normal ≥ 43%**	46.85 %	45.64 %

Table 5. Velopharyngeal area before and after pharyngeal flap.

	Pre-PF	Post-PF
Inhale	1000 cm^2	0.536 cm^2
Exhale	1000 cm^2	0.752 cm^2

By creating a soft tissue bridge connecting the soft palate to the posterior pharyngeal wall, speech can be improved significantly, but the velopharyngeal area can be compromised significantly (Table 5, Fig. 11). At the

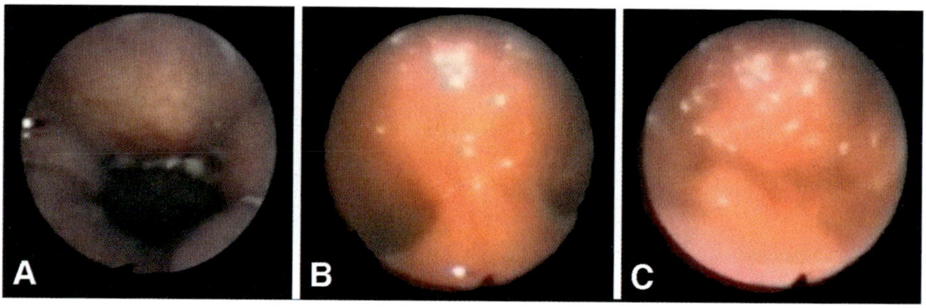

Figure 11. Examples of a pharynx. *A:* Without pharyngeal flap. *B-C:* With pharyngeal flap.

same time, the flap may compromise breathing during sleep because it drastically reduces the pharyngeal cross-sectional area (Fig. 11).

Campos and associates (2016) reported that 77% of middle-aged adults with pharyngeal flap presented obstructive sleep apnea (OSA), while 60% of patients without pharyngeal flap also presented OSA. Therefore, the authors suggested that pharyngeal flap may be one of many causes of OSA in these patients.

All these findings led the LOP to develop a new line of study regarding OSA. Ongoing efforts are investigating questions such as: is it possible that all patients with cleft may have higher risk to develop OSA? How significant is the influence of the anatomical changes over OSA? How greatly can orthognathic surgery or pharyngeal flap influence the development of OSA?

CONCLUSIONS

Patients with cleft lip and palate show anatomical malformations that have a direct influence on physiological functions (e.g., speech and breathing) from birth until the end of growth. Rehabilitation therapies may improve their esthetics, skeletal balance and breathing, but speech is the most compromised function after growth. Therefore, it is mandatory to have an interdisciplinary team to determine the best treatment planning for these patients.

REFERENCES

Campos LD, Trindade-Suedam IK, Sampaio-Teixeira AC, Yamashita RP, Lauris JR, Lorenzi-Filho G, Trindade IE. Obstructive sleep apnea following pharyngeal flap surgery for velopharyngeal insufficiency: A prospective polysomnographic and aerodynamic study in middle-aged adults. Cleft Palate Craniofac J 2016;53(3):e53-e59.

Capelozza Filho L, Normando AD, da Silva Filho OG. Isolated influences of lip and palate surgery on facial growth: Comparison of operated and unoperated male adults with UCLP. Cleft Palate Craniofac J 1996;33(1): 51-56.

Dalston RM. Acoustic assessment of the nasal airway. Cleft Palate Craniofac J 1992;29(6):520-526.

Freitas JA, Trindade-Suedam IK, Garib DG, Neves LT, Almeida AL, Yaedú RY, Oliveira TM, Soares S, Lauris Rde C, Yamashita RP, Trindade AS Jr, Trindade IE, Pinto JH. Rehabilitative treatment of cleft lip and palate: Experience of the Hospital for Rehabilitation of Craniofacial Anomalies/ USP (HRAC/USP): Part 5. Institutional outcomes assessment and the role of the Laboratory of Physiology. J Appl Oral Sci 2013;21(4):383-390.

Fukushiro AP, Trindade IE. Nasal airway dimensions of adults with cleft lip and palate: Differences among cleft types. Cleft Palate Craniofac J 2005;42(4):396-402.

Mars M, Houston WJ. A preliminary study of facial growth and morphology in unoperated male unilateral cleft lip and palate subjects over 13 years of age. Cleft Palate J 1990;27(1):7-10.

Trindade IE, Bertier CE, Sampaio-Teixeira AC. Objective assessment of internal nasal dimensions and speech resonance in individuals with repaired unilateral cleft lip and palate after rhinoseptoplasty. J Craniofac Surg 2009;20(2):308-314.

Trindade IE, Castilho RL, Sampaio-Teixeira AC, Trindade-Suedam IK, Silva-Filho OG. Effects of orthopedic rapid maxillary expansion on internal nasal dimensions in children with cleft lip and palate assessed by acoustic rhinometry. J Craniofac Surg 2010;21(2):306-311.

Trindade IE, Yamashita RP, Suguimoto RM, Mazzottini R, Trindade AS Jr. Effects of orthognathic surgery on speech and breathing of subjects with cleft lip and palate: Acoustic and aerodynamic assessment. Cleft Palate Craniofac J 2003;40(1):54-64.

Trindade-Suedam IK, Castilho RL, Sampaio-Teixeira AC, Araújo BM, Fukushiro AP, Campos LD, Trindade IE. Rapid maxillary expansion increases internal nasal dimensions of children with bilateral cleft lip and palate. Cleft Palate Craniofac J 2016;53(3):272-277.

Yoshida H, Nakamura A, Michi K, Wang GM, Liu K, Qiu WL. Cephalometric analysis of maxillofacial morphology in unoperated cleft palate patients. Cleft Palate Craniofac J 1992;29(5):419-424.

CERVICAL SPINE ANGLES, CRANIOCERVICAL POSTURE, NECK LENGTH AND OROPHARYNGEAL AIRWAY ANALYSES OF OBSTRUCTIVE SLEEP APNEA PATIENTS IN BOTH SUPINE AND UPRIGHT POSITIONS: A RETROSPECTIVE THREE-DIMENSIONAL IMAGING STUDY

Sundaralingam Prem Premaraj, Brian Luong, Kim Sung, Thyagaseely Sheela Premaraj

ABSTRACT

INTRODUCTION: Obstructive sleep apnea (OSA) is one of the most common sleep-related breathing disorders. Previous airway studies of OSA subjects have relied largely on two-dimensional (2D) radiographs. The purpose of this study was to use three-dimensional (3D) imaging to the analyze the relationships among cervical spine angles, craniocervical posture, cervical spine length and the oropharyngeal airway volume in OSA patients in both the supine and upright positions. **METHODS:** Twenty-eight OSA subjects with 3D imaging were included. Airway dimensions, craniocervical posture, spine angles and spine length were assessed using Dolphin® 11.8. Correlation analyses were performed to detect associations among the recorded and measured variables. Mean differences were determined between the supine and upright subjects. **RESULTS AND CONCLUSIONS:** Significant associations were found: positive associations between Apnea-Hypopnea Index (AHI) and age, craniocervical posture and airway volumes, craniocervical posture and cervical vertebrae C1-C2 spinal angle, and spine length and airway volumes. Negative associations were found between craniocervical posture and body mass index (BMI), C2-C3 and C1-C4 angle and age. McGregor and McRae angles were found to be correlated. Subjects in the supine position had significantly smaller oropharyngeal airway dimensions than subjects in the upright position. Craniocervical extension was correlated positively with increased BMI and negatively correlated with airway volumes; however, spinal angles were not. Subjects in the supine position demonstrated smaller airway volumes than upright subjects. Subject positioning and posture are important consideration during the evaluation of OSA.

KEY WORDS: obstructive sleep apnea, cervical spine angle, cone-beam computed tomography (CBCT)

INTRODUCTION

Sleep-disordered breathing (SDB) is a common disorder characterized by shallow breaths or pauses in breathing during sleep. The umbrella category of SDB includes two types of sleep apnea: central sleep apnea (CSA) and obstructive sleep apnea (OSA). OSA is more common and occurs when there is momentary, but repeated, collapse of the airway during respiration. OSA is estimated to affect between 9 to 28% of adults and is more common in males than females (Young et al., 1993). The prevalence of OSA is 23% in females and 49% in males (Heinzer et al., 2015). The risk for developing OSA increases with age and body mass index (BMI; Kapur et al., 2010).

OSA can affect the quality of life dramatically by impairing neurocognitive performance and inducing excessive daytime sleepiness. Beyond these daytime functional impairments, there may be severe comorbidities such as cardiovascular and neurovascular sequelae (Hirsch Allen et al., 2015).

The gold standard for diagnosis of OSA is polysomnography (PSG). Due to its high cost, home sleep apnea testing (HSAT) is an acceptable and valid diagnostic alternative in appropriate candidates for this type of testing (Ayappa et al., 2008; Driver et al., 2011; Cairns et al., 2014).

Two- and three-dimensional (2D, 3D) imaging studies using lateral cephalograms or cone-beam computed tomography (CBCT) and multidetector computed tomography (MDCT) provide an excellent window into the anatomy and physiology of OSA. Studies have shown that there is a high probability of severe OSA with an airway less than 52 mm^2 and low probability if the airway is greater than 110 mm^2 (Lowe et al., 1986). OSA occurs during sleep when subjects are in the supine position. To date, only one study has used CBCT to assess changes in the pharyngeal airway volume of five OSA patients in the supine and upright positions (Camacho et al., 2014). In the supine position, total airway volume (TV) and cross-sectional areas along the length of the pharyngeal airway were significantly smaller (Camacho et al., 2014).

The etiology of OSA is multi-factorial and complex. Besides obesity, craniofacial abnormalities and cervical spine pathologies have been associated with OSA. Limited studies have examined conditions that affect the posterior airway structures (e.g., in cervical spine abnormalities). Several cervical spine abnormalities and pathologies have been associated with OSA (Khan et al., 2014). Studies investigating cervical spine morphology and OSA found a high prevalence of cervical spine fusion in these patients. Osteochondromas, or benign tumors of the spine occurring in the cervical spine, can encroach on the pharyngeal space to cause OSA. These pathologies change the delicate balance in the soft or hard tissues surrounding the upper airway, leading to its collapsibility and subsequent obstruction.

Imaging studies have identified the factors that affect the airway patency in OSA. These factors include positional changes, craniocervical (head) posture, cervical spine angles and neck length. Historically, these studies have relied largely on lateral cephalograms, which did not capture the airway dimensions fully. Advanced 3D imaging studies are superior, but limited in numbers.

Few published studies have examined the association between cervical spine angles and OSA. In a cephalometric study of relationship between cervical spine angles and sleep apnea severity, Dobson and coworkers (1999) found an abnormally convex (kyphotic) curvature of the occiput and upper cervical spine, with the greatest flexion observed in the most severe OSA subjects. Indirect studies have examined the association between cervical spine fusion, disruption of the natural neck alignment and OSA. One study found that the Apnea-Hypopnea Index (AHI)—an indicator for sleep apnea severity—increased 5-10x post-spinal fusion surgery (Guilleminault et al., 2003). Reduction in head and neck flexion-extension after spinal fusion was attributed to the development of OSA. For example, a 10° increase in head flexion resulted in a 37% reduction in posterior airway space (PAS; Ota et al., 2011).

Head posture, or craniocervical posture, also has been associated with OSA. Deviation from a natural head posture in OSA subjects was found as an adaptive mechanism to increase the patency of the airway (Solow et al., 1993, 1996). In a series of cephalometric studies, they found that a forward head position with craniocervical extension increased the lower oropharyngeal airway dimensions. A systematic review identified only one MRI study, which found a positive correlation between head extension and increased hypopharyngeal airway volume (Gurani et al., 2016).

An increase in neck circumference has been associated with OSA (Davies et al., 1992); however, few studies have examined the relationship between neck length and OSA. The few existing studies have reported conflicting conclusions. One study found that a smaller clinical neck length, measured from the hyoid bone to the jugular notch, is associated with increased snoring (Han et al., 2015). Conversely, Kim and associates' CBCT study (2011) showed that an increased neck length was highly predictive of OSA. The authors hypothesized that a longer, but smaller mean cross-sectional area of the airway, is more susceptible to collapse.

There are major weaknesses with prior studies that examined the associations among cervical spine angles, head posture, neck length, upper airway space and sleep apnea severity. Most of those studies relied heavily on 2D radiographs (e.g., whole body x-rays or lateral cephalograms) to make linear and angular measurements. However, 2D radiographic studies do not capture the airway fully in all three spatial planes. The relationship between cervical spine angles, head posture and sleep apnea severity could be studied better using 3D imaging. In addition, these studies utilized radiographs of patients taken in the upright position and failed to show the true anatomic and physiologic relationship between the airway space and the soft and hard tissue during episodes of obstruction, when subjects were in the supine position for sleep.

The body of scientific knowledge on head posture, cervical spine angles and neck length in sleep apnea patients generally is lacking. More research on this subject could be important to improve sleep apnea diagnosis and produce predictable sleep apnea treatment outcomes.

The purpose of the study described below is to build on the limited existing knowledge regarding the effects of craniocervical morphology (cervical spine angles and head posture) and body positioning (supine *versus* upright) on the oropharyngeal airway in OSA subjects using 3D imaging techniques.

MATERIALS AND METHODS

The University of Nebraska Medical Center (UNMC) Institutional Review Board (IRB) approved this study protocol prior to initiating the study. Subjects with both a completed sleep study—either by polysomnogram (PSG) or home sleep study (HST) with a MDCT or CBCT—were included in this retrospective evaluation. Both PSG and HST diagnostic methods produced an AHI score for evaluation of sleep apnea severity.

Three clinic sites (UNMC Sleep Medicine Clinic; The Craniofacial Pain Center of Nebraska [CFP]; and Pioneer Greens Dentistry [PGD]) met the inclusion criteria and were invited to participate in this study.

A total of 221 subjects from three clinics (186 from UNMC, 22 from CFP and 13 from PGD) initially were screened. Subjects with a history of surgical treatment for OSA (e.g., maxillomandibular advancement, uvulopalatopharyngoplasty and/or other hard-soft tissue therapies), cervical spine surgeries (e.g., spinal fusion, disc replacement and/or other spine-related surgeries), cervical spine diseases or injuries (e.g., severe degenerative diseases, scoliosis, herniated discs) and other hard-soft tissue pathologies (e.g., cancer) were excluded from this study. Radiographs with low image quality or inadequate field of view of the airway also were excluded. After applying the exclusion criteria, the final sample size totaled 28 subjects (eleven from UNMC, fourteen from CFP and three from PGD).

Image Acquisition

Radiographic scans varied by clinic. All UNMC Sleep Medicine Clinic radiographs were taken either with the GE Lightspeed Pro or GE V CT (GE Healthcare, Chicago, IL). The voxel size of the head and neck CTs was 0.48 mm. The CT head typically captured the image from the vertex of the cranium to approximately cervical vertebrae 2. The CT neck typically captures the image from above sella to the carina of the trachea, the ridge at the base of the trachea that separates the openings of the right and left main bronchi.

Although the protocol was standardized, one of a possible 18 trained radiology technicians in the radiology department performed the image acquisitions. Subjects were instructed to lie down in the supine position with their head resting gently on a towel without any head positioner. Subjects were instructed to adjust themselves into the most comfortable position before scanning. During scanning, which took about 10 to 15 seconds, they were instructed to hold still to prevent motion artifact.

All radiographic scans from the CFP were taken with the i-CAT 17-19 CBCT (Imaging Sciences International LLC, Hatfield, PA) with a field of view of 23x17 cm and a voxel size of 0.3 mm. Subjects were instructed to stand upright and look straight ahead as if looking into a mirror. They were adjusted to have Frankfort horizontal plane parallel to the floor, if possible.

All radiographic scans from the PGD Clinic were taken with the Galileos GAX5 CBCT (Sirona Dental, Long Island City, NY) with a field of view of 15x15 cm and a voxel size of 0.15 mm. Subjects were instructed to stand upright with relaxed back and shoulders and to look straight as if looking into the distance. Subjects then bit into a bite block to fix the head position. All scans were exported in the DICOM file format for analysis in Dolphin 3D imaging software version 11.8 (Dolphin Imaging and Management Solutions, Chatsworth, CA).

Both examiners (BL and KS) were blinded to the subjects' age, gender, BMI and AHI values during measurements of the airway parameters, craniocervical posture, cervical spine angles and cervical spine length.

Airway Measurement Using Dolphin 3D imaging

To identify the airway, all scans first were oriented in the midsagittal plane using the incisive canal and the second cervical vertebrae (C-2) as guiding landmarks. The radiolucent airway was visualized easily due to its contrast with the surrounding radiopaque soft and hard tissues.

To ensure reproducibility of the measurements, the oropharyngeal airway boundaries were chosen based on easily identifiable hard- and soft-tissue landmarks (Guijarro-Martinez and Swennen, 2013). Table 1 lists in detail a modified description of the superior, inferior, anterior and posterior landmarks used in this study. The TV is partitioned further into the upper (uOP) and lower (lOP) oropharyngeal airway volume (TV = uOP + lOP). The upper and lower oropharyngeal airway are separated by a horizontal plane, parallel to the horizontal aspect of the radiograph, extending from the base of the soft palate and uvula back to the adjacent cervical vertebrae (Fig. 1). The lOP is the difference between TV and uOP.

Measurement of airway volume using Dolphin 3D software consisted of three steps. First, the boundaries for the total, uOP and lOP were set according to the landmarks described above. Second, a number of "seeds" were placed within the selected area of interest.

Table 1. Cephalometric landmarks for the borders of the oropharyngeal airway.

	Boundaries	Plane	Landmarks
Total volume (TV)	Superior	Sagittal	Posterior nasal spine (PNS) to the tip of the odontoid process
	Inferior		Tip of the epiglottis to the posterior border of the adjacent cervical vertebrae along a plane parallel to the horizontal border of the film
	Anterior		PNS to the tip of the epiglottis
	Posterior		Posterior border of the cervical vertebrae at the level of epiglottis to the tip of the odontoid process
Upper oropharyngeal volume	Superior	Sagittal	PNS to the tip of the odontoid process
	Inferior		The base of the uvula to the posterior border of the adjacent cervical vertebrae along a plane parallel to the horizontal border of the film
	Anterior		PNS to the base of the uvula
	Posterior		Posterior border of the cervical vertebrae at the level of uvula to the tip of the odontoid process
Lower oropharyngeal volume	Superior	Sagittal	The base of the uvula to the posterior border of the adjacent cervical vertebrae along a plane parallel to the horizontal border of the film
	Inferior		Tip of the epiglottis to the posterior border of the adjacent cervical vertebrae along a plane parallel to the horizontal border of the film
	Anterior		Base of the uvula to the tip of the epiglottis
	Posterior		Posterior border of the cervical vertebrae at the level of epiglottis to the tip of the odontoid process

Third, the threshold value was adjusted using the interactive threshold interval technique (El and Palomo, 2010). The airway volumes then were calculated automatically (Fig. 2). To calculate the minimum cross-sectional area (MCA), the same boundaries in the prior steps were maintained. The upper and lower limits were selected and the MCA tool was activated.

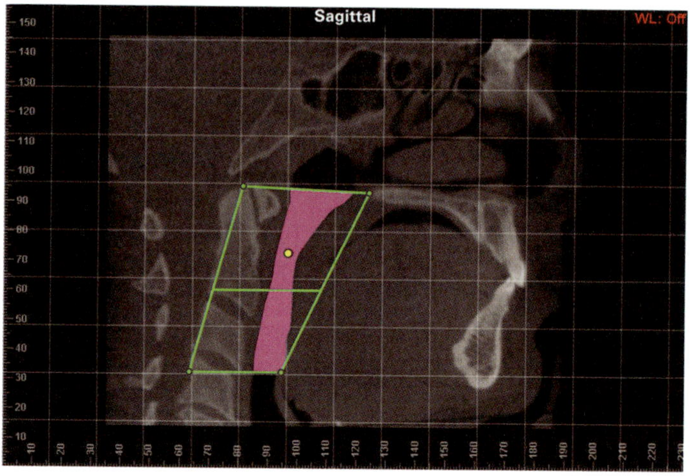

Figure 1. Delineation between upper and lower oropharyngeal airway spaces in Dolphin software. A horizontal line parallel to the horizontal border of the film at the level of the base of the uvula separates the upper and lower oropharyngeal airway.

Cervical Spine Angle Measurements Using Dolphin 3D imaging

In the multi-planar reconstruction view, all radiographs were oriented to the midsagittal plane of the cervical spine using the odontoid process of C2 in the coronal slice and the incisive canal in the sagittal slice (Fig. 2, *top*). Four angular measurements (C1-C2, C2-C3, C3-C4 and C1-C4) were made using the horizontal lines bordering the superior and inferior borders of each cervical spine vertebrae (Fig. 3). Cervical angles were assigned the standard kinematic nomenclature in a right-handed Cartesian coordinate system, with lordotic (extension) angles denoting the "-" sign and kyphotic (flexion) angles denoting the "+" sign (Jackson et al., 1993).

Craniocervical Posture Measurements Using Dolphin 3D imaging

Using the same mid-sagittal view in the cervical spine angle analysis, three craniocervical posture measurements were made. The first two measurements utilized the Rocabado analysis (de Oliveira et al., 2012), which consists of an angular and linear measurement. The angular measurement consisted of the McGregor line—which is the line drawn between the PNS and the most inferior point of the occipital bone—to the odontoid plane. The linear measurement consisted of a line from the base of the occipital bone to the mid-point of the posterior arch of first

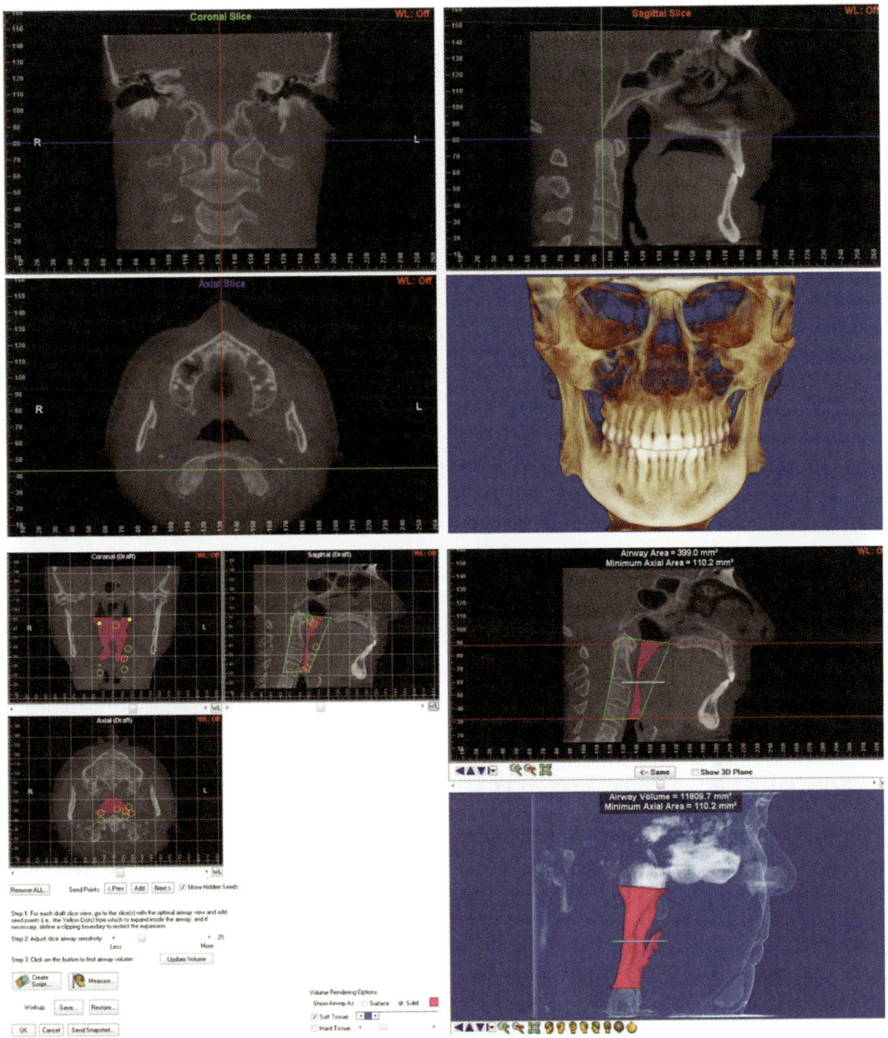

Figure 2. Determination of the TV, uOP and lOP in Dolphin Software. The mid-sagittal plane is set using the four-panel view to visual the odontoid process and incisive canal (*top*). The airway volume and MCA tools are located all in one area (*bottom*).

cervical vertebrae (C1). A third measurement, the McRae line—which extends from basion to opisthion (posterior border of foramen magnum—to odontoid plane angle, was added because PNS was not present in several radiographic scans (Fig. 4).

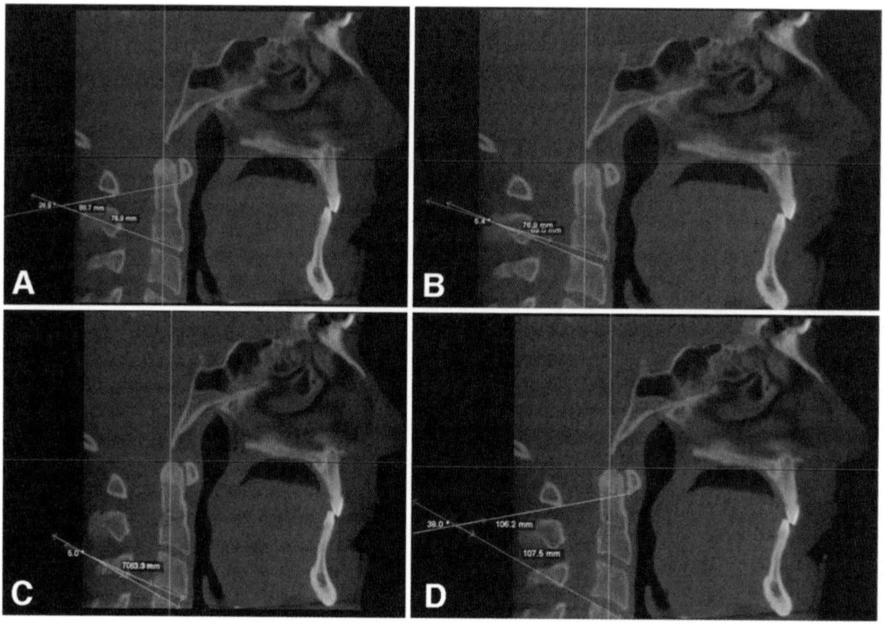

Figure 3. Cervical spine angles analysis. Spinal angles C1-C2 (*A*), C2-C3 (*B*), C3-C4 (*C*) and C1-C4 (*D*) are determined by drawing horizontal tangent lines to the superior and inferior borders of each respective cervical vertebrae and measuring the angles between each line.

Cervical Spine Length Using Dolphin 3D imaging

Using the same midsagittal view in the previous analyses, the cervical spine length was measured from the tip of the odontoid process of C2 to the midpoint, along the inferior border of third cervical vertebrae (C3).

Statistical Analysis and Measurement Error

Fisher exact test was used for the comparison of categorical data due to small sample size. For continuous variables, if data are distributed normally, the t-test was used for comparison. If data were not distributed normally, a non-parametric method was used. Spearman correlation was used for describing the monotonic relationship among numerical variables. Data were analyzed on SAS®9.4 (SAS Institute Inc., Cary, NC).

Intra- and inter-examiner reliability tests were performed to assess the reproducibility of identifying the hard- and soft-tissue landmarks used in each measurement. For intra-examiner reliability, ten subjects

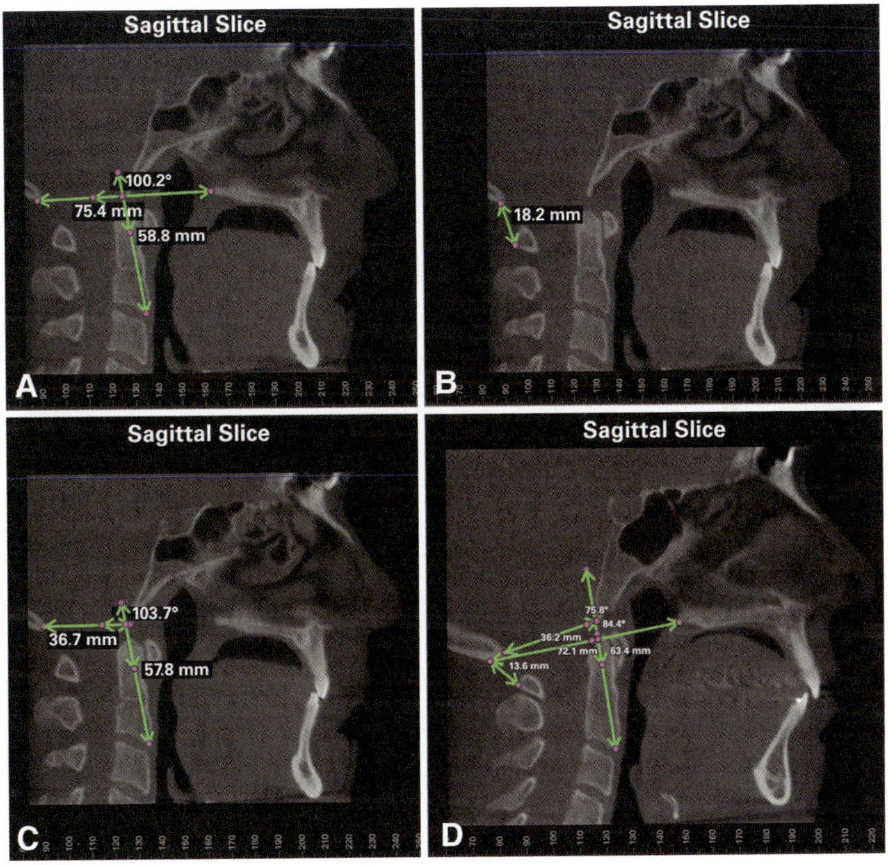

Figure 4. Craniocervical posture analysis. *A:* Craniocervical angle is measured from the intersection between McGregor line (PNS-occipital base) and odontoid plane (tip of C2 to anterior-inferior point). *B:* Atlas-Occiput (AO) length is measured from the base of the occipital bone to the posterior arch of Atlas (C1). *C:* The McRae line is drawn from basion to opisthion. *D:* Both the McRae and McGregor line angles are measured in all scans when possible.

from the three clinics (three from UNMC, six from CFP and one from PGD) were selected randomly two weeks after initial evaluation for repeated measurements of all airway, craniocervical posture, cervical spine angle and cervical spine length measurements. For inter-examiner reliability, two examiners (BL and KS) independently repeated all measurements on the same ten subjects. The Pearson correlation coefficient was calculated for each variable.

Substantial to almost perfect agreement was noted in both the intra- and inter-examiner reliability. The mean intra-examiner reliability was 0.87 and ranged between 0.54 and 1.00. The mean inter-examiner reliability was 0.82 and ranged between 0.21 and 1.00.

RESULTS

Twenty-eight subjects (17 females, 11 males) were included in this study. The mean age of the "normal" (N = 7) sleep apnea group was 39.7 years, the "mild" (N = 9) group 47.6 years, the "moderate" (N = 7) 59.3 and the "severe" (N = 5) group 49.2 years. The mean BMI of the normal group was 31.6, mild group 25.3, moderate 29.1 and severe 30.8. No correlation was observed between AHI and BMI in the present study. A statistically significant difference ($p < 0.05$) in AHI score was observed between males and females (Fig. 5). A weak statistical correlation was noted for sleep apnea severity and age (Fig. 6).

No significant associations were observed between age, gender, BMI and the following airway variables: TV, MCA, location of MCA, uOP and lOP.

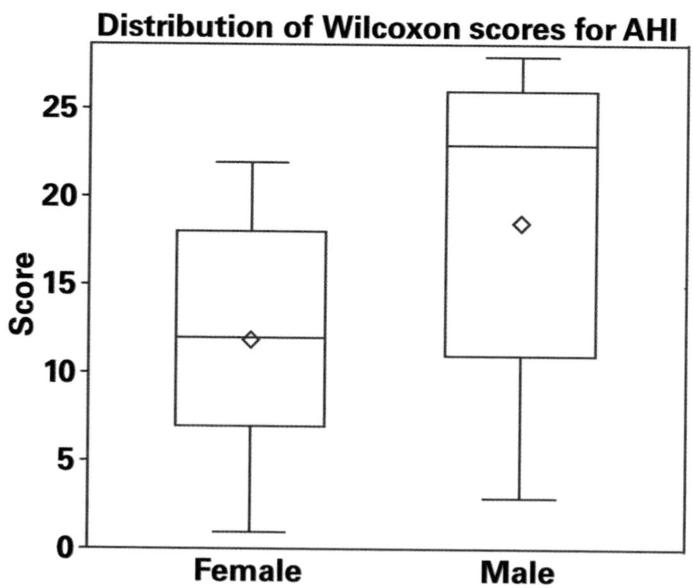

Figure 5. Gender difference in AHI score.

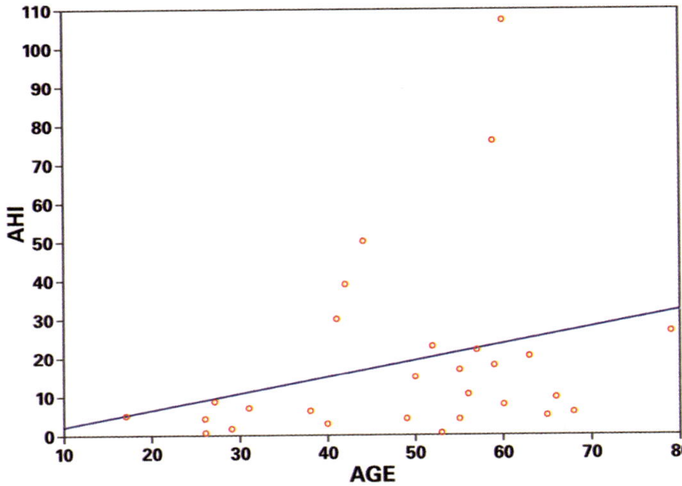

Figure 6. Association between age and AHI.

Craniocervical Posture and BMI

A significant negative correlation was noted between the McGregor-Odontoid plane angle (McG-OP) and BMI (Fig. 7). A significant negative correlation was noted between the McRae-Odontoid Plane angle (McR-OP) and BMI. A significant negative correlation was observed between AO length and BMI.

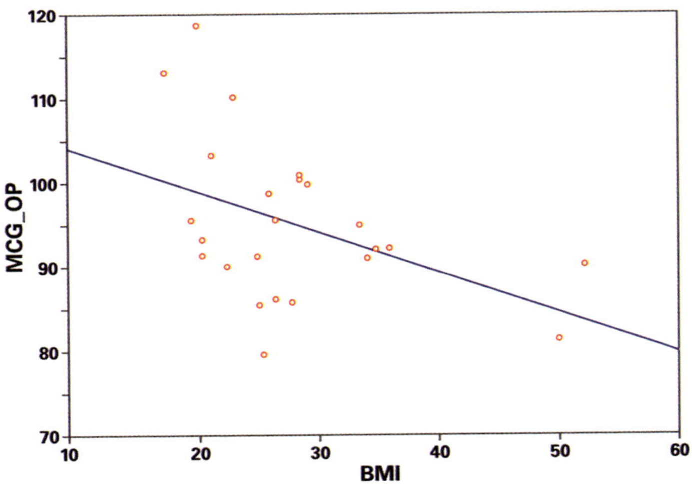

Figure 7. Association between craniocervical posture (McG-OP) and BMI.

Cervical Spine Angles and Age

A significant negative correlation was noted between age and C2-C3 angle as well as C1-C4 angle.

Cervical Spine Neck Length and Gender

A statistically significant difference in C2-C3 length was observed between male (mean 55.5 + 4.5 mm) and female (50.9 + 3.6mm) subjects (Fig. 8).

Craniocervical Posture Variables and AHI Score

No significant relationship observed between craniocervical posture variables and AHI score in the present study.

Craniocervical Posture and Airway Volume

Total airway volume, MCA and uOP volume were correlated positively with the McG-OP (Figs. 9-11). lOP was correlated with both the McG-OP and McR-OP.

Association between McG_OP and McR_OP Angles

The McG_OP and McR_OP were correlated nearly perfectly (Fig. 12).

Cervical Spine Angles and AHI

No significant associations were noted between cervical spine angles (C1-C2, C2-C3, C3-C4 and C1-C4) and AHI.

Cervical Spine Angles and Airway Volume

No significant associations were noted between cervical spine angles (C1-C2, C2-C3, C3-C4 and C1-C4) and the following airway dimensions: TV, minimum area, location of MCA, uOP and lOP.

Cervical Spine Length (C2-C3) and Airway Volume

A significant but weak positive association was noted between spine length and TV. No significant association was noted between spine length and MCA. A positive correlation was noted between spine length and location of MCA. For every unit increase of C2-C3 length, the odds of

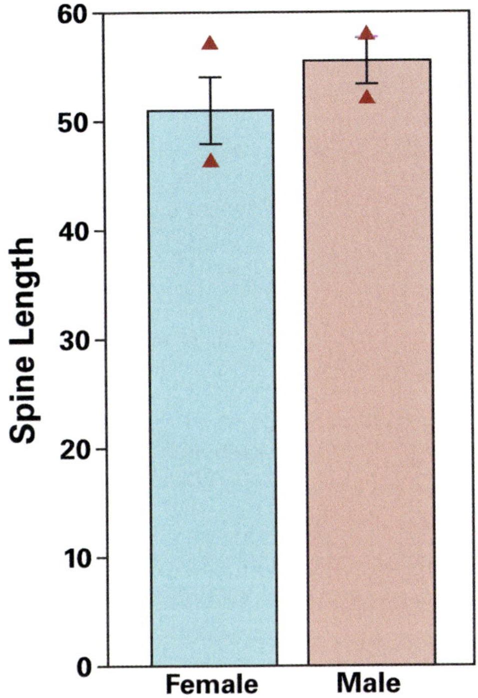

Figure 8. Gender differences in cervical spine length.

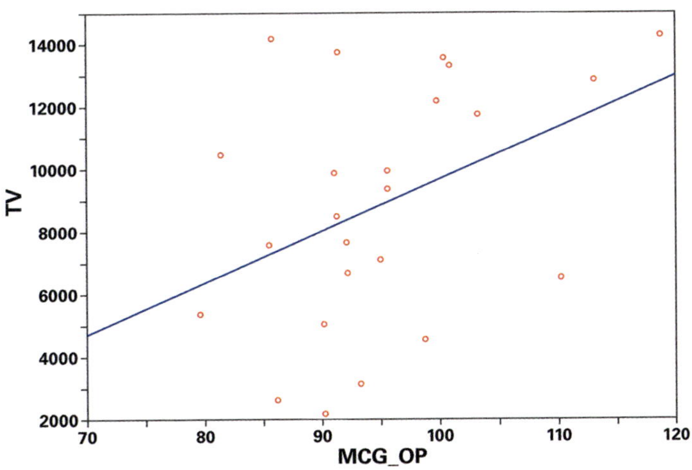

Figure 9. Association between craniocervical posture and TV.

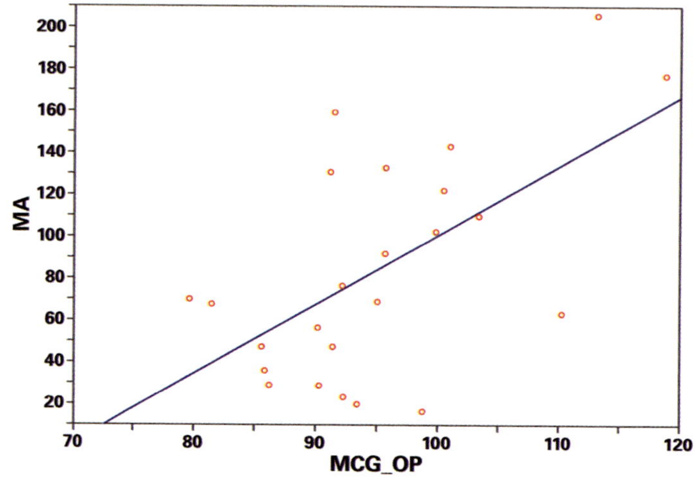

Figure 10. Association between craniocervical posture and minimum area.

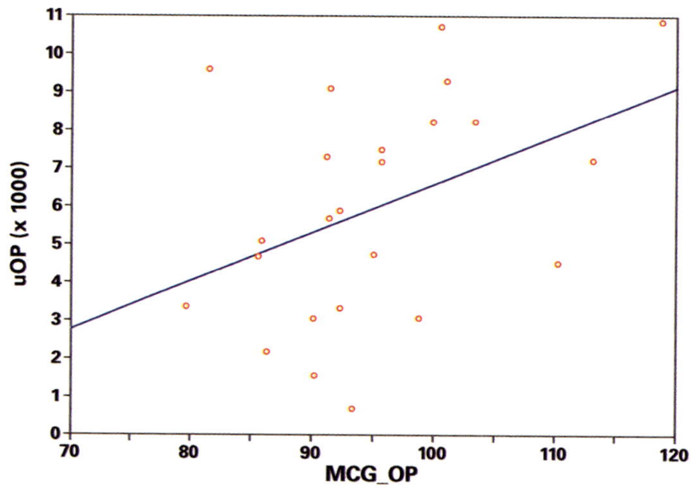

Figure 11. Association between craniocervical posture and uOP.

retro-lingual (RL) to retro-uvular (RU) is expected to increase 1.767 with 95% confidence interval. A significant positive association was noted between uOP and spine length. However, no relationship was noted between lOP and cervical spine length.

Craniocervical Posture and Cervical Spine Angles

A significant positive association was noted between McG_OP and C1-C2 angles (Fig. 13). Similarly, a significant positive association was

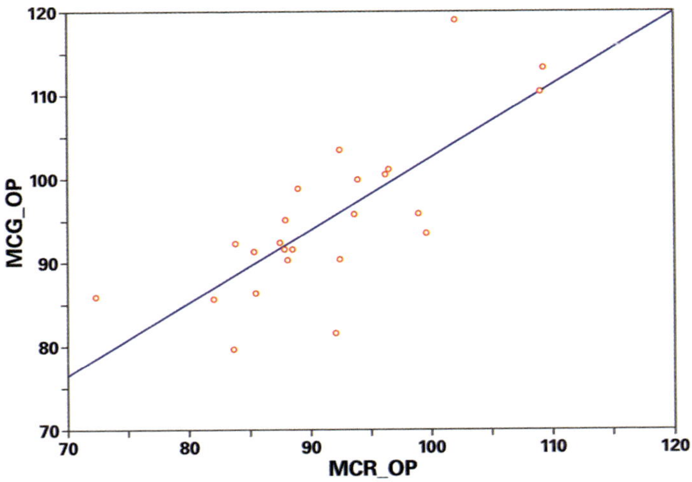

Figure 12. Association between McG and McR.

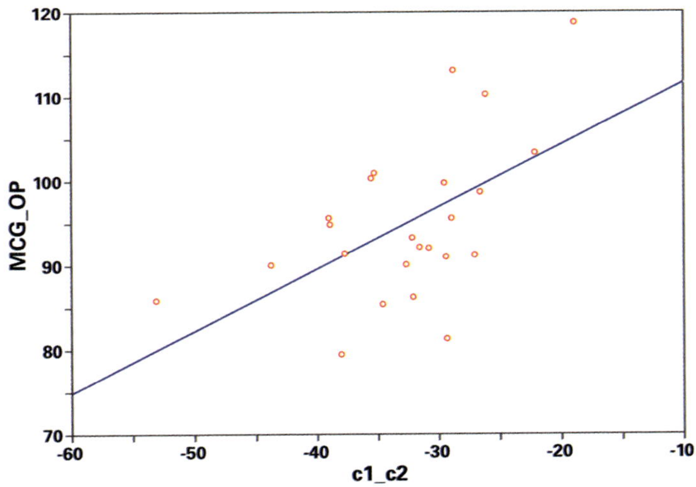

Figure 13. Association between craniocervical posture and C1-C2 spinal angle.

observed between McR_OP and C1-C2 angles. The McG_OP is correlated with AO length.

Imaging Position and Airway Volume Parameters

A significant mean difference of 3,677.0 mm^3 in TV was noted between the images taken in the supine and upright position. A significant mean difference of 2,893.1 mm^3 in uOP was noted between the images taken in the supine and upright positions (Fig. 14). The difference in lOP

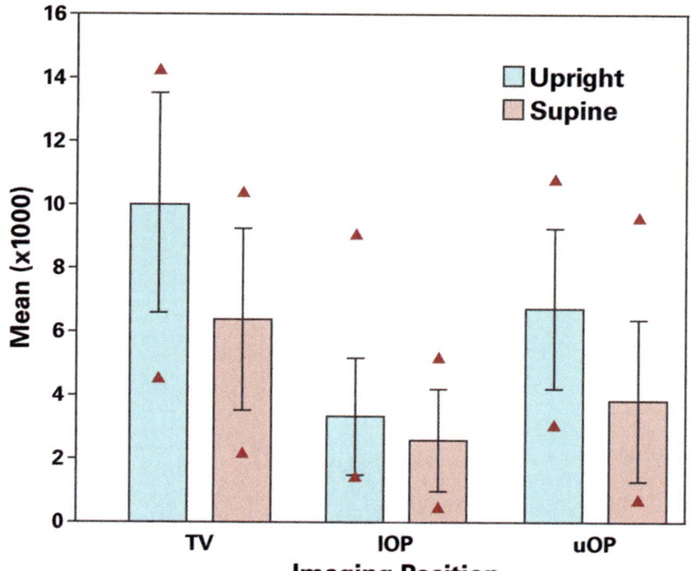

Figure 14. Airway volume and imaging position.

between the images taken in the supine and upright positions was not significant statistically (Fig. 14). A significant mean difference of 50.7 mm^2 in MCA was noted between images taken in the supine and upright positions (Fig. 15). No significant difference in the location of the MCA was noted between images taken in the supine and upright positions. In the images taken in the supine position, 91% had minimum areas in the retro-uvula *versus* 9% in the retro-lingual area. Meanwhile in the upright position, 71% had minimum areas in the retro-uvula *versus* 29% in the retro-lingual area (Fig. 16).

Imaging Position and Craniocervical Posture

There were no statistically significant differences in the measurements of McG_OP, McR_OP and AO length between images taken in the supine and upright positions (Fig. 17).

Imaging Position and Cervical Spine Angles

No statistically significant differences were seen in the following spine angles measured in the images taken in the supine and upright positions: C1-C2, C2-C3, C3-C4 and C1-C4 (Fig. 18).

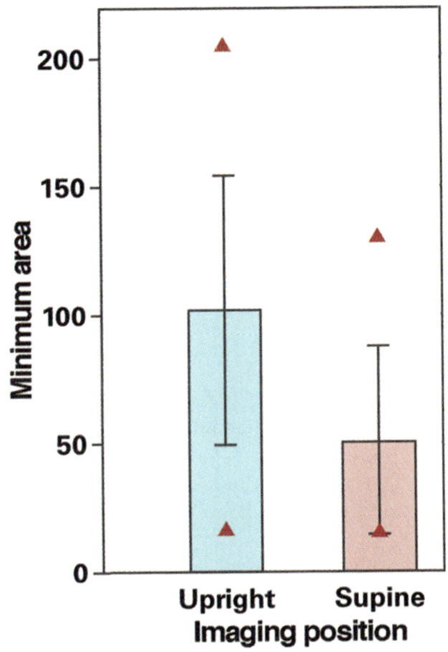

Figure 15. Minimum area of the airway and imaging position.

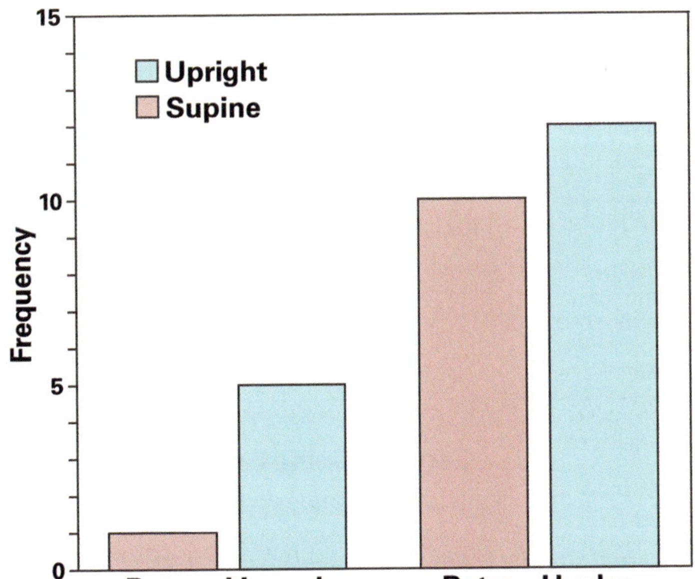

Figure 16. Location of minimum area of the airway and imaging position.

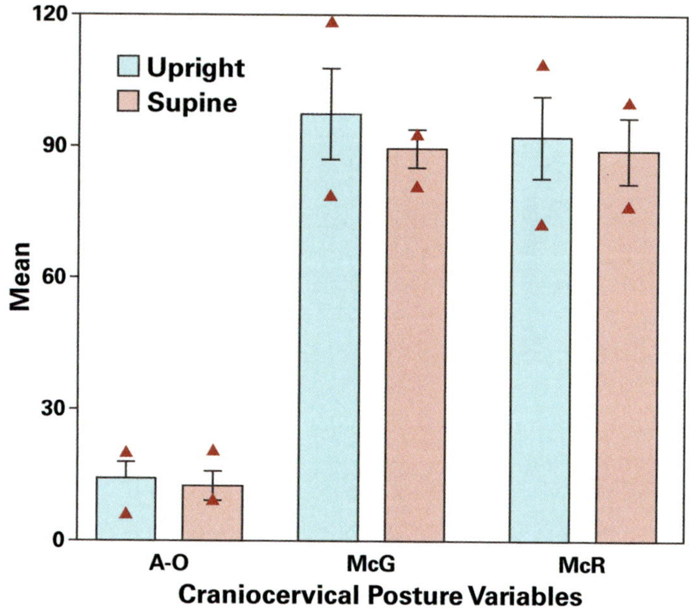

Figure 17. Imaging position and craniocervical posture.

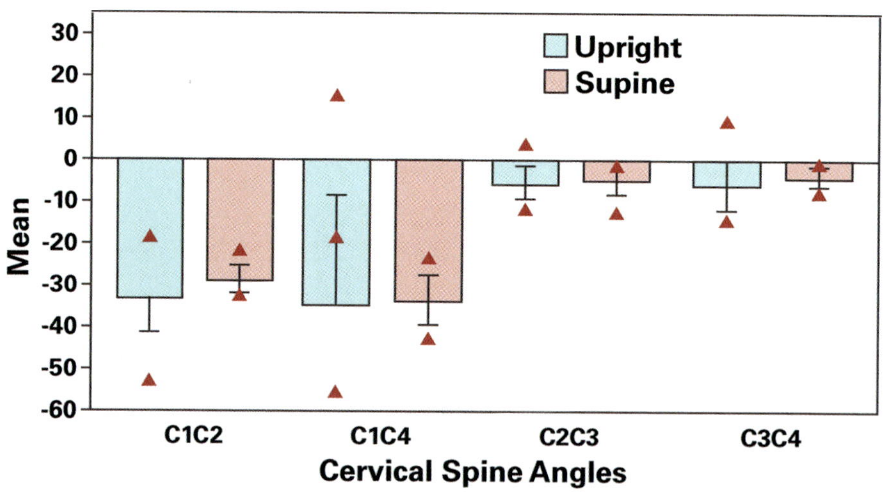

Figure 18. Imaging position and cervical spine angles.

Imaging Position and Cervical Vertebrae (C2-C3) Neck Length

No significant difference in C2-C3 neck length was noted between images taken in the supine position and upright positions (Fig. 19).

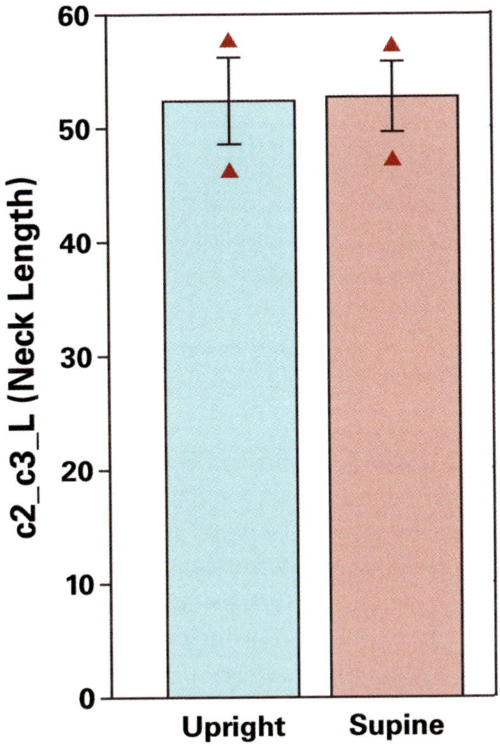

Figure 19. Imaging position and neck length.

DISCUSSION

There was a significant difference in AHI scores between the males and female subjects. This finding is similar to previous findings on gender differences in the prevalence of OSA. Both the Wisconsin and Swiss sleep studies reported that the prevalence of OSA in males are about 2-3x higher than females (Young et al., 1993, 1997, 2002, 2009; Punjabi, 2008).

There was a weak statistical correlation between AHI and age in our study. This correlates with previous studies that showed that the prevalence of OSA increases with age (Kapur, 2010) due to increased fat deposition around the pharynx, lengthening of the soft palate and changes in parapharyngeal structures (Malhotra et al., 2006).

Our study failed to show any statistically significant association between BMI and AHI, though a positive trend was observed in the data.

Quan and associates (1997) demonstrated a strong association between BMI and AHI. The Sleep Heart Health Study showed that a weight gain of 10 kg can confer a 2.5x increase in the chances of increase the AHI by 15/hour.

No differences in cervical spine angles or craniocervical posture were noted in the supine *versus* upright position in the present study. Regarding positional changes in cervical spine angles, a 5° decrease in cervical spine angle has been shown in the supine position, which the author attributed to gravitational forces (Martensen, 2015). Jun and colleagues' study (2014) found an increased cervical lordosis in the upright position. No studies have examined changes in head posture in the supine position.

Our study showed a significant difference in TV, uOP and MCA between upright and supine positions. Subjects in the supine position exhibited a significantly smaller TV compared to the subjects in the upright position. The MCA was smaller significantly in the supine group compared to the upright group. These findings agreed with the study by Camacho and coworkers (2014) that indicated a significant decrease in the total upper airway volume from 14,100 to 9,500 mm^2. The MCA also decreased from 120 to 30 mm^2.

A significant negative correlation was noted between the C2-C3 and C1-C4 angles with age. This finding implies that the cervical spine alignment becomes more curved inwardly (lordotic) with age. Due to a lack of normative data, controversy exists regarding whether the cervical spine becomes more lordotic or kyphotic (straightens) with age (Kim et al., 2014). It is believed that with age and disc degeneration, the spine becomes more kyphotic, whereas others believe the spine becomes more lordotic to maintain gaze (Park et al., 2013). Our results should be interpreted with caution for two reasons. First, the sample size is small (n = 28). Second, among the subjects who were in the supine imaging position, a few patients had their necks supported by a pillow during CT scans.

Significant positive association was noted between the McG and McR angles and the C1-C2 angle. This is not surprising since McG and McR are measurements of head extension and flexion. The atlanto-axial joint of C1-C2 allows mostly for head rotation, but also some head flexion. This implies that the upper head-neck joints involved in head flexion-extension are related.

To date, no CBCT study examining the relationship between head posture and OSA subjects is reported. The Solow studies showed a significant increase in the anteroposterior dimension of the lower oropharyngeal airway with craniocervical extension (Solow et al., 1993, 1996). Our study found a significant positive correlation between craniocervical angulation and TV, uOP, lOP and MCA. These data imply that airway volume increases with head flexion, not extension. These results conflict with findings from a previous cephalometric and 3D imaging study (Muto et al., 2002), which reported a 10° increase that resulted in an approximate 4 mm increase in airway space. Gurani and associates (2016) reported an MRI study that showed head extension resulted in an increased hypopharyngeal airway volume.

These disagreements might be due to differences in the way the radiographs were taken. Subjects who had MDCTs taken in the supine position had a small pillow placed under the neck, which could have tilted the head. However, patients were instructed to adjust their head and/or pillow into a more comfortable position before the acquisition of images. Likewise, the CBCT imaging subjects were instructed to be in the natural head position/posture; however, it is unclear if all subjects were positioned this way consistently.

Interestingly, significant negative associations were noted between McG_OP angle, McR_OP angle, AO length and BMI. These results imply that head extension is correlated highly with BMI. For every unit increase in BMI, 0.5° increase in head extension was noted. These results tend to agree with those proposed by Solow and colleagues (1996). Craniocervical extension from the natural head position is an adaptive mechanism used in OSA subjects, who typically have high BMI, as a way to increase airway patency.

The normal values for the Rocabado parameters for head extension-flexion is 96 to 106° or 4 to 9 mm. Our study showed that the subjects in supine position were in an extended head position (89.5° ± 4.3°), whereas the subjects in upright position exhibited a normal position (97.3° ± 10.2°). Although these differences were not significant statistically, the differences could be related to patient positioning. Moreover, supine subjects had higher AHI and BMI values and lower airway volumes, which could imply that these patients were in the compensated extended head position to maintain airway patency (Hellsing et al., 1987; Solow et al., 1996; Piccin et al., 2016).

Almost all prior cephalometric studies examining head posture have used variations of the Solow or Rocabado analyses (Solo et al., 1993, 1996; de Oliveira et al., 2012) and have depended largely on a wide field of view and the presence of craniofacial structures (e.g., nasion, sella and PNS). For subjects with MDCTs that have limited field of view, a modified method from the Rocabado analysis using the McRae line, instead of the McGregor line, was used in the present study. A nearly perfect correlation was found between McGregor and McRae angles (0.77).

A significant difference in C2-C3 length was noted between males and females. This difference is expected since statural differences between genders have been established well in the literature, as males on average are taller (Graber and Swain, 1985). Furthermore, studies that quantified the differences in neck geometry found that most anthropometric parameters were significantly smaller in females compared to males. Female C3-C7 vertebrae were smaller in the anteroposterior direction and were weaker than male necks in both flexion and extension (Vasavada et al., 2008).

Furthermore, other studies found that males have longer upper airway length and proposed that these gender differences could explain partially the predisposition of men to OSA (Malhotra et al., 2002; Ronen et al., 2007). Significant but weak positive association was noted between spine length and TV. In addition, a significant positive association was noted between the uOP and spine length. Our finding partially agrees with Kim and associates (2011), who found that longer neck length did not result in increased uOP.

Another interesting finding was that for every unit increase in spine length, there is a 1.767x increase in the odds of the minimum constriction area shifting from the retrolingual to the retrouvula area. Camacho and colleagues (2014) noted that the most common site of constriction was located in the retropalatal (uvula) area. To our knowledge, there is no study specifically examining the relationship between neck length and the location of the minimum area.

Study Limitations

The first major limitation of this study is a small sample size of 28 subjects. In general, the smaller the sample size, the more noise is seen in the results. For example, we expected to see a correlation between AHI and BMI; no statistical significance was detected in this dataset, however, which could be due to the small sample size.

The second major limitation is the heterogeneity of the subject pool, which automatically introduces variability in the data. Subjects were recruited from three different clinic sites, each with different demographics and radiology protocol. First, there are major differences between the UNMC and CFP-PGD subjects. UNMC hospital patients had CT scans for other health conditions that were unrelated to sleep apnea. These patients likely presented with more complex medical histories and potentially significant medical comorbidities compared to the CFP-PGD patients. Secondly, the radiology protocols varied among the three clinics. The UNMC radiology department has 18 radiology technicians who might have slightly different routines for capturing CTs. The CFP and PGD clinics used different CBCT machines that have different head positioners.

The third major limitation is inherent in the use of CBCTs and MDCTs to scan the airway. While both imaging techniques have been validated for use in sleep apnea studies, they represent snapshot images of a dynamic process that occurs during sleep. Other imaging techniques (e.g., the four-dimensional [4D] MDCT [Wagnetz et al., 2010] or the drug-induced sleep endoscopy [DISE]) are equipped better at capturing airway volumes and location of MCA in real time.

CONCLUSIONS

To date, no CBCT studies have examined the relationship among craniocervical posture, cervical spine angles and the upper pharyngeal airway space in OSA subjects. Our study showed that craniocervical posture could impact the airway dimension significantly. Cervical spine angles, in the absence of spine pathology, have little-to-no impact on the airway space. However, the atlanto-axial joint, which allows mostly for head rotation and some flexion-extension, may impact the upper airway space. Furthermore, our study showed that airway dimensions are decreased significantly in the supine position, which typically is the sleeping position. For clinicians who frequently order MDCTs, we proposed using the McRae line for evaluating head posture in limited view neck CTs. Altogether, these findings may help providers better assess the clinical characteristics of OSA patients.

REFERENCES

Ayappa I, Norman RG, Seelall V, Rapoport DM. Validation of a self-applied unattended monitor for sleep disordered breathing. J Clin Sleep Med 2008;4(1):26-37.

Cairns A, Wickwire E, Schaefer E, Nyanjom D. A pilot validation study for the NOX T3(TM) portable monitor for the detection of OSA. Sleep Breath 2014;18(3):609-614.

Camacho M, Capasso R, Schendel S. Airway changes in obstructive sleep apnoea patients associated with a supine *versus* an upright position examined using cone beam computed tomography. J Laryngol Otol 2014;128(9):824-830.

Davies RJ, Ali NJ, Stradling JR. Neck circumference and other clinical features in the diagnosis of the obstructive sleep apnoea syndrome. Thorax 1992;47(2):101-105.

de Oliveira LB, Cajaíba F, Costa W, Rocabado M, Lazo-Osório R, Ribeiro S. Comparative analysis of assessment of the craniocervical equilibrium through two methods: Cephalometry of Rocabado and cervical range of motion. Work 2012;41(Suppl 1):2563-2568.

Dobson GJ, Blanks RHI, Boone WR, McCoy HG. Cervical angles in sleep apnea patients: A retrospective study. J Vertebral Subluxation Res 1999;3(1):1-15.

Driver HS, Pereira EJ, Bjerring K, Toop F, Stewart SC, Munt PW, Fitzpatrick MF. Validation of the MediByte® type 3 portable monitor compared with polysomnography for screening of obstructive sleep apnea. Cana Respir J 2011;18(3):137-143.

El H, Palomo JM. Measuring the airway in 3 dimensions: A reliability and accuracy study. Am J Orthod Dentofacial Orthop 2010;137(4 Suppl): S50.e51-e59.

Graber TM, Swain BF. *Orthodontics: Current Principles and Techniques.* 6th ed. St. Louis: C.V. Mosby Co. 2017.

Guijarro-Martínez R, Swennen GR. Three-dimensional cone beam computed tomography definition of the anatomical subregions of the upper airway: A validation study. Int J Oral Maxillofac Surg 2013;42(9): 1140-1149.

Guilleminault C, Li KK, Philip P, Kushida CA. Anterior cervical spine fusion and sleep disordered breathing. Neurology 2003;61(1):97-99.

Gurani SF, Di Carlo G, Cattaneo PM, Thorn JJ, Pinholt EM. Effect of head and tongue posture on the pharyngeal airway dimensions and morphology in three-dimensional imaging: A systematic review. J Oral Maxillofac Res 2016;7(1):e1.

Han TS, Oh MK, Kim SM, Yang HJ, Lee BS, Park SY, Lee WJ. Relationship between neck length, sleep, and cardiovascular risk factors. Korean J Fam Med 2015;36(1):10-21.

Heinzer R, Vat S, Marques-Vidal P, Marti-Soler H, Andries D, Tobback N, Mooser V, Preisig M, Malhotra A, Waeber G, Vollenweider P, Tafti M, Haba-Rubio J. Prevalence of sleep-disordered breathing in the general population: The HypnoLaus study. Lancet Respir Med 2015;3(4):310-318.

Hellsing E, McWilliam J, Reigo T, Spangfort E. The relationship between craniofacial morphology, head posture and spinal curvature in 8, 11 and 15-year-old children. Eur J Orthod 1987;9(4):254-264.

Hirsch Allen AJM, Bansback N, Ayas NT. The effect of OSA on work disability and work-related injuries. Chest 2015;147(5):1422-1428.

Jackson BL, Harrison DD, Robertson GA, Barker WF. Chiropractic biophysics lateral cervical film analysis reliability. J Manipulative Physiol Ther 1993;16(6):384-391.

Jun HS, Chang IB, Song JH, Kim TH, Park MS, Kim SW, Oh JK. Is it possible to evaluate the parameters of cervical sagittal alignment on cervical computed tomographic scans? Spine (Phila Pa 1976) 2014;39(10): E630-E636.

Kapur VK. Obstructive sleep apnea: Diagnosis, epidemiology, and economics. Respir Care 2010;55(9):1155-1167.

Khan A, Than KD, Chen KS, Wang AC, La Marca F, Park P. Sleep apnea and cervical spine pathology. Eur Spine J 2014;23(3):641-647.

Kim EJ, Choi JH, Kim YS, Kim TH, Lee SH, Lee HM, Shin C, Lee SH. Upper airway changes in severe obstructive sleep apnea: Upper airway length and volumetric analyses using 3D MDCT. Acta Otolaryngol 2011;131(5):527-532.

Kim HJ, Lenke LG, Oshima Y, Chuntarapas T, Mesfin A, Hershman S, Fogelson JL, Riew KD. Cervical lordosis actually increases with aging and progressive degeneration in spinal deformity patients. Spine Deform 2014;2(5):410-414.

Lowe AA, Gionhaku N, Takeuchi K, Fleetham JA. Three-dimensional CT reconstructions of tongue and airway in adult subjects with obstructive sleep apnea. Am J Orthod Dentofacial Orthop 1986;90(5):364-374.

Malhotra A, Huang Y, Fogel R, Lazic S, Pillar G, Jakab M, Kikinis R, White DP. Aging influences on pharyngeal anatomy and physiology: The predisposition to pharyngeal collapse. Am J Med 2006;119(1):72.e9-e14.

Malhotra A, Huang Y, Fogel RB, Pillar G, Edwards JK, Kikinis R, Loring SH, White DP. The male predisposition to pharyngeal collapse: Importance of airway length. Am J Respir Crit Care Med 2002;166(10):1388-1395.

Muto T, Takeda S, Kanazawa M, Yamazaki A, Fujiwara Y, Mizoguchi I. The effect of head posture on the pharyngeal airway space (PAS). Int J Oral Maxillofac Surg 2002;31(6):579-583.

Ota M, Neo M, Aoyama T, Ishizaki T, Fujibayashi S, Takemoto M, Nakayama T, Nakamura T. Impact of the O-C2 angle on the oropharyngeal space in normal patients. Spine (Phila Pa 1976) 2011;36(11):e720-e726.

Park MS, Moon SH, Lee HM, Kim SW, Kim TH, Lee SY, Riew KD. The effect of age on cervical sagittal alignment: Normative data on 100 asymptomatic subjects. Spine (Phila Pa 1976) 2013;38(8):e458-e463.

Piccin CF, Pozzebon D, Scapini F, Corrêa EC. Craniocervical posture in patients with obstructive sleep apnea. Int Arch Otorhinolaryngol 2016; 20(3):189-195.

Punjabi NM. The epidemiology of adult obstructive sleep apnea. Proc Am Thorac Soc 2008;5(2):136-143.

Quan SF, Howard BV, Iber C, Kiley JP, Nieto FJ, O'Connor GT, Rapoport DM, Redline S, Robbins J, Samet JM, Wahl PW. The Sleep Heart Health Study: Design, rationale, and methods. Sleep 1997;20(12):1077-1085.

Ronen O, Malhotra A, Pillar G. Influence of gender and age on upper-airway length during development. Pediatrics 2007;120(4):e1028-e1034.

Solow B, Ovesen J, Nielsen PW, Wildschiødtz G, Tallgren A. Head posture in obstructive sleep apnoea. Eur J Orthod 1993;15(2):107-114.

Solow B, Skov S, Ovesen J, Norup PW, Wildschiødtz G. Airway dimensions and head posture in obstructive sleep apnoea. Eur J Orthod 1996; 18(6):571-579.

Vasavada AN, Danaraj J, Siegmund GP. Head and neck anthropometry, vertebral geometry and neck strength in height-matched men and women. J Biomech 2008;41(1):114-121.

Wagnetz U, Roberts HC, Chung T, Patsios D, Chapman KR, Paul NS. Dynamic airway evaluation with volume CT: Initial experience. Can Assoc Radiol J 2010;61(2):90-97.

Young T, Evans L, Finn L, Palta M. Estimation of the clinically diagnosed proportion of sleep apnea syndrome in middle-aged men and women. Sleep 1997;20(9):705-706.

Young T, Palta M, Dempsey J, Peppard PE, Nieto FJ, Hla KM. Burden of sleep apnea: Rationale, design, and major findings of the Wisconsin Sleep Cohort study. WMJ 2009;108(5):246-249.

Young T, Palta M, Dempsey J, Skatrud J, Weber S, Badr S. The occurrence of sleep-disordered breathing among middle-aged adults. N Engl J Med 1993;328(17):1230-1235.

Young T, Peppard PE, Gottlieb DJ. Epidemiology of obstructive sleep apnea: A population health perspective. Am J Respir Crit Care Med 2000;165(9):1217-1239.

THE EFFECTS OF TONGUE, HYOID AND PHARYNGEAL AIRWAY SPACE ON CRANIOFACIAL GROWTH

Yoon-Ji Kim, Youngjun Kim, Laehyun Kim, Jae-Jun Ryu

ABSTRACT

Craniofacial growth is influenced by the surrounding organs' functions such as breathing, swallowing and speaking. The Functional Matrix Theory, which was suggested by Melvin Moss in 1968, hypothesized that the growth of the craniofacial skeleton occurs in response to the functional needs of the adjacent soft tissues in which it is embedded; these soft tissues are known as *functional matrices*. According to the Functional Matrix Theory, the growth of the maxilla and the mandible occurs due to an enlargement of the nasal and oral cavities. This theory was followed later by other studies, which indicated that there may be a direct relationship between the skeletal unit and the functional matrices; this relationship is caused by the signaling of adjacent bone cells through mechanical receptors and the cellular network.

Malocclusion is the result of an interplay between innate genetic factors and external environmental factors. The local environmental factors that affect jaw development occur as a result of the specific muscular functions of the jaws. Therefore, with regard to orthodontic diagnosis, an assessment of the possible functional factors is important to identify the etiology of malocclusion and establish an appropriate treatment plan.

KEY WORDS: tongue, hyoid, pharyngeal airway, facial pattern, malocclusion

---

INTRODUCTION

The tongue is an organ that is situated in the oral cavity and is composed of both intrinsic and extrinsic muscles (Table 1). These muscles apply force constantly to the surrounding structures (e.g., the pharyngeal airway, the maxillomandibular complex, the dentition). Therefore, the position and function of the tongue are important etiological factors in both skeletal and dental malocclusions (Adesina et al., 2013; Ihan Hren and Barbic, 2016).

Table 1. Extrinsic and intrinsic muscles of the tongue.

Extrinsic muscles	Intrinsic muscles
Genioglossus muscle	Superior longitudinal muscle
Hyoglossus muscle	Inferior longitudinal muscle
Styloglossus muscle	Transverse lingual muscle
Palatoglossus muscle	Vertical lingual muscle

THREE-DIMENSIONAL (3D) ANALYSIS OF THE POSITION AND VOLUME OF THE TONGUE

Until recently, only two-dimensional (2D) lateral cephalometric analysis of the tongue and its effect on malocclusions were available. There is a significant correlation between the area of the tongue, measured using lateral cephalograms and three-dimensional (3D) tongue volume (Liégeois et al., 2009); because the tongue is a 3D structure, however, 2D analysis cannot provide fully accurate information regarding the tongue's position and volume.

Many approaches, including the direct impression of the tongue in a protruded position, conventional computed tomography, ultrasonograms and magnetic resonance images (MRIs), have been used to assess the volume of tongue (Tamari et al., 1991*a,b*; Iida-Kondo et al., 2006; Liégeois et al., 2009; Shigeta et al., 2011; Ihan Hren and Barbic, 2016). Following the advent of cone-beam computed tomography (CBCT), attempts have been made to analyze the effects of the tongue's position and volume on craniofacial and dental patterns.

CBCT Image Acquisition and Reorientation

To carry out accurate volumetric analysis of the position and volume of the tongue, CBCT must be performed in natural head position with the patient's tongue and lips also in a relaxed position, while the occlusion is in maximal intercuspation. Because the tongue is a highly mobile organ, Graber and colleagues (1997) suggested that patients should be asked to swallow before undergoing the radiologic examination to find the tongue's relaxed position. As in all 3D volumetric analyses, CBCTs should be reoriented using a reference plane—either horizontal or midsagittal—to allow for an accurate analysis.

Tongue Positions

To assess tongue position, the distance from the dorsum of the tongue and tongue tip to the palate is measured. Hyoid position also can be assessed by measuring the anteroposterior and vertical distance from the most anterosuperior point of the hyoid to an anatomical reference point (e.g., anterior nasal spine [ANS]). An example of tongue and hyoid position measurement is shown in Figure 1.

Tongue Volume

To measure the volume of the tongue, the tongue must be segmented as a 3D volumetric object. 3D models of the patient's tongue can be reconstructed using various segmentation software programs (Fig. 2). As the tongue is composed of many intrinsic and extrinsic muscles, its posterior and inferior borders cannot be defined clearly in radiographic images. Therefore, a bony anatomical landmark is needed to define the tongue's inferior and posterior borders and allow standardized volumetric analysis.

The hyoid bone can be used as a reference to determine the posterior and inferior borders of the tongue because it is located at the base of the tongue (Fig. 3); notably, the hyoglossus muscle originates from the hyoid. It should be noted, however, that position of the hyoid bone varies according to the position of the maxilla and the mandible (Bibby and Preston, 1981; Chang, 1987; Tallgren and Solow, 1987; Adamidis and Spyropoulos, 1992; Abu Allhaija and Al-Khateeb, 2005). These variations should be taken into consideration.

TONGUE POSITIONING IN PATIENTS WITH NORMAL SKELETAL PATTERNS

The tongue is a very mobile organ, which means that its position is highly variable. This variability is reflected in the available statistical data distribution, which indicates that the variables relating to the tongue positions are not distributed normally. Based on the tongue position acquired from adult Class I malocclusion patients, the estimated mean tongue positions in 10 points on the dorsum of tongue have been derived using a statistical model (Table 2). In the sagittal view, the dorsum of the tongue is closer to the palate posteriorly. In the frontal view, a slight depression in the middle of the tongue can be observed.

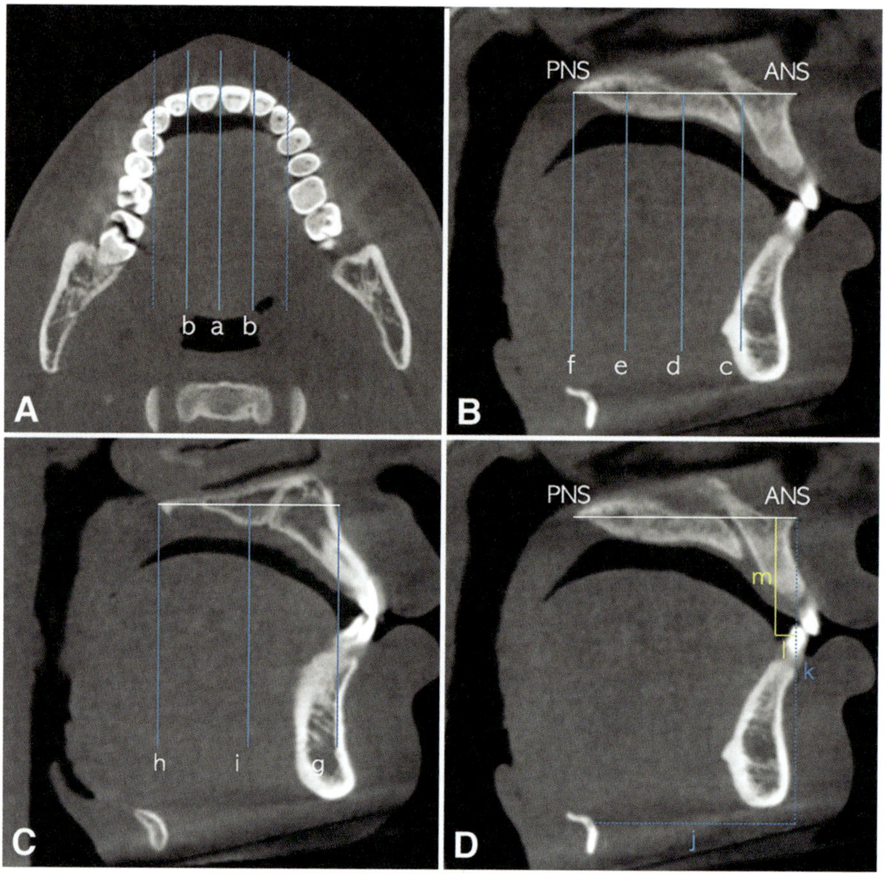

Figure 1. Measurement of tongue position. Cone-beam computed tomography (CBCT) scans were re-oriented using the midsagittal reference plane (MSRP), constructed using ANS, posterior nasal spine (PNS) and crista galli as the reference plane. A plane perpendicular to the MSRP, which passed through ANS and PNS, was defined as the palatal plane and was used as a horizontal reference plane. *A:* Two additional planes running parallel to the MSRP and bisecting the line drawn from the MSRP to the most prominent points of the palatal cusps of the right and left first maxillary molars were constructed for tongue position measurements; these were defined as the lateral planes. *B-C:* In each plane, the shortest distances from the dorsum of the tongue to the palatal gingiva were measured. The average of the right and left values was used to denote the position of the tongue in the lateral plane. The horizontal and vertical positions of the most anterosuperior point of the hyoid bone and tongue tip were measured on the MRSP.

(*Figure 1* continued) *D:* Linear distances from the palatal plane and a plane perpendicular to the palatal plane and passing through the ANS were used as the reference planes for measuring the vertical and horizontal positions of the tongue tip and hyoid, respectively.

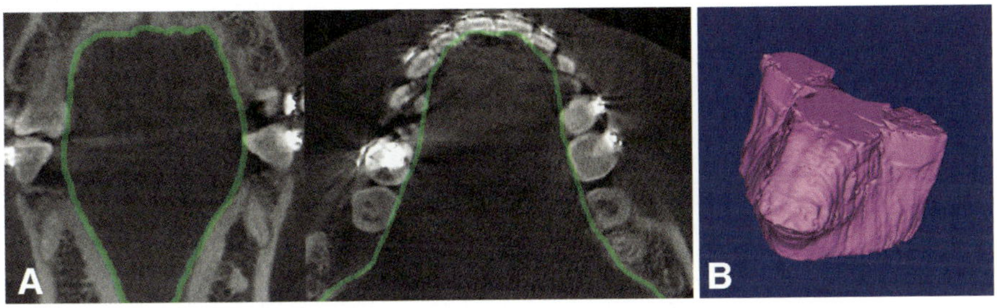

Figure 2. *A.* The 3D model of the tongue of each patient can be reconstructed using segmentation. The 2D boundaries are defined interactively on the axial, sagittal and coronal image slices by the user. *B.* The 3D model of the subject's tongue then is reconstructed from the segmented 2D slice images.

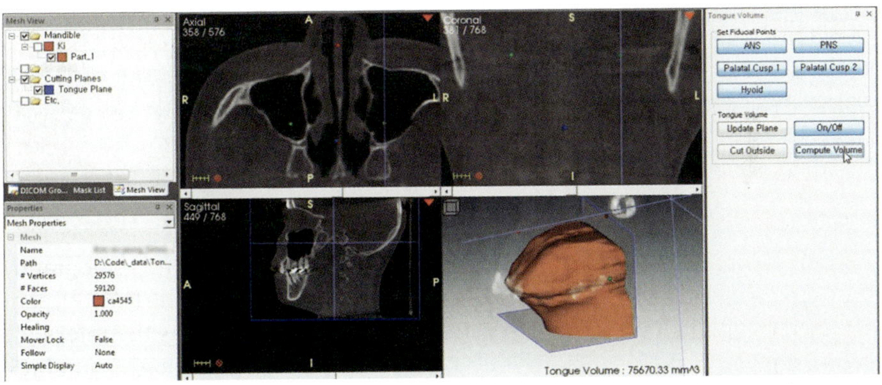

Figure 3. After the modeling procedure, five anatomical points—ANS, PNS, the palatal cusp tips of the maxillary first molars (*right* and *left*) and the anterosuperior-most point of the hyoid—were identified on the CBCT image slices for re-orientation and standardization of the inferior and posterior borders of the tongue using the software developed by the Korea Institute of Science and Technology. Subsequently, the horizontal and vertical planes that crossed the most anterosuperior point of the hyoid were used to crop the tongue's volume. The 3D volume of the cropped tongue model was calculated in cubic millimeters using the software.

Table 2. Estimation of the average tongue posture by linear mixed model analysis.

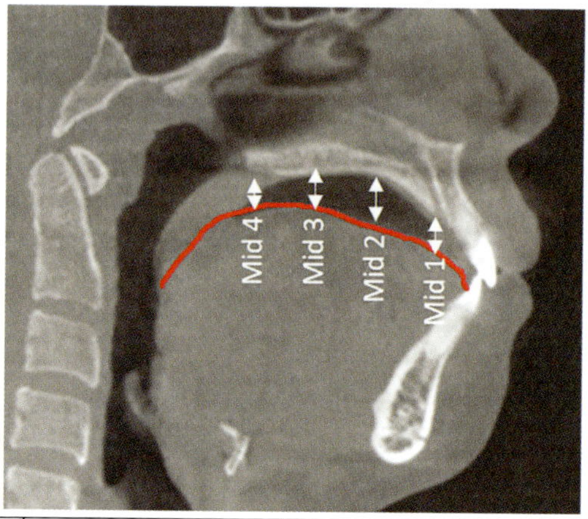

	Estimate (mm)	Standard error	P
Mid 1*	9.3	0.4	< 0.0001
Mid 2*	8.5	0.5	< 0.0001
Mid 3*	5.6	0.4	< 0.0001
Mid 4*	2.8	0.4	< 0.0001

* = In the midsagittal reference plane, the mid 1 = the shortest distance from the dorsum of the tongue to the posterior border of the incisive canal orifice; the mid 4 = the shortest distance to the PNS; and the mid 2 and 3 = distances to the palate measured in points that trisect the line formed by the posterior border of the incisive canal orifice and the PNS.

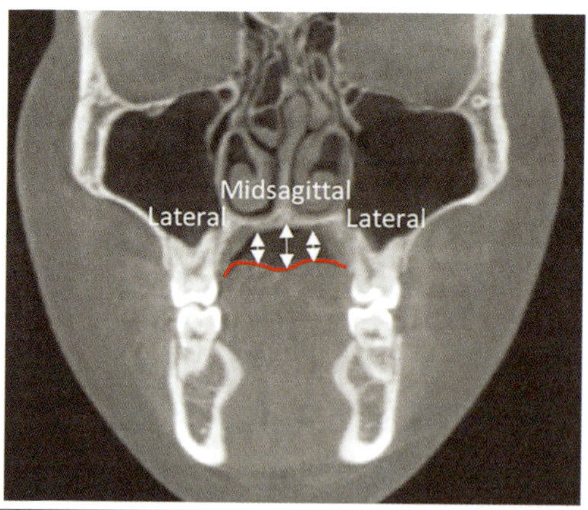

	Estimate (mm)	Standard error	P
Lateral **	6.1	0.4	< 0.0001
MSRP **	7.0	0.4	< 0.0001

Table 2 (continued) ** = Midsagittal reference plane (MSRP), constructed by the ANS, PNS and crista galli was used as the reference plane. Two additional planes that were parallel to the MSRP and bisecting the line drawn from the MSRP to the most prominent points of the palatal cusps of the right and left first maxillary molars were defined as the lateral planes.

Table 3. Correlation between tongue and hyoid. Lat_ant = mean tongue position measured at the most anterior point in the right and left lateral planes. Lat_mid = mean tongue position measured at the midpoint of the anterior and posterior point. Lat_post = mean tongue position measured at the most posterior point. Hyoid_x (mm) = horizontal distance from the antero-superior-most point of the hyoid bone to the line perpendicular to the palatal plane and passing through the ANS. Hyoid_y (mm) = vertical distance from the anterior-most point of the hyoid bone to the palatal plane (ANS-PNS). Tongue tip_x (mm) = horizontal distance from the tongue tip to the ANS on a line parallel to the palate. Tongue tip_y (mm) = vertical distance from the tongue tip to the ANS on a line that is perpendicular to the palate. * = $P < 0.05$. ** = $P < 0.001$.

	Hyoid x		**Hyoid y**	
	Correlation coefficient	P	Correlation coefficient	P
Mid 1	0.353*	0.017	0.248	0.101
Mid 2	0.046	0.766	0.331*	0.026
Mid 3	-0.047	0.761	0.349*	0.019
Mid 4	-0.058	0.706	0.377*	0.011
Lat_ant	0.118	0.438	0.241	0.111
Lat_mid	0.109	0.475	0.328	0.028
Lat_post	0.220	0.146	0.302*	0.044
Tongue tip_x	0.291	0.053	0.404**	0.006
Tongue tip_y	0.065	0.670	-0.111	0.466

Relative to the tongue, the hyoid position is more stable and shows a normal distribution. Because there is a significant correlation between the positions of the tongue and the hyoid, the position of the hyoid can be used to analyze the differences in various types of malocclusion. The vertical position of the hyoid bone is highly correlated, especially with the positions of the middle and posterior tongue (Table 3).

One example of how tongue position can affect the development of a malocclusion arises from the habit of thumb sucking, in which the low positioning of the tongue leads to a collapse of the upper arch due to the imbalance of pressure between the tongue and the cheeks

(Haryett et al., 1967). The effect of tongue size on mandibular prognathism is a controversial subject. Macroglossia has been suggested to be one of the contributing factors leading to mandibular prognathism (Xin et al., 2015; Ihan Hren and Barbic, 2016); however, other studies have concluded that there is no correlation between tongue size and mandibular prognathism (Yoo et al., 1996; Sutherland et al., 2014). Lower incisor crowding also has been associated with tongue size (Uysal et al., 2013). Additionally, an increased tongue length—measured from the hyoid bone to the tongue tip using lateral cephalograms—has been observed in patients with a bimaxillary protrusion (Adesina et al., 2013).

Using lateral cephalograms, Abu Allhaija and Al-Khateeb (2005) measured tongue position at the point below the middle of the soft palate and concluded that there was no difference in the position of the dorsum of the tongue in patients with Class I, II and III skeletal malocclusions. According to another study by Primozic and associates (2013), lower tongue position and greater mouth floor volume were observed more frequently in Class III patients than in Class I patients.

TONGUE POSITION AND VOLUME ON CRANIOFACIAL GROWTH: 3D STUDIES

A greater tongue volume is associated with mandibular prognathism and tongue size has been found to be proportionate to the degree of mandibular prognathism. A large tongue also has been observed in those with hyperdivergent growth patterns. The multi-variate linear regression analysis of the craniofacial morphology in patients with different malocclusions are displayed in Table 4.

Low tongue positions in both the posterior and anterior part of the tongue also were correlated with hyperdivergent growth patterns. The overbite depth indicator (ODI), which is defined as the sum of the palatal plane angle and the AB line to mandibular plane angle (Kim, 1974), has shown to be the strongest vertical skeletal indicator to observe significant correlations with tongue measurements. A shallow overbite also has been associated with a large and low-postured tongue.

FACIAL GROWTH PATTERNS

The effect that respiratory function has on craniofacial growth has been studied for decades and most clinicians now understand that

Table 4. Estimated regression coefficients of the final multi-variate linear regression model for the tongue position and volume on lateral cephalometric measurements. ANB = Point A-Nasion-Point B, or angle between the N-A and N-B line. APDI = anteroposterior dysplasia indicator. ODI = overbite depth indicator. SN-GoGn = angle between the SN plane and Go-Gn line. Hyoid_x (mm) = horizontal distance from antero-superior-most point of hyoid bone to the line perpendicular to palatal plane and passing through ANS. Hyoid_y (mm) = vertical distance from anterior-most point of hyoid bone to the palatal plane (ANS-PNS). Tongue tip_y (mm) = vertical distance from tongue tip to the ANS on a line that is perpendicular to the palate. Mid 4 (mm) = vertical distance from dorsal surface of the tongue to the palate on a line that passes the PNS in the midsagittal plane.

Dependent variable	Independent variable	B	Standard error	T	P
ANB	Intercept	20.4	30.4	0.7	0.506
	Hyoid_x	0.3	0.1	5.6	< 0.001
	Hyoid_y	0.1	0.1	1.8	0.073
	Volume	-3.4	3.1	-1.1	0.282
	Tongue tip_y	-0.1	0.1	-1.1	-0.260
	Mid 4	-0.2	0.1	-2.4	0.023
APDI	Intercept	-83.7	63.0	-1.3	0.191
	Hyoid_x	-0.9	0.1	-8.0	< 0.001
	Hyoid_y	-0.4	0.1	-2.7	0.009
	Volume	21.2	6.5	3.3	0.002
	Tongue tip_y	-0.1	0.2	-0.5	0.628
	Mid 4	0.6	0.2	3.4	0.001
ODI	Intercept	217.3	76.0	2.9	0.006
	Hyoid_x	0.3	0.1	2.3	0.025
	Hyoid_y	0.7	0.2	4.3	< 0.001
	Volume	-16.5	7.8	-2.1	0.040
	Tongue tip_y	-0.7	0.2	03.9	< 0.001
	Mid 4	-0.5	0.2	-2.6	0.013
SN-GoGn	Intercept	50.9	73.2	0.7	0.491
	Hyoid_x	0.3	0.1	2.4	0.022
	Hyoid_y	-0.2	0.2	-1.1	0.294
	Volume	-2.9	7.5	-0.4	0.707
	Tongue tip_y	0.5	0.2	2.7	0.010
	Mid 4	-0.2	0.2	-1.1	0.288
Overbite	Intercept	23.4	24.4	1.0	0.342
	Hyoid_x	0.1	0.0	1.4	0.166
	Hyoid_y	0.1	0.1	2.1	0.042
	Volume	-2.4	2.5	-0.9	0.348
	Tongue tip_y	-0.2	0.1	-3.2	0.003
	Mid 4	0.0	0.1	-0.4	0.696

respiratory function is highly relevant in orthodontic diagnosis and treatment plans. Angle (1907) observed that a Class II division 1 malocclusion is associated with mouth breathing and with an obstruction of the upper pharyngeal airway. Clinical features relating to impaired breathing have been observed by researchers (Ricketts, 1968; Linder-Aronson, 1979; McNamara 1981; Vig, 1998) who indicated that the main characteristics leading to upper airway obstruction are adenoids and tonsil hypertrophy, crossbite, increase in lower facial height, vertical maxillary excess, skeletal open bite, narrow external nares and tongue thrusting.

EFFECT OF THE PHARYNGEAL AIRWAY ON FACIAL GROWTH PATTERNS

According to a recent 3D study of pharyngeal airway volumes, Kim and colleagues (2010) reported that the Class II pre-adolescent children have lower airway volumes; these volumes were measured from the anterior nasal cavity and nasopharynx to the epiglottis. In adults, skeletal Class III patients showed greater airway volume in the superior pharyngeal airway above the level of the PNS (Hong et al., 2011).

TONGUE POSITION AND PHARYNGEAL AIRWAY SPACE IN CLASS I PATIENTS

The superior pharyngeal airway space is surrounded by soft tissues, with the posterior part of the tongue serving as the anterior border of the airway. Therefore, as with tongue volume, pharyngeal airway volume also varies greatly among individuals. In Table 5, airway volume showed significant correlation with tongue position in the posterior part, whereas the minimal cross-sectional area (MCA) was correlated with the anterior part of the tongue's position, including the horizontal position of tongue tip. Lower tongue positions were associated with a greater airway volume.

According to linear regression analysis, tongue positioning in the midsagittal plane as well as a MCA affected the airway volume (Table 6). Lower tongue positions and greater airway minimal cross-sectional area were associated with the greater airway volumes, which is similar to the results of the correlation analysis. However, the effect of the MCA was much greater than tongue position that was measured in the midsagittal plane.

Table 5. Correlation analysis of the upper airway volume, MCA and tongue positions. * = P < 0.05. ** = P < 0.001.

	Airway volume	**Minimal cross-sectional area**
Mid 1	0.053	-0.107
Mid 2	0.081	-0.117
Mid 3	0.313**	0.052
Mid 4	0.298**	0.169
Lat_ant	0.042	-0.273**
Lat_mid	0.267**	0.013
Lat_post	0.224*	0.199*
Tongue tip_x	0.005	-0.242*
Tongue tip_y	0.159	-0.043

Table 6. Regression analysis of the airway volume measured from the palatal plane to the plane that crosses the most superior point of the epiglottis.

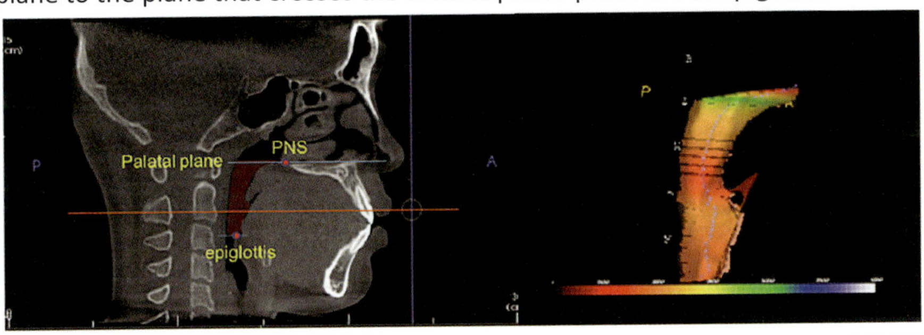

Dependent variable	**Independent variable**	**B**	β	**P**
Airway volume	(Constant)	1.913		< 0.001
	Minimal cross-sectional area	0.003	0.577	< 0.001
	Mid 3	0.022	0.240	< 0.001

CONCLUSIONS

3D analysis of the head using CBCT technology has led to a better understanding of the effects that the tongue and pharyngeal airway volume has on craniofacial growth patterns. Although volumetric assessment of the tongue and pharyngeal airway may be more reliable than using 2D images, it should be noted that tongue is a very mobile organ and varies greatly from person to person. Therefore, the tongue positioning and size assessed *via* radiographic images should be analyzed carefully. As possible associations of tongue and pharyngeal airway with the facial growth patterns have been suggested, further studies to validate their impact on growth may be warranted.

REFERENCES

Abu Allhaija ES, Al-Khateeb SN. Uvulo-glosso-pharyngeal dimensions in different anteroposterior skeletal patterns. Angle Orthod 2005;75(6): 1012-1018.

Adamidis IP, Spyropoulos MN. Hyoid bone position and orientation in Class I and Class III malocclusions. Am J Orthod Dentofacial Orthop 1992;101(4):308-312.

Adesina BA, Otuyemi OD, Kolawole KA, Adeyemi AT. Assessment of the impact of tongue size in patients with bimaxillary protrusion. Int Orthod 2013;11(2):221-232. [In English and French.]

Angle EH. *Treatment of Malocclusion of the Teeth: Angle's System.* S.S. White Dental Manufacturing Company 1907.

Bibby RE, Preston CB. The hyoid triangle. Am J Orthod 1981;80(1):92-97.

Chang YI. A radiographic study of the hyoid bone position in malocclusion. Korean J Orthod 1987;17(1):7-13.

Graber TM, Rakosi T, Petrovic AG. *Dentofacial Orthopedics with Functional Appliances.* St. Louis, MO:Mosby 1997.

Haryett RD, Hansen FC, Davidson PO, Sandilands ML. Chronic thumb-sucking: The psychologic effects and the relative effectiveness of various methods of treatment. Am J Orthod 1967;53(8):569-585.

Hong JS, Oh KM, Kim BR, Kim YJ, Park YH. Three-dimensional analysis of pharyngeal airway volume in adults with anterior position of the mandible. Am J Orthod Dentofacial Orthop 2011;140(4):e161-e169.

Ihan Hren N, Barbic U. Tongue volume in adults with skeletal Class III dentofacial deformities. Head Face Med 2016;12:12.

Iida-Kondo C, Yoshino N, Kurabayashi T, Mataki S, Hasegawa M, Kurosaki N. Comparison of tongue volume/oral cavity volume ratio between obstructive sleep apnea syndrome patients and normal adults using magnetic resonance imaging. J Med Dent Sci 2006;53(2):119-126.

Kim YH. Overbite depth indicator with particular reference to anterior open-bite. Am J Orthod 1974;65(6):586-611.

Kim YJ, Hong JS, Hwang YI, Park YH. Three-dimensional analysis of pharyngeal airway in preadolescent children with different anteroposterior skeletal patterns. Am J Orthod Dentofacial Orthop 2010;137(3):306.e301-e311.

Liégeois F, Albert A, Limme M. Comparison between tongue volume from magnetic resonance images and tongue area from profile cephalograms. Eur J Orthod 2010;32(4):381-386.

Linder-Aronson S. Respiratory function in relation to facial morphology and the dentition. Br J Orthod 1979;6(2):59-71.

McNamara JA Jr. Influence of respiratory pattern on craniofacial growth. Angle Orthod 1981;51(4):269-300.

Moss ML. A theoretical analysis of the functional matrix. Acta Biotheor 1968;18(1):195-202.

Moss ML. The functional matrix hypothesis revisited: 1. The role of mechanotransduction. Am J Orthod Dentofacial Orthop 1997*a*;112(1):8-11.

Moss ML. The functional matrix hypothesis revisited: 2. The role of an osseous connected cellular network. Am J Orthod Dentofacial Orthop 1997*b*;112(2):221-226.

Moss ML, Salentijn L. The primary role of functional matrices in facial growth. Am J Orthod 1969;55(6):566-577.

Primozic J, Farcnik F, Perinetti G, Richmond S, Ovsenik M. The association of tongue posture with the dentoalveolar maxillary and mandibular morphology in Class III malocclusion: A controlled study. Eur J Orthod 2013;35(3):388-393.

Ricketts RM. Forum on the tonsil and adenoid problem in orthodontics: Respiratory obstruction syndrome. Am J Orthod 1968;54(7):495-507.

Shigeta Y, Ogawa T, Ando E, Clark GT, Enciso R. Influence of tongue/ mandible volume ratio on oropharyngeal airway in Japanese male patients with obstructive sleep apnea. Oral Surg Oral Med Oral Pathol Oral Radiol Endod 2011;111(2):239-243.

Sutherland K, Schwab RJ, Maislin G, Lee RW, Benedikstdsottir B, Pack AI, Gislason T, Juliusson S, Cistulli PA. Facial phenotyping by quantitative photography reflects craniofacial morphology measured on magnetic resonance imaging in Icelandic sleep apnea patients. Sleep 2014;37(5): 959-968.

Tallgren A, Solow B. Hyoid bone position, facial morphology and head posture in adults. Eur J Orthod 1987;9(1):1-8.

Tamari K, Murakami T, Takahama Y. The dimensions of the tongue in relation to its motility. Am J Orthod Dentofacial Orthop 1991*a*;99(2):140-146.

Tamari K, Shimizu K, Ichinose M, Nakata S, Takahama Y. Relationship between tongue volume and lower dental arch sizes. Am J Orthod Dentofacial Orthop 1991*b*;100(5):453-458.

Uysal T, Yagci A, Ucar FI, Veli I, Ozer T. Cone-beam computed tomography evaluation of relationship between tongue volume and lower incisor irregularity. Eur J Orthod 2013;35(5):555-562.

Vig KW. Nasal obstruction and facial growth: The strength of evidence for clinical assumptions. Am J Orthod Dentofacial Orthop 1998;113(6): 603-611.

Xin N, Tao W, Ashwin DV, Jinlin S. Establishment of integral biomechanical balance in the correction of tongue source skeletal dentomaxillofacial open bite deformities. J Craniofac Surg 2015;26(2):473-476.

Yoo E, Murakami S, Takada K, Fuchihata H, Sakuda M. Tongue volume in human female adults with mandibular prognathism. J Dent Res 1996;75(12):1957-1962.

DISORDERED BREATHING IN THE PERINATAL PERIOD INDUCES BONE AND METABOLIC INJURY

Eung-Kwon Pae

ABSTRACT

Disrupted breathing patterns subject the infant to repetitive exposure to normoxic followed by hypoxic conditions (i.e., intermittent hypoxia [IH]). Emerging evidence indicates that IH exposure exerts significant injury to IH-prone body systems. The extent of such injury is described poorly, despite the well-outlined prevalence of such breathing patterns in premature and post-natal infants. We examined in rat models how the incidence of perinatal IH influences adverse outcomes in long bones and the mandible. We treated post-natal day 0 (or P0) ~ P2 rat pups with IH for periods as short as one hour and then maintained the pups in normal ambient air. Tibia and mandibular hardness and elasticity significantly declined three weeks after IH exposure. Reduced mineralization and alkaline phosphatase appeared, in addition to other molecular changes (e.g., decreased *collagen 1* and *Runx2* expression in bone and in cultured osteoblasts treated under IH). Deficiencies also appeared in metabolic systems. Zinc levels declined in pancreatic beta cells, together with levels of the zinc uptake transporter (ZIP8), presumably from oxidative stress from IH. Disturbed zinc balance from diminished ZIP8 levels in pancreatic beta cells after IH challenge likely contributes to diabetogenic effects of increased blood glucose and decreased insulin levels, as well as disturbed glucose sensitivity. Exposure to IH through common appearances of disturbed breathing in the human neonate may exert similar consequences to the deleterious outcomes to bone density and pancreatic function found in these animal models.

KEY WORDS: diabetes, osteoporosis, perinatal intermittent hypoxia, zinc homeostasis

---

INTRODUCTION

Pediatric obstructive sleep apnea (OSA) is recognized as a distinctive type of OSA relative to OSA in adults, yet perinatal-disordered breathing differs from pediatric OSA in epidemiology (Lumeng and Chervin, 2008) and etiopathophysiology (Harvey et al., 1999). Its adverse outcomes may differ from other presentations. Bone mineral density in males with severe perinatal OSA symptoms frequently is compromised (Hamada et al., 2016). Because a low bone mass in such patients also is common, a previous review suggests that abnormal metabolic consequences are associated with hypoxia and disrupted circadian rhythm (Drager et al., 2010; Swanson et al., 2015).

First of all, we need to know how teeth move in perinatal intermittent hypoxia (IH) treated animals. If one acknowledges that a brief IH exposure at an early age induces such problems in alveolar bone as described above, dentists must be alerted to the possibility of OSA at the first patient interview.

Sleep disturbances can influence neural regulation on energy homeostasis and autonomic nerve balance (Weiss et al., 2015). Particularly, increased sympathetic nerve tone due to damage in the hypothalamus (Hunt et al., 2016), brainstem (Semenza and Prabhakar, 2015) and cerebellum (Pae et al., 2005*a*, 2011) may result in osteoporosis and high muscle tone (Pae et al., 2005*b*). Most studies investigated cause-effect relationships between OSA and bony outcomes using adult humans and full-grown animals. However, our current discussion will focus on processes that occur in the bone in post-natal animals after a brief intermittent hypoxic challenge during early life.

PERINATAL IH EVENTS

Prevalence of Perinatal IH Events in Humans is Unknown, But Presumably High

The perinatal period includes pre- and post-natal periods. Prenatal IH usually results from maternal exposure to IH events during the gestational period (Louis et al., 2012; Xu et al., 2014; Rice et al., 2015). Therefore, the prevalence of OSA in obese women during pregnancy could be translated into prevalence of pre-natal IH. This relationship is known to constitute approximately 5% of total pregnancies (Antony et al., 2014*b*).

Maternal obesity and OSA are associated strongly with neonatal outcomes such as pre-term birth or neonatal intensive care unit admission (Olivarez et al., 2011; Bin et al., 2016). Post-natal breathing problems are common in neonates with prematurity (Centers for Disease Control, 2017), infants who experience difficult delivery (Eichenwald et al., 2016; Tapia et al., 2016) and neonates with breathing difficulty for unclear reasons (Herlenius and Kuhn, 2013). Prevalence of post-natal IH could be as high as 10% of the general population worldwide (Beck et al., 2010).

Effects of Adverse Outcomes in Neonates May Last Long

A sudden post-natal collapse in healthy newborn infants is rare, but such an outcome often is dreadful (Herlenius and Kuhn, 2013). Hypoxic events or oxygen desaturation during sleep in infants to a mother with adverse maternal outcomes commonly are observed (Deschamp and Daftary, 2017). Complications at birth signified by a low APGAR scores at five minutes often require prolonged medical attention beyond the immediate post-natal period. Type-1 diabetes (T1D) is considered an associated post-natal outcome.

The origins of T1D remain unclear. A consensus has emerged that three conditions exist to initiate this chronic medical condition:

1. Genetic susceptibility;
2. A timely trigger; and
3. Subsequent exposure to an antigen.

Of these conditions, fewer than 10% of genetically susceptible individuals progress into the disease (Knip et al., 2005); therefore, the role of the other two conditions and environmental factors appear to play major roles in the prevalence of T1D. Autoimmune antibodies that develop in most patients are a consequence of exposure to antigens. However, our knowledge on environmental factors that trigger autoimmune responses has been limited. Based on animal studies and a compilation of case reports with infant hypoxic events, it could be conjectured that IH exposure during the immediate post-natal period may be a key environmental factor triggering an autoimmune response.

Case F.B.

F.B.'s case study provides an example of neonatal adverse outcomes of early IH insult. T1D mellitus is known to be an autoimmune disease that has multi-faceted triggering factors. However, environmental factors recently have been suggested as a contributing cause. We suggest a new subtype of T1D that is triggered by IH post-natally.

Medical History. F.B. was born on April 28, 2004 as a full-term infant with her umbilical cord wrapped around her neck. She showed the following APGAR scores after the severe hypoxia from the birth complication (Table 1); other details from her medical history are presented in Table 2.

Table 1. Apgar scores for F.B.

	1 minute	**5 minutes**
Heart rate	2	2
Respiratory effort	2	2
Reflex	2	2
Muscle tone	2	2
Skin color	0	1

Table 2. Medical history details of note for F.B.

DATE	**REPORTED INFORMATION**
04/28/04	Birth of F.B.
07/02/04	Pediatrician visit noted "a very fussy baby"
01/03/05	Consulted by pediatrician for "snoring" during sleep
01/08/05-01/13/05	Disturbed sleep reported
03/22/10	Severe snoring reported with episodes like apnea, grumpy morning and meltdown
05/06/10	Adenoidectomy performed for chest pain and palpitation
07/23/10	Epigastric pain reported
12/27/10	"Meltdown" appeared on chart
03/03/11	Tardiness reported at school
04/24/11	PSG performed with these results: oxygen nadir 89%, desat O2 (SaO2 lower than 4%) 70 events, 64 arousals (per 8 hours), AHI = 9, more stage 2 than 3, REM 13.5%, sleep efficiency 95%
05/25/11	Tonsillectomy with turbinate removal performed
02/14/12	"Epic meltdown" reported which mother related to breathing problems
09/04/12	First (pre-operative) sleep study indicated the patient had OSA and established candidacy for adenoidectomy

Table 2. (Continued)

02/05/13	Diabetes suspected in association with mood issue
02/21/13	Headaches with low blood sugar; puffiness noted around eyes
01/13/14	Diabetes diagnosed with A1C = 13%, GAD65 = 2.3, IAA < 0.4, glucose 150 mg/dL
01/22/14	SOAP note with average blood sugar 306 (check 4x/day at finger tips), insulin 4x/day (Lantus Solostar Pen 100 units/mL SubQ) before bed; Humalog 100 units/mL SubQ 3x before meals; mood and anger issue since age four charted
02/26/14	T1D diagnosed
04/17/14	Attended pump class based on HbA1c at 7.5%
04/22/14	Palate expansion performed
05/29/14	Visited Stanford Children's Hospital Pediatric Pulmonary and Sleep Clinic
06/19/14	Visited Madison Clinic for Pediatric Diabetes, University of California-San Francisco: blood sugar me asured 7x/day; 159 mg/dL on average, fluctuating between 36-453 mg/dL
06/20/14	PSG performed at Stanford Children's Hospital with these results: sleep efficiency 90.7%, OSA 26 episodes in total (with the longest episode at 28 seconds), desat O2 nadir 92%, end tidal CO2 28.1 ~ 44.3%; CPAP prescribed based on the record
07/10/14	Used insulin pump from Madison Clinic at 110 mg/dL
07/17/14	Madison Clinic details: HbA1c 7.1%, random glucose 233 mg/dL, blood glucose 10x/day fluctuating 410-175-47
10/03/14	Stanford Children's Hospital CPAP 6-8 hours/day in addition to palate expander

PERINATAL DISORDERED BREATHING AND DIABETES MELLITUS

Obtaining Diabetic Pups

Our laboratory has been using rat pups for an animal model of disrupted breathing in human infants (Pae et al., 2011). The length of pregnancy in Sprague Dawley rats is approximately 22 days. Timed-pregnant animals were purchased and housed separately for a week before parturition. Immediately after delivery between post-natal day 0 (P0) and P2, we maintained newborn pups in a hypoxic chamber with the mother under IH condition (oxygen level in the chamber fluctuating between approximately 10% and 21%, balanced with nitrogen every four minutes) for one hour (Pae et al., 2013, 2014).

After a one-time one-hour exposure, the animals were maintained in ambient air with control animals. These pups showed diabetogenic symptoms and signs characterized by significantly high blood glucose with significantly low (< 50%) insulin levels when examined three weeks later (Fig. 1). We observed an abnormal tendency in glucose intolerance as well (Fig. 2). All these outcomes occurred without changes in islet mass and beta-cell counts (Fig. 3).

Animal and In Vitro Models

Our perinatal rat model consistently exhibited T1D-like symptoms after short-term IH exposure. Our protocol used one- or five-hour exposures at P0-P2. In contrast to other rodent models that principally show T2D symptoms (Carreras et al., 2012; Polak et al., 2013) because our perinatal model uniquely provided a means to investigate etiopathophysiology following disordered breathing in perinatal animals (i.e., how the condition may have developed from events at an early age). For *in vitro* studies, we utilized primary islets obtained from developing rats.

PERINATAL APNEA AND BONE DEVELOPMENT

Numerous reports have related disturbed breathing to risk for perinatal outcomes (Antony et al., 2014*a*; Xu et al., 2014; Rice et al., 2015). The focus of the studies, however, has been the effect of maternal OSA on the fetus during gestation. Most studies examining effects of post-natal-disturbed breathing focused on perinatal outcomes of apnea of prematurity and, thus, did not assess effects of true post-natal apnea. Post-natal apnea is very common (approximately 9.6%) in premature newborns (Brockmann, 2016; Tapia et al., 2016); it also is common in full-term infants.

Although we do not know underlying mechanisms, perinatal apnea (or IH) episodes disrupt bone remodeling processes (Swanson et al., 2015) and result in osteoporosis in long bones, as well as in the mandible (Kim et al., 2016). One plausible causal mechanism associated with osteoporosis resulting from apnea is oxidative stress that damages neural

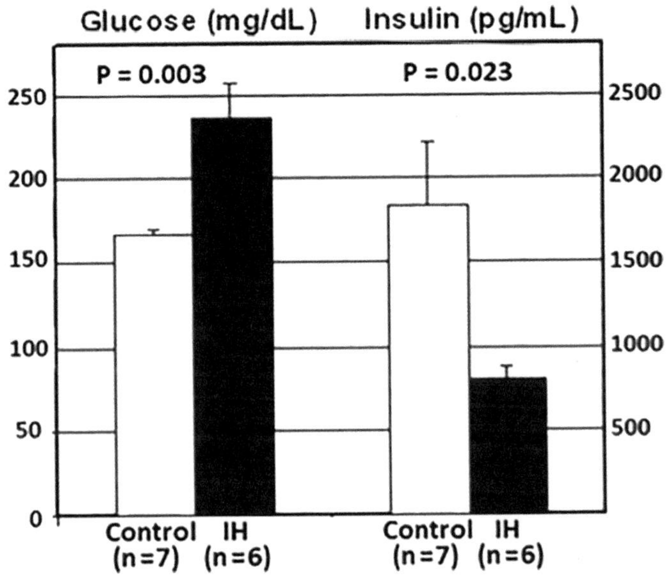

Figure 1. Comparison of serum glucose and insulin levels of IH treated *versus* control animals. The differences of both blood and insulin are significant statistically. Referenced from Pae et al., 2013.

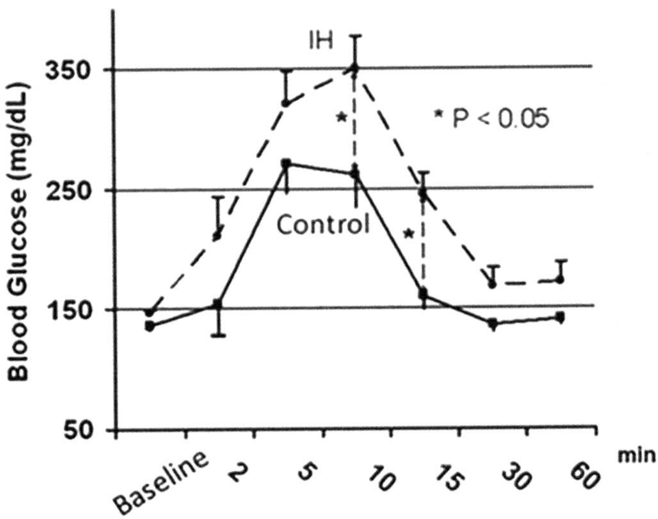

Figure 2. Glucose tolerance tests (GTT). GTT results show a significant difference at 10- and 15-minute points. At the 30-minute point, the glucose level returns to the baseline in control pups. Referenced from Pae et al., 2013.

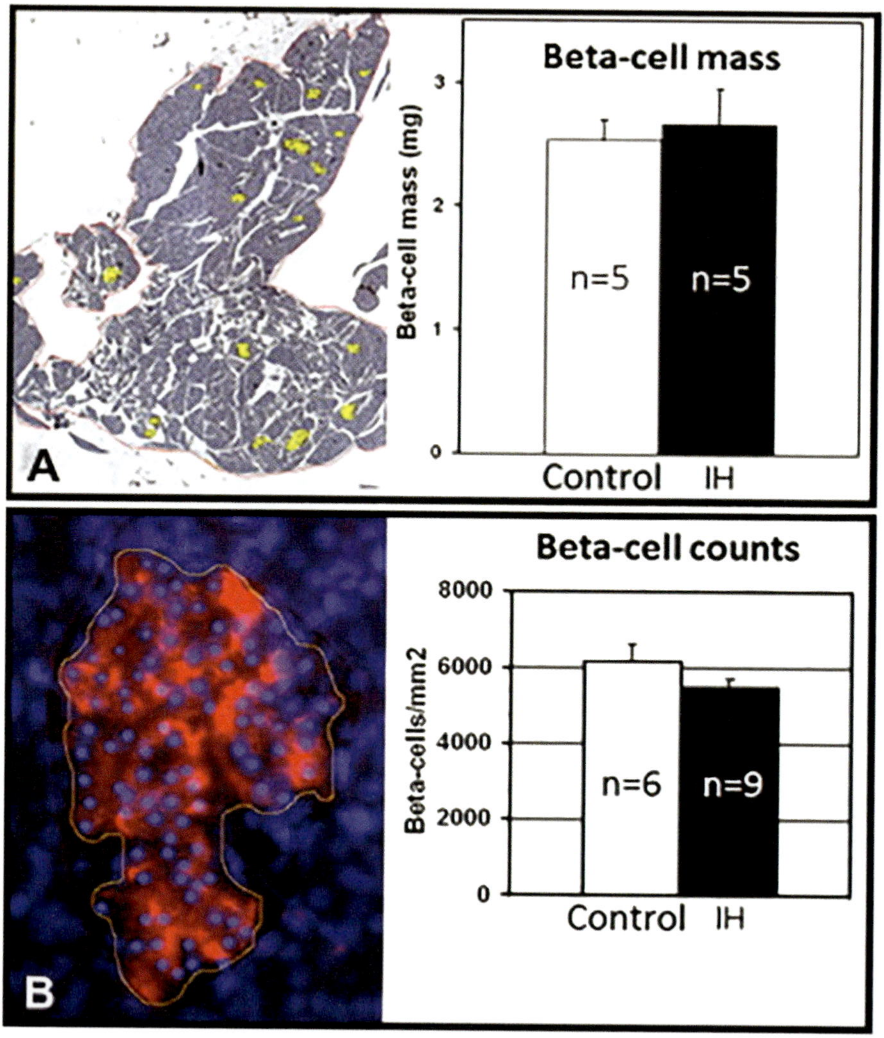

Figure 3. Beta cell mass and counts in IH-treated and control animals. *A:* Islets are identified by yellow stain. The difference in beta cell mass was not significant statistically. *B:* The number of beta cells was counted in an islet demarcated by the orange line. The difference in beta cell count was not significant statistically. Referenced from Pae et al., 2013.

pathways modulated by orexin. Orexin is an upstream neurotransmitter that regulates multiple basic physiologic functions including sleep-wake control, food intake and energy expenditure. Orexin pathways in apnea might be damaged by oxidative stress. One system regulated by orexin is a central neural pathway associated with leptin, which regulates food intake (Schwartz et al., 1998).

A more likely possibility, however, is that the autonomic nervous system is damaged by IH insults, a well-known consequence of adult OSA, as shown by a large body of MRI structural and functional studies. The injury including both gray matter and fiber damage appears especially in autonomic regulatory areas (e.g., the insula, hypothalamus, cingulate and ventral lateral medulla; Macey et al., 2002; Harper et al., 2012). Purkinje cells in the cerebellar cortex were damaged significantly after a five-hour IH challenge (Pae et al., 2005*a*). Because the cerebellar cortex provides a dampening effect on extremes of autonomic outflow (e.g., sympathetic drive for blood pressure), damaged cerebellar Purkinje cells will lead to an inability to reduce extremely elevated sympathetic tone. Periosteal development is regulated by sympathetic outflow and excessive sympathetic tone will modify bone development.

One more concern is that perinatal apnea induces diabetogenic processes (Pae et al., 2013, 2014; Kim et al., 2016). Our reports demonstrate that a transient IH challenge for one hour to immediately-postnatal rat pups elevates blood glucose levels and significantly decreases insulin levels three weeks later (Fig. 1). This set of changes accompanies glucose intolerance as well (Fig. 2). Interestingly, this dysfunctional process is not associated with abnormal islet mass or a decreased beta cell count. We suspect that this diabetogenic dysfunction of beta cells results in a fracture-prone and brittle bone condition.

A SUGGESTED LINK CONNECTS OSA WITH DIABETES AND TO OSTEOPOROSIS

The processes leading to development of OSA remain unknown; we understand the pathophysiology of OSA, but not the mechanisms underlying its onset. Our contribution to understanding the pathology is that brief IH challenges during the perinatal stage induce diabetes and fragile bones. These two outcomes of perinatal IH may occur independently, or be associated with the onset mechanisms of OSA.

A Link Connects Perinatal IH to Diabetes-like Symptoms

Among the cells residing in many exocrine tissues containing zinc in our body, pancreatic beta cells use zinc as a basic ingredient in the production of insulin. Insulin molecules require zinc ions for maturation and secretion in a functional form of the hexamer. Zinc ions are centered in the insulin hexamer like a metal chelator holding the insulin molecule together (Dodson and Steiner, 1998). Therefore, if zinc is depleted in the cytoplasm of beta cells, insulin cannot be assembled and secreted.

Zinc in beta cells also is needed to overcome oxidative stress as an antioxidant. Because beta cells handle byproducts containing s-s chains during the insulin manufacturing process, the intracellular zinc concentration must be supplied constantly by intracellular organelles or extracellular space *via* zinc transporters such as ZIP8 (or Zrt- and Irt-like protein from *Slc39a8* gene), a zinc uptake transporter (Pae and Kim, 2014).

Zinc influx and efflux processes are regulated tightly because intracellular zinc levels are critical in numerous body functions and physiologies (Fukada et al., 2013; Hojyo and Fukada, 2016). Zinc deficiency is prevalent worldwide and in children more than adults (Prasad, 2012). Further, infants at birth have a small reservoir of zinc (Ackland and Michalczyk, 2016); thus, zinc deficiency results in growth retardation (Kambe et al., 2015), immune failure (Maywald and Rink, 2015, 2016) and impaired cognitive functions (Sandstead, 2013). However, the role of zinc in diabetogenic effects is not understood well.

We reported that a lack of ZIP8, an iron and zinc transporter, as a consequence of brief post-natal IH exposure can result in reduction of insulin secretion (Pae et al., 2014). Chabosseau and Rutter's study (2016) reviewed a role of zinc in pancreatic islets, yet discussion was limited to zinc transporter 8 (ZnT8) transporters which efflux zinc ions to the cytosol or extracellular space. In contrast, we demonstrated that an IH insult decreases ZIP8 expression in the beta cell membrane, independent of insulin production processes. A lack of ZIP8 inhibits zinc influx from the extra-cellular space. In our studies, the level of "zinc drop" in cytosol after IH insults appeared less than the drop incurred by siRNA-mediated ZIP8 (siZIP8); however, the degree of decrease was significant (P = 0.005), compared to controls in our rat study (Pae et al., 2014). The question arises: what type of diabetes do those characteristics exhibited by our rodent model manifest, T1D or T2D? Although inconclusive, we suggest that the disturbance of zinc homeostasis due to perinatal IH results in a mild form of T1D-like symptoms in the rat.

Connecting Diabetes-like Symptoms to Osteoporosis

Although the underlying mechanisms still are obscure (Maddaloni et al., 2017; Yeap et al., 2017), numerous studies report osteoporosis or subnormal bone density in patients with diabetes (Thraikill et al., 2005; Leidig-Bruckner et al., 2014; Hough et al., 2016). Our rodent model demonstrates that a short-term IH insult induces diabetes-like symptoms *via* zinc perturbation in pancreatic beta cells. Recently, zinc signaling draws much attention from many directions. Among them, Hojyo and Fukada's review (2016) exhibited a significant contribution of ZIP13 and ZIP14 to bone and systemic growth.

We demonstrated that IH insults disturbed zinc regulation in bone. Our results clearly show that the level of ZIP8 transporter in bones decreased significantly as the zinc level declines. This outcome paralleled other measurements (e.g., Runx2, Collagen-1, osteocalcin, alkaline phosphatase and bone minerals; see Fig. 3 in Kim et al., 2016). Particularly, the tibia and the mandible harvested from IH-treated animals showed significant fragility in bone strength and low elasticity (Fig. 4).

Observations from the *in vitro* studies on osteoblasts paralleled those of *in vivo* evaluations as well. We confirmed that a short-term IH exposure may result in fragile bone conditions *via* a diabetogenic process. However, we also confirmed that IH exposure could result in a similar outcome when beta cells *in vitro* were exposed to IH (Fig. 5). This demonstration would be the first report exhibiting a direct cause-effect relationship between IH and bone fragility.

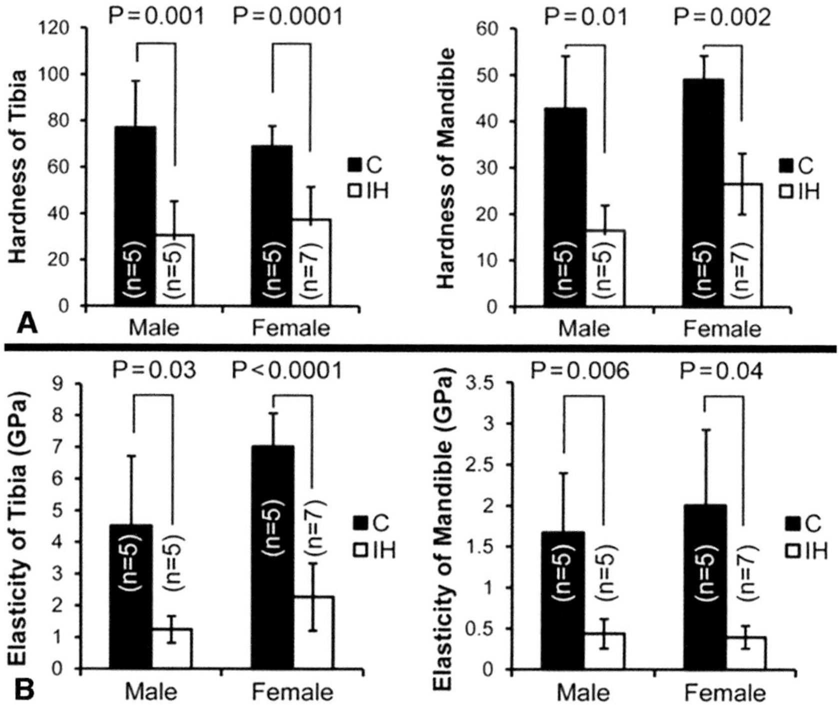

Figure 4. Comparisons of dynamic hardness and elasticity of a long bone (tibia) and the mandible between male and female control and IH animals. Closed (C) represents the control group (five males and five females) and open (IH) represents the intermittent hypoxia group (five males and seven females). *A:* Hardness in males (control tibia 76.9±20.20 *versus* IH tibia 30.5±14.78; control mandible 42.6±11.41 *versus* IH mandible 16.5±5.39) and in females (control tibia 68.9±8.84 *versus* IH tibia 37.4±14.18; control mandible 49.0±5.11 *versus* IH mandible 26.6±6.57). *B:* Elasticity of male control tibia 4.53±2.19 *versus* IH 1.25±0.42 and control mandible values of 1.67±0.73 *versus* IH mandible 0.44±0.18. Control female tibia values were 7.02±1.05 *versus* IH tibia 2.28±1.06 and female control mandible values were 2.01±1.92 *versus* IH mandible of 0.40±0.14. Referenced from Kim et al., 2016.

→ Figure 5. Comparisons between control and IH-treated osteoblasts. IH-treated human osteoblasts show decreased expression of alkaline phosphatase (*A*) and mineralization (*B*), assessed by Alizarin S Red. Runx2, Collagen-1 and ZIP8 protein contents (*C*) assayed using Western blots. Osteocalcin levels (*D*) quantified from

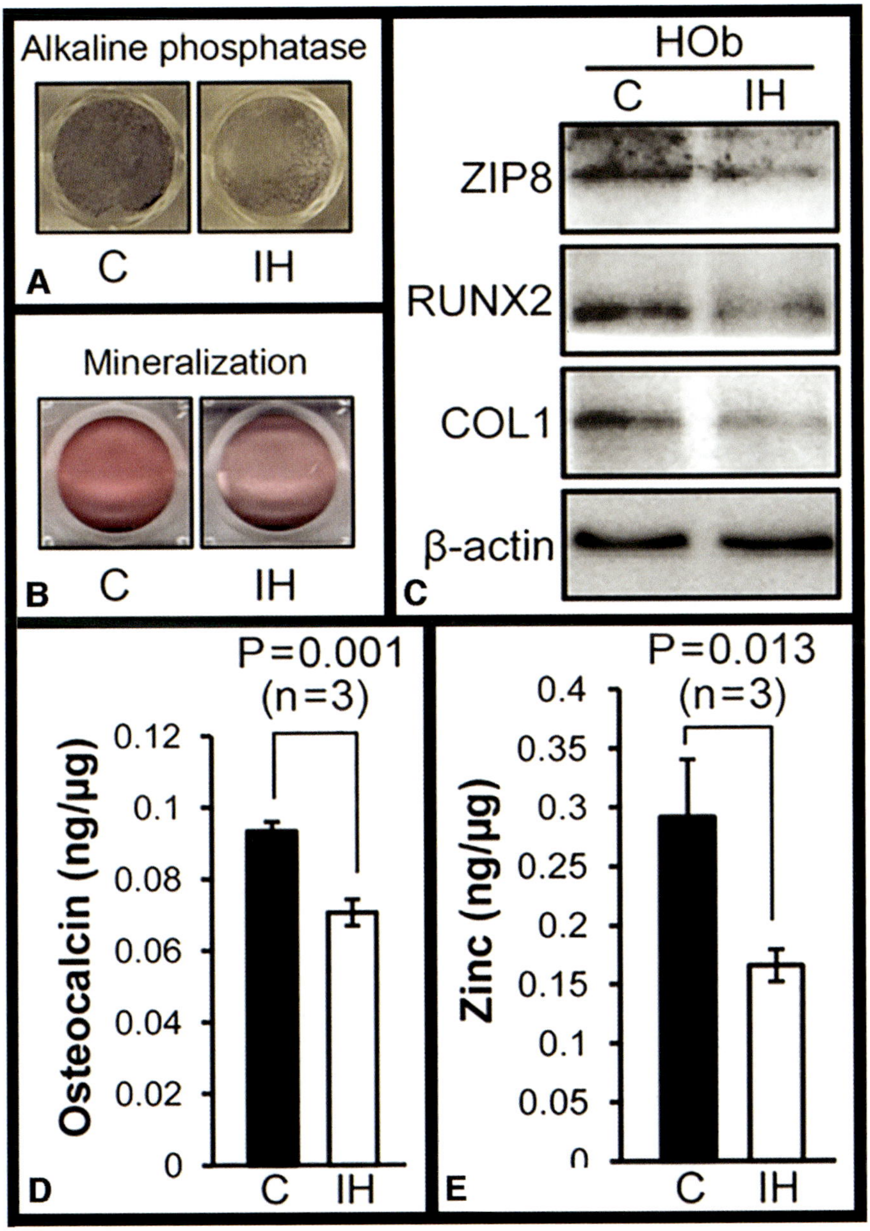

cell lysates using ELISA assays show a significant decline with IH. Zinc contents (*E*), estimated in whole-cell lysates obtained from three sets of osteoblasts culture dishes (n = 3), decreased after IH treatment. Referenced from Kim et al., 2016.

CONCLUSIONS

Clinical studies show that zinc deficiency could result in diabetic conditions. However, whether zinc supplementation improves insulin homeostasis remains unclear (Ruz et al., 2016; de Carvalho et al., 2017). Estimating bone quality clinically in patients with diabetes is difficult.

Despite multiple studies on animals and human infants, we still have insufficient data to conclude whether perinatal IH results in notable long-term growth defects in craniofacial areas.

REFERENCES

Ackland ML, Michalczyk AA. Zinc and infant nutrition. Arch Biochem Biophys 2016;611:51-57.

Antony KM, Agrawal A, Arndt ME, Murphy AM, Alapat PM, Guntupalli KK, Aagaard KM. Association of adverse perinatal outcomes with screening measures of obstructive sleep apnea. J Perinatol 2014*a*;34(6):441-448.

Antony KM, Agrawal A, Arndt ME, Murphy AM, Alapat PM, Guntupalli KK, Aagaard KM. Obstructive sleep apnea in pregnancy: Reliability of prevalence and prediction estimates. J Perinatol 2014*b*;34(8):587-593.

Beck S, Wojdyla D, Say L, Betran AP, Merialdi M, Requejo JH, Rubens C, Menon R, Van Look PF. The worldwide incidence of preterm birth: A systematic review of maternal mortality and morbidity. Bull World Health Organ 2010;88(1):31-38.

Bin YS, Cistulli PA, Ford JB. Population-based study of sleep apnea in pregnancy and maternal and infant outcomes. J Clin Sleep Med 2016; 12(6):871-877.

Brockmann PE. When does the risk for obstructive sleep apnea start? The importance of perinatal factors. Sleep 2016;39(4):721-722.

Carreras A, Kayali F, Zhang J, Hirotsu C, Wang Y, Gozal D. Metabolic effects of intermittent hypoxia in mice: Steady *versus* high-frequency applied hypoxia daily during the rest period. Am J Physiol Regul Integr Comp Physiol 2012;303(7):R700-R709.

Centers for Disease Control, 2017. https://www.cdc.gov/Reproductive-Health/MaternalInfant Health/PretermBirth.htm?

Chabosseau P, Rutter GA. Zinc and diabetes. Arch Biochem Biophys 2016;611:79-85.

de Carvalho GB, Brandão-Lima PN, Maia CS, Barbosa KB, Pires LV. Zinc's role in the glycemic control of patients with type 2 diabetes: A systematic review. Biometals 2017;30(2):151-162.

Deschamp A, Daftary A. A newborn infant with oxygen desaturation during sleep. Chest 2017;151(1):e17-e20.

Dodson G, Steiner D. The role of assembly in insulin's biosynthesis. Curr Opin Struct Biol 1998;8(2):189-194.

Drager LF, Jun JC, Polotsky VY. Metabolic consequences of intermittent hypoxia: Relevance to obstructive sleep apnea. Best Pract Res Clin Endocrinol Metab 2010;24(5):843-851.

Eichenwald EC;Committee on Fetus and Newborn, American Academy of Pediatrics. Apnea of prematurity. Pediatrics 2016;137(1).

Fukada T, Hojyo S, Furuichi T. Zinc signal: A new player in osteobiology. J Bone Miner Metab 2013;31(2):129-135.

Hamada S, Ikezoe K, Hirai T, Oguma T, Tanizawa K, Inouchi M, Handa T, Oga T, Mishima M, Chin K. Evaluation of bone mineral density by computed tomography in patients with obstructive sleep apnea. J Clin Sleep Med 2016;12(1):25-34.

Harper RM, Kumar R, Macey PM, Ogren JA, Richardson HL. Functional neuroanatomy and sleep-disordered breathing: Implications for autonomic regulation. Anat Rec (Hoboken) 2012;295(9):1385-1395.

Harvey JM, O'Callaghan MJ, Wales PD, Harris MA, Masters IB. Aetiological factors and development in subjects with obstructive sleep apnoea. J Paediatr Child Health 1999;35(2):140-144.

Herlenius E, Kuhn P. Sudden unexpected post-natal collapse of newborn infants: A review of cases, definitions, risks, and preventive measures. Transl Stroke Res 2013;4(2):236-247.

Hojyo S, Fukada T. Roles of zinc signaling in the immune system. J Immunol Res 2016;2016:6762343.

Hough FS, Pierroz DD, Cooper C, Ferrari SL; IOF CSA Bone and Diabetes Working Group. Mechanisms in endocrinology: Mechanisms and evaluation of bone fragility in type 1 diabetes mellitus. Eur J Endocrinol 2016;174(4):R127-R138.

Hunt NJ, Russell B, Du MK, Waters KA, Machaalani R. Changes in orexinergic immunoreactivity of the piglet hypothalamus and pons after exposure to chronic post-natal nicotine and intermittent hypercapnic hypoxia. Eur J Neurosci 2016;43(12):1612-1622.

Kambe T, Fukue K, Ishida R, Miyazaki S. Overview of inherited zinc deficiency in infants and children. J Nutr Sci Vitaminol (Tokyo) 2015;61 Suppl:S44-S46.

Kim G, Elnabawi O, Shin D, Pae EK. Transient intermittent hypoxia exposure disrupts neonatal bone strength. Front Pediatr 2016;4:15.

Knip M, Veijola R, Virtanen SM, Hyöty H, Vaarala O, Akerblom HK. Environmental triggers and determinants of type 1 diabetes. Diabetes 2005;54(Suppl 2):S125-S136.

Leidig-Bruckner G, Grobholz S, Bruckner T, Scheidt-Nave C, Nawroth P, Schneider JG. Prevalence and determinants of osteoporosis in patients with type 1 and type 2 diabetes mellitus. BMC Endocr Disord 2014; 14:33.

Louis J, Auckley D, Miladinovic B, Shepherd A, Mencin P, Kumar D, Mercer B, Redline S. Perinatal outcomes associated with obstructive sleep apnea in obese pregnant women. Obstet Gynecol 2012;120(5):1085-1092.

Lumeng JC, Chervin RD. Epidemiology of pediatric obstructive sleep apnea. Proc Am Thorac Soc 2008;5(2):242-252.

Macey PM, Henderson LA, Macey KE, Alger JR, Frysinger RC, Woo MA, Harper RK, Yan-Go FL, Harper RM. Brain morphology associated with obstructive sleep apnea. Am J Respir Crit Care Med 2002;166(10): 1382-1387.

Maddaloni E, D'Eon S, Hastings S, Tinsley LJ, Napoli N, Khamaisi M, Bouxsein ML, Fouda SM, Keenan HA. Bone health in subjects with type 1 diabetes for more than 50 years. Acta Diabetol 2017;54(5):479-488.

Maywald M, Rink L. Zinc homeostasis and immunosenescence. J Trace Elem Med Biol 2015;29:24-30.

Maywald M, Rink L. Zinc supplementation induces CD4+CD25+Foxp3+ antigen-specific regulatory T cells and suppresses IFN-γ production by upregulation of Foxp3 and KLF-10 and downregulation of IRF-1. Eur J Nutr 2017;56(5):1859-1869.

Olivarez SA, Ferres M, Antony K, Mattewal A, Maheshwari B, Sangi-Haghpeykar H, Aagaard-Tillery K. Obstructive sleep apnea screening in pregnancy, perinatal outcomes, and impact of maternal obesity. Am J Perinatol 2011;28(8):651-658.

Pae EK, Ahuja B, Kim M, Kim G. Impaired glucose homeostasis after a transient intermittent hypoxic exposure in neonatal rats. Biochem Biophys Res Commun 2013;441(3):637-642.

Pae EK, Chien P, Harper RM. Intermittent hypoxia damages cerebellar cortex and deep nuclei. Neurosci Lett 2005*a*;375(2):123-128.

Pae EK, Kim G. Insulin production hampered by intermittent hypoxia *via* impaired zinc homeostasis. PLoS One 2014;9(2):e90192.

Pae EK, Wu J, Nguyen D, Monti R, Harper RM. Geniohyoid muscle properties and myosin heavy chain composition are altered after short-term intermittent hypoxic exposure. J Appl Physiol (1985) 2005*b*;98(3):889-894.

Pae EK, Yoon AJ, Ahuja B, Lau GW, Nguyen DD, Kim Y, Harper RM. Perinatal intermittent hypoxia alters γ-aminobutyric acid: A receptor levels in rat cerebellum. Int J Dev Neurosci 2011;29(8):819-826.

Polak J, Shimoda LA, Drager LF, Undem C, McHugh H, Polotsky VY, Punjabi NM. Intermittent hypoxia impairs glucose homeostasis in C57BL6/J mice: Partial improvement with cessation of the exposure. Sleep 2013; 36(10):1483-1490.

Prasad AS. Discovery of human zinc deficiency: 50 years later. J Trace Elem Med Biol 2012;26(2-3):66-69.

Rice JR, Larrabure-Torrealva GT, Luque Fernandez MA, Grande M, Motta V, Barrios YV, Sanchez S, Gelaye B, Williams MA. High risk for obstructive sleep apnea and other sleep disorders among overweight and obese pregnant women. BMC Pregnancy Childbirth 2015;15:198.

Ruz M, Carrasco F, Sánchez A, Perez A, Rojas P. Does zinc really "metal" with diabetes? The epidemiologic evidence. Curr Diab Rep 2016; 16(11):111.

Sandstead HH. Human zinc deficiency: Discovery to initial translation. Adv Nutr 2013;4(1):76-81.

Schwartz MW. Orexins and appetite: The big picture of energy homeostasis gets a little bigger. Nat Med 1998;4(4):385-386.

Semenza GL, Prabhakar NR. Neural regulation of hypoxia-inducible factors and redox state drives the pathogenesis of hypertension in a rodent model of sleep apnea. J Appl Physiol (1985) 2015;119(10):1152-1156.

Swanson CM, Shea SA, Stone KL, Cauley JA, Rosen CJ, Redline S, Karsenty G, Orwoll ES. Obstructive sleep apnea and metabolic bone disease: Insights into the relationship between bone and sleep. J Bone Miner Res 2015;30(2):199-211.

Tapia IE, Shults J, Doyle LW, Nixon GM, Cielo CM, Traylor J, Marcus CL; Caffeine for Apnea of Prematurity – Sleep Study Group. Perinatal risk factors associated with the obstructive sleep apnea syndrome in school-aged children born preterm. Sleep 2016;39(4):737-742.

Thrailkill KM, Liu L, Wahl EC, Bunn RC, Perrien DS, Cockrell GE, Skinner RA, Hogue WR, Carver AA, Fowlkes JL, Aronson J, Lumpkin CK Jr. Bone formation is impaired in a model of type 1 diabetes. Diabetes 2005;54(10):2875-2881.

Weiss JW, Tamisier R, Liu Y. Sympathoexcitation and arterial hypertension associated with obstructive sleep apnea and cyclic intermittent hypoxia. J Appl Physiol (1985) 2015;119(12):1449-1454.

Xu T, Feng Y, Peng H, Guo D, Li T. Obstructive sleep apnea and the risk of perinatal outcomes: A meta-analysis of cohort studies. Sci Rep 2014;4: 6982.

Yeap BB, Davis WA, Peters K, Hamilton EJ, Rakic V, Paul Chubb SA, Davis TME. Circulating osteocalcin is unrelated to glucose homoeostasis in adults with type 1 diabetes. J Diabetes Complications 2017;31(6):948-951.

CBCT EVALUATION OF VOLUMETRIC CHANGES IN THE UPPER AIRWAY FOLLOWING REPOSITIONING THE MANDIBLE TO CENTRIC RELATION

Thyagaseely Sheela Premaraj, Jason Robert Scott, Mary E. Burns, Sundaralingam Prem Premaraj

ABSTRACT

Oral appliances for treating obstructive sleep apnea (OSA) could alter pharyngeal airway dimensions and morphology. The purpose of this retrospective study was to determine the volumetric and minimum cross-sectional area (MCA) changes within the pharyngeal airway space after positioning the mandibular condyles in centric relation (CR) with the maxillary anterior guided orthotics (MAGO). Thirty-one subjects presented with symptoms of upper airway obstruction and sleep-related breathing abnormalities with centric relation-centric occlusion (CR-CO) discrepancies. They were treated with MAGO and evaluated using cone-beam computed tomography (CBCT) scans to assess pharyngeal airway volumetric changes. Airway measurements included total volume (TV), nasopharyngeal volume (NV), oropharyngeal volume (OV) and MCA. The location of the MCA was recorded in reference to the second and third cervical vertebrae and to the occlusal plane. A paired two-sample T-test was performed for statistical significance. F-test was performed to determine the variability for gender and age with all measurements. TV and OV did not show any significant differences following MAGO therapy. NV showed a statistically significant difference after treatment ($p = 0.05$). Pre- and post-treatment differences in MCA were not significant statistically ($p = 0.22$). MCA was located inferior to the occlusal plane in 90% of the subjects. Positioning the condyles in CR significantly increased the NV, but did not increase the TV or OV necessarily. MCA was not affected by the treatment. The majority of patients had the location of MCA below the occlusal plane. Further studies are needed to assess the clinical efficacy of MAGO in relation to pharyngeal airway patency.

KEY WORDS: centric relation (CR), maxillary anterior guided orthotics (MAGO), pharyngeal airway, minimum cross-sectional area (MCA), oral appliances

---

INTRODUCTION

Over the past two decades, public health awareness of obstructive sleep apnea (OSA) and other sleep disorders has increased markedly. Recent data from a Swiss study estimates the prevalence of OSA to be 23% in females and 49% in males (Heinzer et al., 2015). The Wisconsin Sleep Cohort (WSC)—named this as its baseline sample size consisted of 1,500 Wisconsin state employees—is an ongoing longitudinal study of the causes and natural history of sleep disorders, particularly sleep apnea; the WSC has been collecting data for over 20 years. Sleep apnea was assessed by polysomnography (PSG) overnight, in-laboratory sleep study. Data from the WSC suggests a prevalence of 24% in men and 9% in women ages 30 to 60 years of age, with an increased prevalence in older adults (Young et al., 1993, 2002; Hla et al., 2002). Furthermore, prevalence data suggests that OSA is as common in the developing world as it is in western countries (Kapur, 2010).

The etiology of OSA is multi-factorial with structural considerations including palatine and lingual tonsillar and adenoid hypertrophy; soft palate enlargement; enlarged tongue size and position; mandibular retrognathism; rheumatoid arthritis; osteophytes; osteochondromas; and cervical spine fusion (Khan et al., 2014). These pathologies contribute to upper airway collapsibility and obstruction. OSA is a condition in which the airway momentarily, but repeatedly, collapses or becomes obstructed during sleep, resulting in shallow breathing or breathing pauses (Garvey et al., 2015).

The complex interaction between tissues of the upper airway, including soft tissues (24 muscles in addition to other soft tissues), cartilaginous tissues and bony structures (Strohl, 1981) are required for optimal functioning (Bogaerts et al., 2012). The maxilla and mandible are related closely to the pharyngeal airway, having many muscle attachments between these two structures. Any positional changes of these bony structures could influence the patency of the pharyngeal airway.

Continuous positive airway pressure (CPAP) currently is the gold standard for the non-surgical treatment of OSA. An effective alternative to CPAP treatment is oral appliance therapy. The American Academy of Sleep Medicine recommends oral appliance therapy for patients with mild to moderate OSA, as well as for patients with severe OSA who are unwilling or unable to use CPAP (Kushida et al., 2006). Oral appliances improve ventilation by airway volume expansion and a reduction in the collapsibility of the soft palate and tongue. Mandibular advancement appliances (MADs), one category of oral appliance therapy, have been shown to be effective at increasing the upper airway volume (Ryan et al., 1999; Johal and Battagel, 2001; Kyung et al., 2005; Suga et al., 2014).

The most orthopedic and musculoskeletally reproducible position of the mandible is known as centric relation (CR; McKee, 2005; Fig. 1), where the mandible is positioned backward and the condyle is seated at its most superior and anterior position in the glenoid fossa. Stabilization splints, also known as splint therapy, maintain the mandible in CR. Splint therapy is used commonly to stabilize the occlusion, musculature and joint in temporomandibular disorders (Glass et al., 1991, 1993). This type of appliance has proven effective in deprogramming masticatory muscles, while stabilizing the temporomandibular joint (TMJ) in CR (Clark, 1984). CR, being the most orthopedic and musculoskeletally stable position, has the potential to influence the muscles associated with the mandible and could have a positive influence on the pharyngeal muscles. However, positioning the mandible backward could impinge the retropharyngeal airway. Therefore, our aim was to investigate the effects of splint therapy on pharyngeal airway dimensions using three-dimensional (3D) cone-beam computed tomography (CBCT).

In this retrospective study, we investigated the effects of positioning the mandible in CR, using maxillary anterior guided orthotics (MAGO). We examined the effects of MAGO on the pharyngeal airway

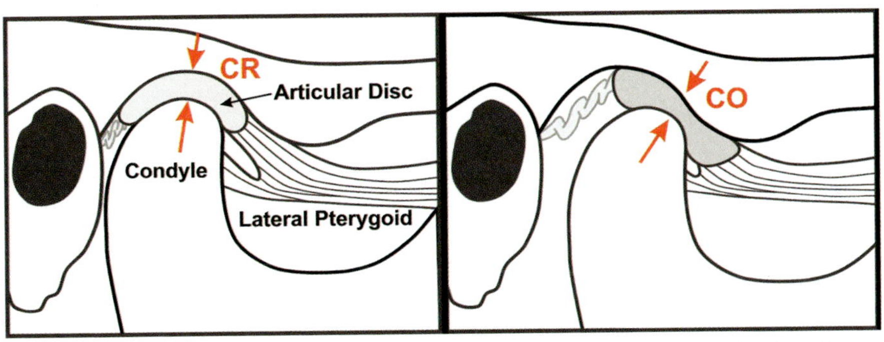

Figure 1. Condylar position in the glenoid fossa in centric occlusion (CO) and centric relation (CR). In CR, the condyle is seated in the most posterior and superior position in the glenoid fossa. Arrow shows the position of the condyle. In CO, the condyle is positioned forward in the glenoid fossa, which is not physiological.

volume and changes in the minimum cross-sectional (MCA) area in patients who had CR and centric occlusion (CR-CO) discrepancies and also reported having sleep related breathing abnormalities (e.g., snoring and/or respiratory effort-related arousals). Our objective was to assess whether positioning the mandible in CR impinges on upper airway volume.

MATERIALS AND METHODS

Patient Information

After obtaining Institutional Review Board approval from the University of Nebraska Medical Center (IRB approval #681-14-EP), records were evaluated for the study. Informed consent was obtained from all subjects. Subjects received a MAGO (Fig. 2) that positioned the mandible in CR. Subjects were patients at a private orthodontic office who presented with signs and symptoms indicative of sleep-related disorders (e.g., snoring, disturbed sleep and arousals); all of them exhibited CR-CO discrepancies. The stabilization splint was placed on the maxillary dentition and adjusted until the condyles were seated fully in CR. This appliance was worn by the patient for 24 hours/day. CBCT scans were taken before delivery of the appliance. The second scan was taken six to seven months after the initiation of appliance therapy. Informed consent was obtained from all subjects. All patients wore the appliance all day and night during this period.

Image Acquisition

All radiographic scans analyzed in this study were taken with Kodak CBCT machine (Carestream Health, Toronto, Canada) with a field view of 18.4 x 20.6 cm and a voxel size of 0.3 mm. All scans were completed at the private office of Dr. Mary Burns in New Hope, PA.

Initial scans were taken on all subjects in maximum intercuspation in the upright position with Frankfort horizontal plane parallel to the floor. Post-treatment scans were taken in CR with the MAGO placed on the maxillary arch. The average time between pre- and post-scans was seven months and eight days. A total of 31 subjects' pre- and post-treatment CBCT scans were evaluated (13 males and 18 females). Many of the subjects received orthodontic treatment following MAGO therapy. Both CBCT scans review and orthodontic therapy were performed by the

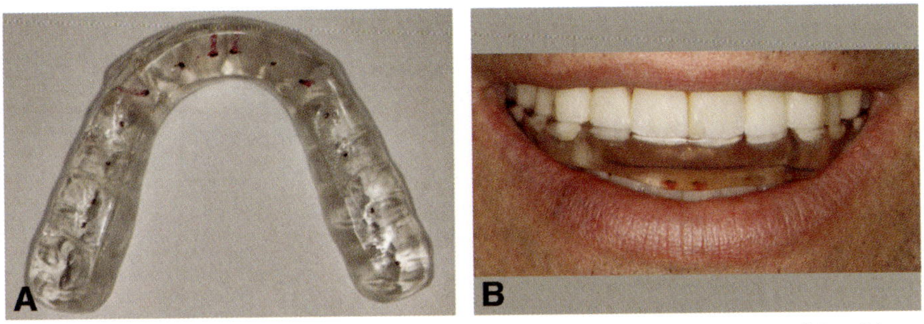

Figure 2. *A:* Maxillary anterior guided orthotic (MAGO). Courtesy of Dr. Mary Burns. *B:* MAGO fitted in a patient. Note the vertical component.

attending orthodontist. The mean subject age at the time of the post-MAGO scan was 46 years and ranged from 19 to 64 years of age.

All scans were acquired in DICOM file format and exported for interpretation. The interpretation of all CBCT images was performed on Anatomage Invivo5 viewing software Version 5.4 (Anatomage, San Jose, CA) licensed to the University of Nebraska Medical Center College of Dentistry.

Volumetric Airway Analysis

All scans were oriented in the sagittal view in the Anatomage software, based on the incisive canal and the second cervical vertebrae (CV2). In preparation for upper airway analysis, several anatomic planes and points were determined to measure different regions of the airway. A plane between the posterior nasal spine (PNS) and the center of sella turcica (S) formed the superior limit, and a plane passing through the inferior border of the third cervical vertebrae (CV3) formed the inferior limit (Fig. 3).

Total upper airway volume was defined as the sum of nasopharyngeal and oropharyngeal volumes as calculated by measuring the space between a horizontal plane at the inferior border of CV3 and a vertical plane connecting the center of S and PNS. Nasopharyngeal volume was defined as the area between a plane parallel with Frankfort horizontal at PNS and a plane passing through PNS and the center of S (Fig. 3). The oropharyngeal volume was defined by the area between a plane parallel with Frankfort horizontal at PNS and a plane parallel with Frankfort horizontal at the inferior border of CV3 (Fig. 3).

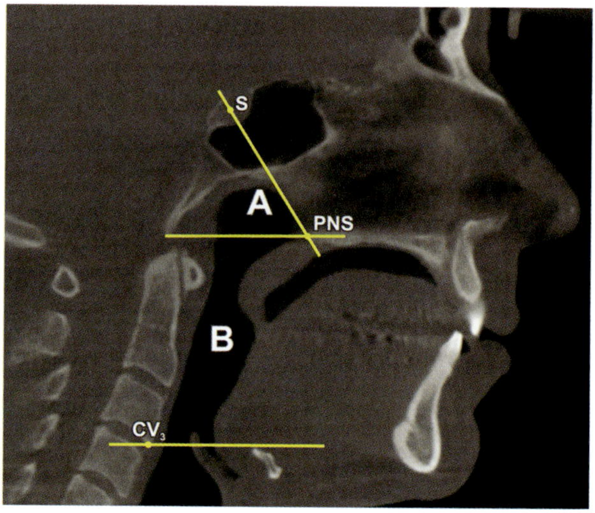

Figure 3. CBCT scan in lateral view. Upper yellow line—a plane connecting posterior nasal spine (PNS) and center of sella (S) representing upper border of the total volume of the pharyngeal airway. Middle yellow line—a plane parallel to Frankfort horizontal at PNS. Lower yellow line—a plane passing through the most inferior border of CV3, parallel to PNS representing the lower boarder of pharyngeal airway. *A* indicates the nasopharyngeal airway volume. *B* indicates the oropharyngeal airway volume. Total upper airway volume was calculated by *A+B*.

Clipping and sculpting tools were used to isolate total upper airway volume and oropharyngeal volume. Clipping initially was performed in gray scale to visualize better the skeletal reference points. The sculpting tool then was used to remove unnecessary soft tissue and skeletal structures from both the right and left sagittal views of the airway (Fig. 4*A*). The lateral walls of the airway then were sculpted in the frontal view (Fig. 4*B*). The volume measurement tool was used to calculate the volume in cubic centimeters (Fig. 4*C*). The nasopharyngeal volume was calculated by subtracting the total volume by the oropharyngeal volume. All scans were measured with a lower threshold value of -1000 and upper value at -603 Hounsfield units (HU). These thresholds values were adapted from a previous study (Hart et al., 2015). Pre- and post-treatment scans were recorded and compared for changes in airway volume for all three regions. All volumetric analyses were performed by a single examiner.

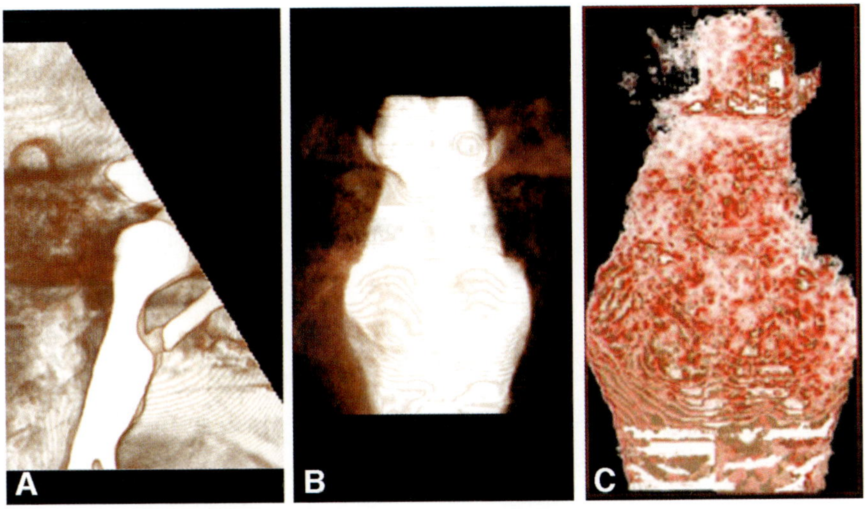

Figure 4. *A:* Sagittal view of total pharyngeal airway volume after initial clipping. Inverse color scale was used to better visualize the airway. Freehand sculpture tool then was used to draw the border of airway, removing unnecessary soft tissue and skeletal structures from both right and left sides. *B:* Frontal view of total volume. Freehand sculpture tool was used to outline the border of airway, removing unnecessary soft tissue and skeletal structures from the lateral boundaries. *C:* Volume rendering of isolated total volume airway with density of -603 HU. Airway volume was calculated in cubic centimeters.

MCA Analysis

MCA is defined as the most constricted portion of the airway. The airway analysis tool from Anatomage Version 5.4 was used to determine the MCA and its location along the total airway, bordered superiorly with a plane between PNS and the center of S, and inferiorly by a plane passing through the inferior border of CV3 and measured in mm^2 (Fig. 5*A-B*). The location of the MCA was observed and recorded in relation to the superior, middle and inferior thirds of the bodies of CV2 and CV3 and the occlusal plane. Pre- and post-treatment MCA and locations were compared.

Intra-examiner Reliability

Intra-observer reliability tests were performed to investigate any potential error in identifying skeletal landmarks and airway volumetric and

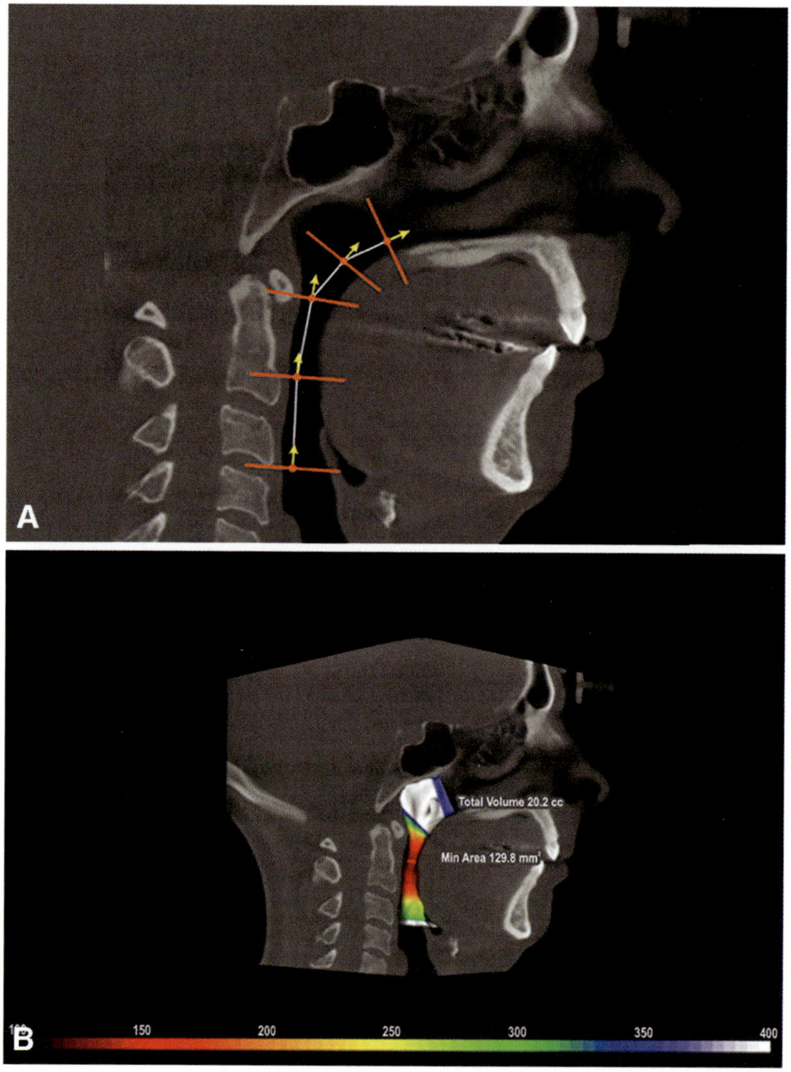

Figure 5. *A:* Images were oriented in the sagittal view using the incisive canal and CV2. Volume rendering tool used to determine the minimum cross-sectional area (MCA) of total airway. *B:* MCA was identified and recorded in mm^2.

MCA measurements. After one month of initial tracings, tracings were repeated. The Pearson correlation coefficient was calculated for each variable.

Statistical Analysis

All volumetric and MCA measurements were recorded for pre- and post-treatment scans. Mean volumetric and MCA measurements also were calculated. A paired two-sample T-test was performed using SPSS (Version 16.0, IBM, Armonk, NY) to determine the significance of change in volume and MCA. An F-test was performed to determine the variability for gender with all measurements. The p-value for all volumetric and area measurements was set at < 0.05.

RESULTS

Pharyngeal Airway Volume Measurements

The average pre- and post-treatment measurements for total volume, nasopharyngeal volume and oropharyngeal volume on all subjects are shown in Figure 6. Error bars represent the upper and lower 95% confidence limits of each measurement. Thirteen of the 31 subjects (42%) showed a decrease in total volume, while 18 (58%) exhibited an increase in total volume. A paired sample t-test showed no significant difference ($p = 0.45$) for total upper airway volume. The nasopharyngeal volume pre- and post-treatment revealed a statistically significant increase in volume after treatment with MAGO ($p = 0.05$). The oropharyngeal volume showed no significant difference between pre- and post-treatment ($p = 0.93$). No differences were found between males and females with all three volumetric measurements of pharyngeal airway.

MCA Measurement

The mean MCA measurements for all subjects pre- and post-treatment are shown in Figure 7. There was no statistical difference found between pre- and post-treatment in MCA values. Error bars represent the upper and lower 95% confidence limits of each measurement. Thirteen of the 31 subjects (42%) showed a decrease in MCA, while 18 (58%) exhibited larger MCAs ($p = 0.22$). The pre-treatment mean MCA was 115.6 mm^2 and the post-treatment mean MCA was 105.5 mm^2. The mean difference was -10.1 mm^2, a difference that was not significant statistically ($p = 0.22$). No significant differences in MCA measurements were found between male and female subjects.

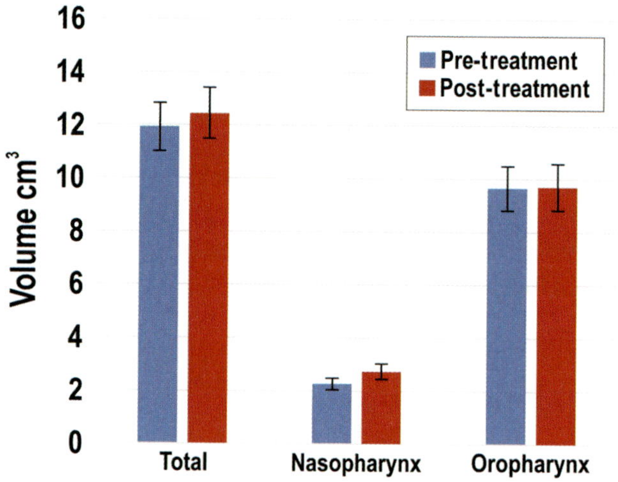

Figure 6. Pre- and post-treatment mean total volume, nasopharynx volume and oropharynx volume. Error bars represent the upper and lower 95% confidence limits of each measurement.

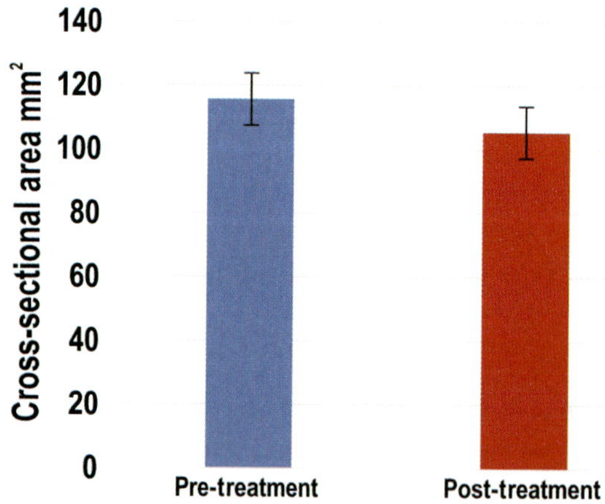

Figure 7. Pre- and post-treatment mean MCA (mm^2). Error bars represent the upper and lower 95% confidence limits of each measurement.

Location of MCA

The location of the MCA was recorded in relation to the superior, middle and inferior thirds of the bodies of CV2 and CV3 and the occlusal plane. These findings are represented in Figure 8.

In 90% of patients, changes in the location of the MCA with the appliance are worth mentioning, the importance of which and the influence of the appliance on this positional change are discussed in the "Discussion" section of this chapter. Differences between pre- and post-treatment MCA locations were found on fourteen of the 31 patients (45%). Half of these subjects showed MCA relocation inferior to the pre-treatment recording, while the other half showed a more superior relocation. Twenty-six subjects (84%) displayed an MCA located at the superior, middle or inferior third of the CV2 body. In addition, the MCA was found to be inferior to the occlusal plane in 28 subjects (90%).

Intra-examiner Reliability

The repeatability of each value was tested by calculating the Pearson correlation coefficient on initial and final measurements one month later, on ten randomly chosen subjects. The average correlation coefficient was 0.986, ranging from 0.960 to 0.998, indicating significant correlation between measurements.

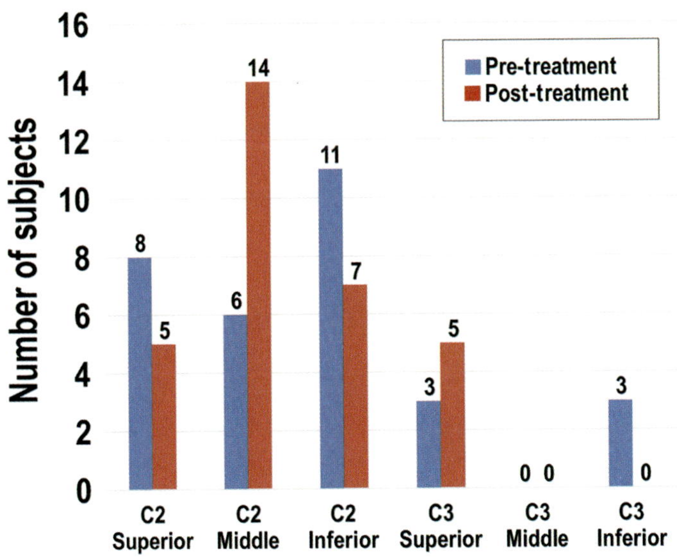

Figure 8. Location of MCA pre- and post-treatment. The location of the MCA was observed and recorded in relation to the superior, middle and inferior thirds of the bodies of CV2 and CV3.

DISCUSSION

Several studies have indicated that a fully seated condylar position in CR is an essential component of a stable and functional occlusion, and that a healthy masticatory muscular position is observed when the condyle is in CR, independent of the occlusion (Schaerer et al., 1967; Bakke and Møller, 1980; Williamson and Lundquist, 1983; Riise et al., 1994; Utt, 1995; Hannam et al., 1997; Crawford, 1999; Cordray, 2006; Okeson, 2015). Okeson (2015) described CR as the most orthopedic and musculoskeletally stable position of the mandible necessary for healthy masticatory system. The positional stability of the TMJ is determined by the muscles attached to the joint like in any other mobile joint.

When a set of muscles are in a physiological state of harmony, it is expected that the neighboring anatomical structures also would achieve physiologic harmony. The pharynx is interconnected with the maxilla and mandible *via* an array of complex musculature. It is conceivable that mandibular positional changes, along with its associated muscular stability, could influence retropharyngeal airway patency.

There is no published literature documenting the relationship between CR and its influence on the posterior pharyngeal airway. Patients with TMD-related symptoms often are treated with splint therapy positioning the mandible in CR. With increasing attention to OSA and an increased prevalence of TMD or related symptoms, treatment with splint therapy needs to be examined in regard to pharyngeal airway volume. The important question is: would splint therapy help or hurt the patient with undiagnosed OSA?

All subjects evaluated in this study presented with symptoms of sleep-related breathing abnormalities, including snoring and/or frequent respiratory effort-related arousals or daytime sleepiness as reported by the patients. All the subjects had CR-CO discrepancies; however, the CR-CO shift was not measured in this population. A small discrepancy of less than 1.0 mm in the vertical or horizontal plane is considered normal and seems not to be a risk factor for temporomandibular disorders (Crawford, 1999). On the other hand, a discrepancy in the vertical or horizontal plane greater than 2 mm should not be ignored.

None of the patients had PSG performed for the diagnosis of OSA. However, all subjects received MAGO therapy for their occlusal discrepancies. The attending orthodontists believed MAGO treatment not only eliminated the CR-CO discrepancies, but also eliminated patient symptoms related to OSA (anecdotal evidence).

The results of this retrospective study indicated there were no statistically significant differences observed in total airway volume or in oropharyngeal airway volume following MAGO therapy ($p = 0.45$, $p = 0.92$, respectively). Fifty-eight percent of the subjects showed an increase in total airway volume after therapy; this increase was not significant statistically. To answer the question of whether this therapy could provide clinical significance in improving breathing-related problems, a larger sample size would have been needed. Interestingly, nasopharyngeal volume showed a significant increase with MAGO treatment ($p = 0.05$). This implies that the MAGO, the splint therapy, did not impinge on the pharyngeal airway, but improved the nasopharyngeal airway volume.

MAGO's contribution to nasopharyngeal airway volume could be due to two reasons: 1) its vertical component opening the jaw (Fig. 2*B*); and 2) the support given to the posterior palate and to the soft palate area, which improved rigidity and preventing collapsibility. These two factors may have contributed to patient's comfort related to breathing and corroborating with the reported patient's experiences. OSA is the more common condition where the airway momentarily but repeatedly collapses or becomes obstructed during sleep, resulting in dyspnea; this is manifested as shallow breathing or breathing pauses, snoring and daytime sleepiness (Garvey et al., 2015).

MAGO may provide support to the posterior palate and its related soft tissues and muscles preventing collapse. No significant differences were found between male and female subjects with all volumetric and MCA measurements. This finding is in agreement with previous studies that found no correlation with gender after examining the frequency, direction and magnitude of condylar changes in CR *versus* CO (Utt et al., 1995).

Even though it is counterintuitive to conclude that positioning the mandible backward could improve or increase pharyngeal airway volume, the observed changes may be attributed to the design of the MAGO appliance. MAGO, as shown in Figure 2*B*, has a vertical thickness that opens the occlusion and separates the jaws, in addition to positioning the mandible backward. This vertical opening of the jaws influences the pharyngeal wall architecture and may increase the airway. In addition, musculoskeletal stable positioning of the mandible in CR may have influenced the airway musculature.

Even though mandibular advancement devices that place the mandible in a protruded position have been shown to be effective at increasing the upper airway volume, their long-term use has negative effects on dentoalveolar structures. Unfortunately, significant advancement of the mandible often results in unwanted dentoalveolar effects, particularly when used over long periods of time. The adverse effects include proclination of mandibular incisors, retroclination of maxillary incisors, molar extrusion and an anteroposterior change in molar relationship (Almeida et al., 2006*a,b*). These sequelae would not be anticipated with MAGO or splint therapy due to its physiologic positioning of the jaws.

This study did not reveal any statistically significant differences after MAGO therapy in MCA of the pharyngeal airway ($p = 0.22$). However, 58% of the patients in this study showed increased MCA after MAGO. In majority of the patients, the location of MCA remained in the same location (55%). Among the patients who showed changes in the location of MCA, half showed a downward movement and the other half showed an upward movement in the pharyngeal airway. The location of MCA is important as the oral appliance used is mainly tooth-borne and its influence on the pharyngeal airway could be due to its proximity, which is in line with/near the occlusal plane. This finding is consistent with that of Ogawa and associates (2007), who found the location of the MCA in patients with OSA was below the occlusal plane in more than 70% of the subjects.

The identification of MCA below the occlusal plane emphasizes the importance of the relationship of this location with regard to oral appliances. Oral appliances may not influence the position of MCA of the pharyngeal airway befitting its location. These appliances are tooth and tissue borne in the oral cavity, changing the position of the mandible while supporting the palatal tissue, especially providing support to the posterior surface of the palate and the musculature. Therefore, while such appliances could influence retroglossal and/or retropalatal spaces, they may not influence the airway space below the occlusal plane.

The MCA and its location is an important parameter pertaining to OSA, breathing and CPAP therapy. When air flows through a constricted area, airflow dynamics could be changed. In patients with OSA, increasing the MCA could be of value in improving breathing or avoiding a reverse airflow by a positive pressure. MAGO did not change the MCA significantly in this retrospective study. We did not measure the exact location of MCA relative to occlusal plane in millimeters. As discussed previously, however, in that the appliance was supported by the occlusal surface of teeth, patients who had the location of the MCA changed below or above their pre-treatment position may have had the MCA located in line with or near the occlusal plane compared to most subjects who did not show changes with MAGO.

The high correlation coefficient between all volumetric and MCA measurements reflect intra-examiner accuracy and ensures reliability for all measured outcomes ($R^2 = 0.098$).

There were shortcomings in this retrospective study. First, we did not have an objective assessment of the sleep-related symptoms. The outcome regarding symptom-relief following appliance therapy with MAGO was not measured objectively; instead, it consisted of anecdotal patient reports. However, the objective of our study was to assess the volumetric and cross-sectional area changes in pharyngeal air way volume after MAGO therapy, and the outcome of the study was not dependent on the AHI index. The AHI before and after treatment would be of immense value regarding MAGO usage in OSA.

Secondly, although CBCT has been proven to be an effective tool in evaluating airway parameters, it represents a static evaluation of a dynamic structure. All scans were taken in an upright position, which is not representative of the airway in the supine position during sleep. CBCT studies have shown that the airway's volumetric and cross-sectional area dimensions are significantly smaller in patients in the supine position compared to an upright position (Camacho et al., 2014).

Therefore, the true nature of the relationship between the airway changes and the relief of symptoms cannot be assessed. In essence, the methodology of the present study did not measure the clinical significance of the volumetric changes observed following MAGO therapy in relieving symptoms related to breathing. Further well-controlled prospective studies utilizing supine CBCT scans are needed to assess the clinical efficacy of splint therapy in reducing the severity and complications that arise from OSA.

CONCLUSIONS

Positioning the mandibular condyles in CR using MAGO does influence posterior pharyngeal airway volume, especially the nasopharyngeal volume. It did not contribute to any significant volumetric changes in total airway or oropharyngeal airway volume and MCA. The MCA was found to be inferior to the occlusal plane in the majority of the subjects. MAGO therapy may not affect patients adversely with OSA according to the outcome of this study. Additional studies are needed to assess the clinical efficacy of MAGO in reducing the severity and complications that arise from OSA. However, reliable index factors to measure the outcome of oral appliance treatment in OSA therapy require more than volumetric and MCA measurements. The clinical efficacy of these appliances in airway patency may be elucidated well with a computational fluid dynamics approach, assessing the air flow pattern for a more reliable outcome in the treatment of OSA.

REFERENCES

Almeida FR, Lowe AA, Otsuka R, Fastlicht S, Farbood M, Tsuiki S. Long-term sequellae of oral appliance therapy in obstructive sleep apnea patients: Part 2. Study-model analysis. Am J Orthod Dentofacial Orthop 2006*a*;129(2):205-213.

Almeida FR, Lowe AA, Sung JO, Tsuiki S, Otsuka R. Long-term sequellae of oral appliance therapy in obstructive sleep apnea patients: Part 1. Cephalometric analysis. Am J Orthod Dentofacial Orthop 2006*b*; 129(2):195-204.

Bakke M, Møller E. Distortion of maximum elevator activity by unilateral tooth contact. Scand J Dent Res 1980;88(1):67-75.

Bogaerts M, Deggoujf N, Huart C, Hupin C, Laureyns G, Lemkens P, Rombaux P, Ten Bosch JV, Gordts F. Physiology of the mouth and pharynx, Waldeyer's ring, taste and smell. B-ENT 2012;8 Suppl 19:13-20.

Camacho M, Capasso R, Schendel S. Airway changes in obstructive sleep apnoea patients associated with a supine *versus* an upright position examined using cone beam computed tomography. J Laryngol Otol 2014;128(9):824-830.

Clark GT. A critical evaluation of orthopedic interocclusal appliance therapy: Design, theory, and overall effectiveness. J Am Dent Assoc 1984;108(3):359-364.

Cordray FE. Three-dimensional analysis of models articulated in the seated condylar position from a deprogrammed asymptomatic population: A prospective study. Part 1. Am J Orthod Dentofacial Orthop 2006;129(5):619-630.

Crawford SD. Condylar axis position, as determined by the occlusion and measured by the CPI instrument, and signs and symptoms of temporomandibular dysfunction. Angle Orthod 1999;69(2):103-115.

Garvey JF, Pengo MF, Drakatos P, Kent BD. Epidemiological aspects of obstructive sleep apnea. J Thorac Dis 2015;7(5):920-929.

Glass EG, Glaros AG, McGlynn FD. Myofascial pain dysfunction: Treatments used by ADA members. Cranio 1993;11(1):25-29.

Glass EG, McGlynn FD, Glaros AG. A survey of treatments for myofascial pain dysfunction. Cranio 1991;9(2):165-168.

Hannam AG, De Cou RE, Scott JD, Wood WW. The relationship between dental occlusion, muscle activity and associated jaw movement in man. Arch Oral Biol 1977;22(1):25-32.

Hart PS, McIntyre BP, Kadioglu O, Currier GF, Sullivan SM, Li J, Shay C. Postsurgical volumetric airway changes in 2-jaw orthognathic surgery patients. Am J Orthod Dentofacial Orthop 2015;147(5):536-546.

Heinzer R, Vat S, Marques-Vidal P, Marti-Soler H, Andries D, Tobback N, Mooser V, Preisig M, Malhotra A, Waeber G, Vollenweider P, Tafti M, Haba-Rubio J. Prevalence of sleep-disordered breathing in the general population: The HypnoLaus study. Lancet Respir Med 2015;3(4):310-318.

Hla KM, Skatrud JB, Finn L, Palta M, Young T. The effect of correction of sleep-disordered breathing on BP in untreated hypertension. Chest 2002;122(4):1125-1132.

Johal A, Battagel JM. Current principles in the management of obstructive sleep apnoea with mandibular advancement appliances. Br Dent J 2001;190(10):532-536.

Kapur VK. Obstructive sleep apnea: Diagnosis, epidemiology, and economics. Respir Care 2010;55(9):1155-1167.

Khan A, Than KD, Chen KS, Wang AC, La Marca F, Park P. Sleep apnea and cervical spine pathology. Eur Spine J 2014;23(3):641-647.

Kushida CA, Morgenthaler TI, Littner MR, Alessi CA, Bailey D, Coleman J Jr, Friedman L, Hirshkowitz M, Kapen S, Kramer M, Lee-Chiong T, Owens J, Pancer JP; American Academy of Sleep. Practice parameters for the treatment of snoring and obstructive sleep apnea with oral appliances: An update for 2005. Sleep 2006;29(2):240-243.

Kyung SH, Park YC, Pae EK. Obstructive sleep apnea patients with the oral appliance experience pharyngeal size and shape changes in three dimensions. Angle Orthod 2005;75(1):15-22.

McKee JR. Comparing condylar positions achieved through bimanual manipulation to condylar positions achieved through masticatory muscle contraction against an anterior deprogrammer: A pilot study. J Prosthet Dent 2005;94(4):389-393.

Ogawa T, Enciso R, Shintaku WH, Clark GT. Evaluation of cross-section airway configuration of obstructive sleep apnea. Oral Surg Oral Med Oral Pathol Oral Radiol Endod 2007;103(1):102-108.

Okeson JP. Evolution of occlusion and temporomandibular disorder in orthodontics: Past, present, and future. Am J Orthod Dentofacial Orthop 2015;147(5 Suppl):S216-S223.

Riise C, Sheikholeslam A. The influence of experimental interfering occlusal contacts on postural activity of the anterior temporal and masseter muscles in young adults. J Oral Rehabil 1982;9(5):419-425.

Ryan CF, Love LL, Peat D, Fleetham JA, Lowe AA. Mandibular advancement oral appliance therapy for obstructive sleep apnoea: Effect on awake calibre of the velopharynx. Thorax 1999;54(11):972-977.

Schaerer P, Stallard RE, Zander HA. Occlusal interferences and mastication: An electromyographic study. J Prosthet Dent 1967;17(5):438-449.

Strohl KP. Upper airway muscles of respiration. Am Rev Respir Dis 1981;124(3):211-213.

Suga H, Mishima K, Nakano H, Nakano A, Matsumura M, Mano T, Yamasaki Y, Ueyama Y. Different therapeutic mechanisms of rigid and semi-rigid mandibular repositioning devices in obstructive sleep apnea syndrome. J Craniomaxillofac Surg 2014;42(8):1650-1654.

Utt TW, Meyers CE Jr, Wierzba TF, Hondrum SO. A three-dimensional comparison of condylar position changes between centric relation and centric occlusion using the mandibular position indicator. Am J Orthod Dentofacial Orthop 1995;107(3):298-308.

Williamson EH, Lundquist DO. Anterior guidance: Its effect on electromyographic activity of the anterior temporal and masseter muscles. J Prosthet Dent 1983;49(6):816-823.

Young T, Palta M, Dempsey J, Skatrud J, Weber S, Badr S. The occurrence of sleep-disordered breathing among middle-aged adults. N Engl J Med 1993;328(17):1230-1235.

Young T, Peppard PE, Gottlieb DJ. Epidemiology of obstructive sleep apnea: A population health perspective. Am J Respir Crit Care Med 2002;165(9):1217-1239.